Abbreviations for trivial names of commonly referenced biochemicals

Abbreviation	Name
MOPS	3-(N-Morpholino)propanesulfonic acid
MSH	Melanotropin (melanocyte-stimulating hormone)
NAD, NAD$^+$, NADH[b]	Nicotinamide-adenine dinucleotide and its oxidized and reduced forms
NADP, NADP$^+$, NADPH[b]	Nicotinamide-adenine dinucleotide phosphate and its oxidized and reduced forms
NMN	Nicotinamide mononucleotide
NMR	Nuclear magnetic resonance
NTP	Nucleoside triphosphate
NTSB	2-Nitro-5-thiosulfobenzoate
P_i	Inorganic orthophosphate
poly(A)	Polyadenylic acid
poly(C)	Polycytidylic acid
poly(G)	Polyguanylic acid
poly(T)	Polythymidylic acid
poly(U)	Polyuridylic acid
POPOP	1,4-bis(5-Phenyl-2-oxazolyl)benzene
PP$_i$	Inorganic pyrophosphate
PPO	2,5-Diphenyloxazole
QAE-cellulose	Quaternary aminoethylcellulose
RNA	Ribonucleic acid or ribonucleate
SAM	S-Adenosylmethionine
SDS	Sodium dodecylsulfate
TCA	Trichloroacetic acid
TDP	Ribosylthymine 5'-diphosphate[a]
TEAE-cellulose	O-(Triethylaminoethyl)cellulose
TEMED	N,N,N',N'-Tetramethylethylenediamine
TFA	Trifluoroacetic acid
THAM	Tris(hydroxymethyl)aminomethane
TMP	Ribosylthymine 5'-monophosphate[a]
TMS-	Trimethylsilyl
TPCK	N-Tosyl-L-phenylalanine chloromethylketone
TPN[b]	Triphosphopyridine nucleotide
Tris	Tris(hydroxymethyl)aminomethane
TTP	Ribosylthymine 5'-triphosphate[a]
UDP	Uridine 5'-diphosphate[a]
UDP-Gal	Uridine diphosphogalactose
UDP-Glc	Uridine diphosphoglucose
UMP	Uridine 5'-monophosphate[a]
UTP	Uridine 5'-triphosphate[a]

[a] The prefix d may be used to represent the corresponding deoxyribonucleoside phosphates, *e.g.*, dATP. The various isomers of adenosine monophosphate may be written 2'-AMP, 3'-AMP, or 5'-AMP (in case of possible ambiguity). A similar procedure may be applied to other nucleoside or deoxyribonucleoside monophosphates.

[b] Note that DPN$^+$-DPNH, NAD$^+$-NADH, and TPN$^+$-TPNH, NADP$^+$-NADPH are in fact paired abbreviations for the same substances in their oxidized and reduced forms, respectively. DPN, NAD, TPN, and NADP are the same as DPN$^+$, NAD$^+$, TPN$^+$, and NADP$^+$

BIOCHEMICAL TECHNIQUES
Theory and Practice

BIOCHEMICAL TECHNIQUES
Theory and Practice

John F. Robyt
Bernard J. White
Iowa State University

Prospect Heights, Illinois

For information about this book, write or call:

Waveland Press, Inc.
P.O. Box 400
Prospect Heights, Illinois 60070
(708) 634-0081

7

PREFACE

History shows that advancements in science depend on the development of new methods and techniques. Biochemistry was born at the turn of the century with Emil Fischer's use of phenylhydrazine to determine the structures of carbohydrates and his determination that proteins were polypeptides made up of amino acids. Also at this time, the Büchner brothers were isolating yeast enzymes that fermented carbohydrates to alcohol, and Michaelis and Menten were studying enzyme kinetics and developing their hypothesis for the way enzymes perform catalysis.

At the beginning of World War II, biochemistry was still in its infancy. Two techniques, however, were being developed that were to put biochemistry on an exponential course. The first of these was the production of radioisotopes and the methods to quantitatively determine them. ^{14}C-labeled compounds became available and were used as quantitative tracers in biochemical experiments. The second technique was the development of chromatography as a means of separating biochemicals that have very similar chemical and physical properties, such as mixtures of carbohydrates or amino acids. In the early 1960s two very important chromatographic techniques appeared: the development by Sober and Peterson of weak ion-exchange celluloses that could be used to fractionate mixtures of proteins under mild conditions and the development by Flodin and Porath of cross-linked dextrans that could be used as molecular sieves to separate macromolecules by molecular weight. Both of these techniques continue to be extensively applied in the study of biochemistry.

The technique of separating charged molecules in an electrical field was first developed in the 1930s by Tiselius. When subjected to an electric current, charged proteins in solution migrated at different rates towards one of the two electrodes and were detected by a complex optical method. For many years this tedious and laborious procedure of solution electrophoresis was the only method available for the quantitative analysis of complex mixtures of proteins. In the late 1950s, however, zone electrophoresis was developed in which the charged molecules were added to a hydrated, porous solid-support and were separated by the passage of current through the support. In the early 1960s, Ornstein and Davis developed polyacrylamide-gel zone electrophoresis, giving biochemistry a simple and powerful electrophoretic technique for separating charged macromolecules.

The polyacrylamide prevented convection and permitted high resolution of very small amounts of proteins that could be easily detected by specific staining methods. This technique is widely used today for the qualitative and quantitative separation and analysis of proteins and nucleic acids.

Combined with the use of radioisotopes, chromatography, and electrophoresis, quantitative spectrophotometric methods were being developed and applied to the measurement of biochemicals. The use of these techniques put biochemistry on a quantitative basis.

Biochemical Techniques presents the fundamental theory and practice used in biochemistry laboratory courses and in biochemical research. Chapter 1 covers the making of measurements and error analysis, the presentation of data in tables and graphs, and the writing of reports. Chapter 2 presents the preparation and properties of solutions and includes pH and buffers. For some students, the contents of Chapters 1 and 2 may already have been covered in other chemistry courses. We have found, however, that the material is usually needed, if for no other reason than as a review. Chapter 3 proceeds with qualitative and quantitative techniques including UV and visible spectrophotometry, fluorimetry, and infrared and nuclear magnetic resonance spectroscopy. Chapter 4 covers chromatographic methods of analysis and is divided into two parts. Part I discusses the general aspects of the many types of chromatography and Part II covers specific applications for separating carbohydrates, amino acids, proteins, nucleic acids, and lipids. Chapter 5 presents electrophoretic methods of separation and places emphasis on zone electrophoresis, especially polyacrylamide- and agarose-gel electrophoresis. Chapter 6 discusses the nature and properties of radioisotopes and techniques for quantitatively measuring radioactivity by gas-ionization, liquid and solid scintillation spectrometry, and autoradiography. Specific methods of radioisotope labeling of biochemical compounds and applications of their use in tracer experiments are presented. Chapter 6 concludes with a discussion of the points that should be considered in designing an experiment using radioisotopes. Chapter 7 gives various, commonly used methods for the qualitative and quantitative determination of carbohydrates, amino acids, proteins, nucleic acids, and lipids. Wherever possible we have given the underlying chemistry and the limitations of the methods. Chapter 8 presents the techniques used in the preparation of different kinds of biochemical substances and includes cell lysis, centrifugation, filtration, concentration, dialysis, and fermentation. Specific methods of preparing organelles, proteins, carbohydrates, nucleic acids, and lipids are given. Within each class, specific examples were chosen to illustrate different sources, methods, and biochemical properties. For example, in the protein class, the preparation of jack bean urease is described for historical reasons as well as for the simplicity of the method of crystallization; yeast alcohol dehydrogenase is chosen to illustrate the combination of several different types of techniques used in protein purification; human salivary α-amylase is presented because it can be readily obtained from a human source and because of its ease of assay and frequent use in studying enzyme catalysis in laboratory courses; *E. coli* alkaline phosphatase and *B. amyloliquefaciens* (BamI) restriction endonuclease were chosen for their use in the study of the structure and sequence of nucleic acids. Chapter 9 discusses the topic of enzymology and includes the theory and application of Michaelis-Menten kinetics, methodology of enzyme assays, and types of inhibition. Chapter 10 presents strategies involved in the determination of the structures of carbohydrates, proteins, nucleic acids, and lipids. The strategies used in the analysis of DNA include cloning techniques using plasmid and phage DNA vectors. Chapter 10 shows how many of the techniques presented in Chapters 1 through 9 can be used in studying modern biochemical problems.

Three important types of information are set apart in the text by placement in boxes: equations that define an important concept or result from a derivation; examples or calculations that emphasize how a particular method or technique can be applied; and directions for carrying out experimental methods or preparations. Specific directions for the use of most instruments and equipment have not been given, as there are a number of models and manufacturers for each instrument, making it an impossible task to give detailed directions for each type. Specific directions for the use of individual instruments are provided by the manufacturer.

Many instructors of biochemistry laboratory courses have developed their own sets of experiments to fit their needs and facilities. *Biochemical Techniques* is not intended to be a laboratory manual but has been designed to be used in conjunction with experiments already developed by individual instructors. The authors have intended to give the students a book that they will keep and use in their chosen careers in the study of biological subjects. Historical aspects of the development of the techniques are given where appropriate and references to classical experiments from primary sources are given to expose the student to the biochemical literature. Besides providing the background to enhance the understanding of experiments conducted in biochemistry laboratory courses, the book provides additional material that the student can use to explore more advanced areas and techniques related to the experiments. Extensive bibliographies are given for each chapter, so the student can study and obtain further details about the experimental methods and techniques.

It is the authors' philosophy that scientists develop by designing and executing their own experiments, rather than just performing experiments. *Biochemical Techniques* provides the background and techniques that students can use in designing and executing their own experiments either in a laboratory course or in their research. For example, ^{14}C-labeled glycogen could be isolated from the liver of an animal that had been given ^{14}C-glucose. The approximate molecular size of the glycogen could be determined using gel permeation chromatography. The labeled glycogen could then be digested with human salivary α-amylase and the products examined by thin layer chromatography and autoradiography. The principles and specific experimental techniques for these experiments can be obtained in Chapter 4, "Chromatographic Techniques"; Chapter 6, "Theory, Measurement, and Use of Radioisotopes"; Chapter 8, "Biological Preparations"; and Chapter 9, "Enzymology". Another combination could involve the isolation and assay of alcohol dehydrogenase (Chapters 3, 7, 8, and 9) and characterization by polyacrylamide gel electrophoresis (Chapter 5), the determination of its molecular weight by gel permeation chromatography (Chapter 4), and the measurement of its kinetics and inhibition (Chapter 9). Using their imaginations, students could equally well develop many other experiments from the procedures given in the text. Perhaps some of these students would develop a new method or technique that would dramatically change the course of biological science.

ACKNOWLEDGMENTS

We wish to thank Iowa State University and the Department of Biochemistry and Biophysics for providing the intellectual atmosphere conducive to writing this book. Thanks also go to Pierre Robitaille, Jay-lin Jane, and Paul Gollnick for reading and commenting on various parts of the manuscript, to Laura Totaro for reading parts of the manuscript and helping with the figures, and to Steve Eklund for his varied and valuable assistance in the preparation of the manuscript.

John F. Robyt
Bernard J. White

CONTENTS

CHAPTER 3
SPECTROSCOPIC METHODS   40

CHAPTER 10
STRUCTURAL ANALYSIS OF BIOLOGICAL MOLECULES   321

CHAPTER 1

ANALYZING AND REPORTING EXPERIMENTAL DATA

In the scientific method, an experiment is designed to answer a question or to add weight and proof to a hypothesis. The experimental method often involves making a measurement of one quantity as another quantity is varied. For example, determining the effect of varying the concentration of a reactant on the initial velocity of. an enzyme-catalyzed reaction is a classical experiment in the study of enzymes. The experiment requires the measurement of the initial velocity or rate of the reaction and thus requires the collection of data. The importance of collecting data and having concrete numbers as results was succinctly expressed by Lord Kelvin (William Thomson):

> When you can measure what you are speaking about and express it in numbers, you know something about it, and when you cannot measure it, when you cannot express it in numbers, your knowledge is of a meagre and unsatisfactory kind. It may be the beginning of knowledge, but you have scarcely in your thought advanced to the stage of a science.

In this first chapter, we discuss the recording and handling of experimental data, error and precision analysis of data, the expression of data in tables and graphs, the preparation of controls and blanks, general rules for nomenclature, and the writing of laboratory or research reports.

1.1 RECORDING AND HANDLING EXPERIMENTAL DATA

1.1.1 Significant Figures

The gathering of experimental data is only a part of laboratory experimentation. Equally important are the handling and interpretation of the acquired data. In the use of experimental data, we must be careful neither to overextend the reliability of the measurement nor to underutilize the accuracy of the data. Guidelines are available for making such judgments. Experimenters should record data with as many digits as can be accurately measured, with the last digit or the smallest increment of the measurement being an estimate of the true value. This is termed the **significant figures** approach. For example, if the amount of lipid extracted from a membrane sample is measured by weighing the dry sample on an analytical balance and the weight is recorded as 1.973 g, the measurement is said to have four significant figures. The last digit, 3, is an estimate of the true value of the number. If the weighing were repeated, a value of 1.972 g or 1.974 g might be expected, assuming that the balance is capable of weighing accurately to one milligram (0.001 g). In this case, the limit to the number of digits to report for the measurement is determined by the accuracy of the balance used. (If a sample is found to weigh 1.970 g, the measurement still contains four significant figures because the zero is a measured digit.)

Another way of looking at the measurement of the weight of the lipid is that it has an accuracy limit of 0.001 g in 1.973 g or 0.001/1.973, which is 1 part in 1973 parts. This can also be expressed as a percentage: $1/1973 \times 100 = 0.05\%$. We can say that the measurement of the weight has an accuracy of $\pm 0.05\%$. In this example, we are considering only the limitations of accuracy, overlooking the limits due to other experimental errors. We will consider the effects of random errors in Section 1.2.

There are established rules for the use of significant figures in calculations involving experimental measurements with different significant figures.

Multiplication: The product of two or more measurements can have only as many significant figures as are contained in the multiplicand that has the fewest significant figures. For example,

5.17 mL $\times$ 1.973 g/mL = 10.20041 g, or 10.2 g rounded off to three significant figures, because although the dry weight (1.973 g) contains four significant figures, the measured volume (5.17 mL), which contains only three, is the controlling number.

Division: The quotient of two measured quantities can have only as many significant figures as are contained in the term with the fewest significant figures. For example,

139.7 μmol $\div$ 2.67 min = 52.322 μmol/min should be rounded off to 52.3 μmol/min: only three significant figures are given because the number of minutes, which has three significant figures, is controlling.

Another way to determine the correct number of significant figures is to examine the variance in each of the two measurements. In the example of division above, the variances are 0.1 μmol/139.7 μmol or 1 part/1397, and 0.01 min/2.67 or 1 part/267. The result should then be the same order of magnitude as the largest variance, or 1/523 and not 1/52322.

Addition or subtraction: The sum or difference of two measurements can be no more accurate than the least accurate of any individual measurement. For a digit to be significant, it must arise from a significant digit in every term of the sum or difference. For example, if we add 1.5 mL to 0.51 mL and obtain 2.01 mL, the final digit, "1," is not significant because the measurement "1.5 mL" does not include a digit in the hundredths place. The answer must be rounded off to 2.0 mL. If, however, the weight of a sample-filled flask is 4.199 g and the weight of the flask is 3.387 g, the weight of the sample will be reported as 0.812 g. Since both subtrahend and minuend are reported to 0.001 g or 1 mg, the answer (i.e., $4.199 - 3.387 = 0.812$) is acceptable as written. Note that the difference contains only three significant figures, whereas each of the measurements contains four.

Two further considerations arise. Should nonsignificant figures be rounded off at each step of a calculation? A reasonable policy is to carry one non-significant digit through a series of calculations and round off to the correct number of significant figures in the final answer. What if the only nonsignificant digit is a 5? For example, 9.92 mmol/mL × 3.00 L = 29.75 mmol. There is no set rule for this situation, so one must choose a consistent policy and round up in all cases, or round down in all cases. Therefore, 29.75 mmol would be rounded up to 29.8 mmol, or rounded down to 29.7 mmol. If the 5 occurs in an intermediate step in a series of calculations, it should be carried through to the last answer and then rounded off.

1.1.2 Scientific Notation

Confusion regarding significant figures can arise in numbers such as the following: 6100 mm, 0.00412 mol, 0.310 g. These numbers, however, can be represented as the product of a number between 1 and 10 and 10 raised to some appropriate power: 6100 would become 6.1×10^3 mm, 0.00412 mol would become 4.12×10^{-3} mol, and 0.310 g would become 3.10×10^{-1} g. In the first two examples, the zeros are assumed to be nonsignificant, only marking the position of the decimal point. In the third example, the zero is still present because it indicates that the measurement is accurate to the third decimal place, or to 0.001 g.

1.1.3 Units

The international system of measurement, *Système International d'Unités*, abbreviated SI, is based on the MKS (meter·kilogram·seconds) system. The base units of SI are given in Table 1-1. Although the kilogram is the base unit of mass, the gram is still commonly used. Other SI units are derived from the base units: for example, the newton (N, which is $kg \cdot m \cdot s^{-2}$), the unit of force, and the joule (J, which is $m^2 \cdot kg \cdot s^{-2}$), the unit of energy. Some non-SI units continue to be used, such as pressure in atmospheres, and temperature on the Celsius scale.

Multiples and submultiples of units are commonly used, and Table 1-2 gives the prefixes that designate multiples and submultiples. Attaching one of these prefixes directly to a base unit, we obtain abbreviations such as milli-second (ms), nanomole (nmol), and micrometer (μm). Table 1-3 lists some of the basic units of measurement in common usage.

TABLE 1-1. **Basic units of the *Système International d'Unités* (SI)**

Quantity	Unit	Abbreviation
Length	meter	m
Mass	kilogram	kg
Time	second	s
Temperature	kelvin	K
Electric current	ampere	A
Amount of substance	mole	mol
Radioactivity	becquerel	Bq

TABLE 1-2. **Prefixes for multiple and submultiple units**

Quantity	Prefix	Abbreviation
10^{12}	tera	T
10^{9}	giga	G
10^{6}	mega	M
10^{3}	kilo	k
1	—	—
10^{-1}	deci	d
10^{-2}	centi	c
10^{-3}	milli	m
10^{-6}	micro	μ
10^{-9}	nano	n
10^{-12}	pico	p
10^{-15}	femto	f
10^{-18}	atto	a

Any measured quantity must include the units of the measurement. Thus, "path length = 10.0" is not acceptable, but "path length = 10.0 cm" is a complete statement. When making numerical calculations, it is a good idea to include units with all numbers and then to convert all multiple and submultiple units to the same base unit. Thus rather than using moles and millimoles or grams and milligrams together in a single calculation, it is advisable to convert all amounts to moles or to millimoles (or to grams or to milligrams) and then perform the calculation, canceling units where appropriate to obtain a meaningful final answer. As an example, let us determine the molar concentration (M) of alcohol dehydrogenase in a solution that is 10 mg/mL. The mass of alcohol dehydrogenase is 150,000 g/mol.

$$\text{concentration} = \frac{10 \text{ mg/mL} \cdot 1 \text{ g}/10^3 \text{ mg}}{1.5 \times 10^5 \text{ g/mol}} = \frac{10 \times 10^{-3} \text{ g/mL}}{1.5 \times 10^5 \text{ g/mol}}$$

$$= 6.7 \times 10^{-8} \frac{\text{mol}}{\text{mL}} \cdot \frac{10^3 \text{ mL}}{\text{L}}$$

$$= 6.7 \times 10^{-5} \text{ mol/L}$$

$$= 6.7 \times 10^{-5} \text{ M}$$

TABLE 1-3. **Basic units of measurement**

Name	Symbol	Definition
Units of length		
meter	m	Basic unit
kilometer	km	10^3 meters
centimeter	cm	10^{-2} meter
millimeter	mm	10^{-3} meter
micrometer	μm	10^{-6} meter
nanometer	nm	10^{-9} meter
angstrom	Å	10^{-10} meter
Units of volume		
liter	L	Basic unit
deciliter	dL	10^{-1} liter
milliliter	mL	10^{-3} liter
microliter	μL	10^{-6} liter
Units of mass		
kilogram	kg	Basic unit
gram	g	10^{-3} kilogram
milligram	mg	10^{-6} kilogram
microgram	μg	10^{-9} kilogram
nanogram	ng	10^{-12} kilogram
Units of time		
second	s	Basic unit
millisecond	ms	10^{-3} second
minute	min	60 seconds
hour	h	60 minutes
day	da	24 hours
month	mo	28–31 days
year	yr	12 months
Units of radioactivity		
becquerel	Bq	1 disintegration per second (dps)
curie	Ci	3.7×10^{10} Bq

1.2 ANALYSIS OF EXPERIMENTAL DATA

1.2.1 Error Analysis and the Estimation of Precision

Error in an experimental measurement is defined as the difference between the experimental value and the true value. In many cases the true value is not known, so an experimenter is concerned about how well the data represent the accepted true value.

If a number of measurements of an experimental quantity, such as the activity of an enzyme or the disintegration rate of a radioactive sample, are very close together and agree in value, they are said to be **precise**. One usually obtains a mean value for a series of measurements and then calculates the **sample deviation**; that is, one determines the deviation of each value in the series from the mean. A mean deviation can then be obtained by calculating the sum of the absolute values of the sample deviations and dividing by the number of measurements:

$$\text{sample deviation} = x_i - \bar{x} \tag{1-1}$$

$$\text{mean deviation} = \bar{d} = \frac{\sum |x_i - \bar{x}|}{n} \tag{1-2}$$

where x_i = the individual sample values and $\bar{x}$ = the mean value of n samples. In the example presented in Table 1-4, five determinations of the initial velocity for an enzyme reaction are measured. The results of this experiment would be reported as 1.35 ± 0.04 μmol of product per minute.

Another term used in estimating precision is the percentage error or percentage deviation, which is calculated as follows:

$$\text{percentage deviation} = \frac{\text{mean deviation}}{\text{mean value}} \times 100$$

Applying this formula to the example of Table 1-4 yields:

$$\frac{0.04}{1.35} \times 100 = 2.96\% = 3.0\%$$

The percentage deviation indicates that the precision of this set of measurements is within 3%. Other expressions of deviation are **standard deviation** and **standard error**. The standard deviation is a statistical measure of the precision of a set of values. Standard deviation from the mean, s, for a sample of data of n observations can be calculated by Equation 1-3.

$$s = \sqrt{\frac{\sum (x_i - \bar{x})^2}{n - 1}} \tag{1-3}$$

TABLE 1-4.   Sample data for calculation of $\bar{x}$ and $\bar{d}$

Initial velocity (μmol product/min)	Deviation from the mean		
x	$x - \bar{x}$	$|x - \bar{x}|$	$(x - \bar{x})^2$
1.31	-0.04	0.04	0.0016
1.35	0.00	0.00	0.0000
1.38	$+0.03$	0.03	0.0009
1.41	$+0.07$	0.07	0.0049
1.28	-0.07	0.07	0.0049
$\sum = 6.73$		$= 0.21$	$= 0.0123$

mean velocity $= \bar{x} = 6.73/5 = 1.35$ μmol of product per minute
mean deviation $= \bar{d} = 0.21/5 = \pm 0.04$

The standard deviation is the square root of the sum of the squares of the sample deviations divided by the number of observations minus 1. This equation for the standard deviation is valid for any series of measurements that is described by the standard normal distribution. In graphical form, the normal distribution curve shows the dispersion of values about the mean value $\bar{x}$. For a single set of sample values, 68.3% of the observed values will occur within the interval $x \pm s$, and 95.5% of the observed values will occur within the interval $x \pm 2s$, as indicated in Figure 1-1.

For the example of the initial velocity measurements presented in Table 1-4, the standard deviation from the mean is calculated as follows:

$$s = \sqrt{\frac{0.0016 + 0.0000 + 0.0009 + 0.0049 + 0.0049}{5 - 1}} = \sqrt{\frac{0.0123}{4}}$$

$$= \sqrt{0.003075} = 0.055$$

and the velocity is 1.35 ± 0.055 μmol product/min. Therefore, it can be stated with 68.3% confidence that the value of any single additional measurement will not be outside this interval. If $2s$ were used as the interval, the value would be 1.35 ± 0.10 μmol/min, with a confidence of 95.5% that the additional measurements of the initial velocity would be within this interval.

The standard deviation of a set of measurements is most useful in relation to the mean sample value. Therefore, the standard deviation is converted to percent **relative standard deviation** (RSD) or **coefficient of variation**, which is expressed as follows:

$$c = \frac{s}{\bar{x}} \times 100 \tag{1-4}$$

where c = coefficient of variation, s = standard deviation, and $\bar{x}$ = mean sample value. From our example, $c = 0.055/1.35 \times 100 = 4.1\%$. This indicates the relative size of the error and allows the experimenter to compare the precision of two different techniques or different steps of an experimental protocol.

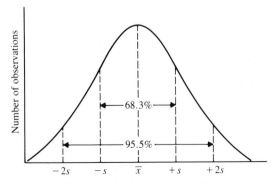

FIGURE 1-1. The normal distribution curve.

1.2.2 Errors and the Rejection of Data

How do we know when an experimental value is a mistake and should be discarded? Obviously if we suspect that a mistake has been made in weighing or pipetting or in making a dilution, we should stop the analysis and discard the data collected for that sample. But, if we gather together a set of data for an experiment without any obvious errors, how do we treat data that appear to be incorrect? First we calculate the mean, $\bar{x}$, and the standard deviation, s. Next, we calculate the deviations from the mean of each value. If the deviation from the mean for any individual measurement is greater than $2s$, the probability of that measurement's being a true member of the set is less than 5%, and there is some justification for asserting that the value is a result not of experimental error but of some kind of mistake. If the deviation from the mean is greater than $3s$, the probability that the measurement is a member of the set is less than 0.3%, and there is considerable justification for discarding the value. For a discussion of other criteria for the rejection of data, see Chapter 3 of Smart's *Combination of Observations*,[1] or a quantitative analysis text, such as Kolthoff et al.[2]

1.2.3 Calculating the Uncertainty of the Result of Mathematical Operations

In the section on error analysis, methods for estimating the error of individual determinations were presented. When these individual determinations, with errors, are further used in calculations, new error estimates will result. The following rules are used for estimating the resulting error for each type of mathematical operation.

Summation: $(a \pm s_a) + (b \pm s_b)$
The uncertainty of $\bar{c}$, where $\bar{c} = \bar{a} + \bar{b}$, is given by:

$$s_c = (s_a^2 + s_b^2)^{1/2}$$

For example, if we were to add 24.8 mg $\pm$ 0.4 mg to 13.6 mg $\pm$ 0.3 mg, the value of the sum and its probable error would be:

$$s_c = [(0.4)^2 + (0.3)^2]^{1/2} = [0.25]^{1/2} = 0.5 \text{ mg}$$

and the sum would be 38.4 $\pm$ 0.5 mg.

Difference: $(a \pm s_a) - (b \pm s_b)$
The uncertainty for $\bar{c} = \bar{a} - \bar{b}$ is given by:

$$s_c = (s_a^2 + s_b^2)^{1/2}$$

Division: $(a \pm s_a) \div (b \pm s_b)$

$$\text{uncertainty in } \bar{c} = s_c = \frac{\bar{a}}{\bar{b}}\left[\left(\frac{s_a}{\bar{a}}\right)^2 + \left(\frac{s_b}{\bar{b}}\right)^2\right]^{1/2}$$

Multiplication: $(a \pm s_a)(b \pm s_b)$

$$\text{uncertainty in } \bar{c} = s_c = \bar{a} \cdot \bar{b}\left[\left(\frac{s_a}{\bar{a}}\right)^2 + \left(\frac{s_b}{\bar{b}}\right)^2\right]^{1/2}$$

TABLE 1-5. Relative error of the mean as a function of the number of measurements

$$s_{\bar{x}} = \frac{s}{\sqrt{N}}$$

N	$s_{\bar{x}}$
1	$1.00s$
2	$0.71s$
3	$0.58s$
4	$0.50s$
5	$0.45s$
6	$0.41s$
10	$0.32s$

1.2.4 The Relation Between Precision and the Number of Determinations

It can be shown that the arithmetical mean, derived from N measurements, is $\sqrt{N}$ times as reliable as a single measurement.

$$\text{estimated average deviation of the mean} = \bar{d}_{\bar{x}} = \frac{\bar{d}}{\sqrt{N}}$$

or

$$\text{estimated standard deviation of the mean} = s_{\bar{x}} = \frac{s}{\sqrt{N}}$$

Precision can be expressed as the standard deviation of a single measurement. The standard deviation of the mean, which indicates the precision of a set of measurements, can be decreased by increasing the number of measurements. As shown in Table 1-5, the standard deviation of the mean $s_{\bar{x}}$ decreases by 29% in going from one to two measurements, and by 42% in going from one to three measurements. In contrast, increasing the number of measurements from four to five yields only 5% increase in precision. Thus, in determining the mean of the experimental value, the precision increases greatly in going from one to three or four measurements, but begins to reach diminishing returns at more than four.

1.3 TABLES AND GRAPHS

Experimental data are best presented in tabular or graphical forms, which offer a concise and rapid method of communicating to the reader the data collected and the experimental results obtained. Both graphs and tables should have explanatory titles and should contain enough information to be intelligible without reference to the text.

1.3.1 Tables

Tables require careful labeling of all data presented. Every row and column should have an appropriate heading or label, and any codes or abbreviations

TABLE I. *Equilibrium binding constants of Triton X-100-solubilized prolactin receptors*
Affinity constants and binding capacity were determined by Scatchard plot analysis of ^{125}I-hGH displacement experiments. The values represent the mean $\pm$ S.D. of three measurements.

Sample	K_a	Capacity
	$\times 10^9 \, M^{-1}$	*pmol/mg protein*
Triton X-100-solubilized membranes	3.5 ± 2.1	6.0 ± 1.4
Concanavalin A-Sepharose-purified receptors	3.9 ± 1.3	39.8 ± 5.1
hGH-Sepharose-purified receptors	3.5 ± 1.0	$20,440 \pm 4,790$

TABLE I.   *Amino acid and heme composition of Cytochrome c-554*

Component	Residues/heme	Residues/molecule[a]
Cys	1.62 ± 0.19	6.5
Asx	3.91 ± 0.24	15.6
Thr	1.76 ± 0.11	7.0
Ser	2.29 ± 0.34	9.2
Glx	6.18 ± 0.43	24.7
Pro	3.35 ± 0.53	13.4
Gly	4.62 ± 0.45	18.5
Ala	4.29 ± 0.58	17.2
Val	1.93 ± 0.21	7.7
Met	1.02 ± 0.10	4.1
Ile	0.34 ± 0.08	1.4
Leu	1.35 ± 0.12	5.4
Tyr	1.13 ± 0.11	4.5
Phe	3.06 ± 0.19	12.2
His	2.39 ± 0.17	9.6
Lys	8.73 ± 0.22	34.9
Arg	1.45 ± 0.06	5.8
Heme		4

[a] Values were calculated assuming a content of 4 hemes and a molecular weight of 25,000.

FIGURE 1-2. Examples of tables from the literature.

should be explained in a footnote to the table. The units of all measurements should be specified. If the condition is the same for all tabulated data, this is indicated at the top of the column.

It is not wise to put too much data into a single table. Use of several tables will emphasize or highlight the important aspects of the data.

Do not include more significant figures than are justified by the accuracy of the determinations (see Section 1.1.1). If mean values are reported, some accepted measure of the range should be indicated, along with the number of individual observations contributing to the mean value. Figure 1-2 presents examples of tables from the literature.

1.3.2 The Use of Powers of 10 in Tables and Figures

Multipliers by powers of 10 (10^3, 10^{-3}, etc.) are sometimes useful in table headings and figures. Because these multipliers may confuse the readers, however,

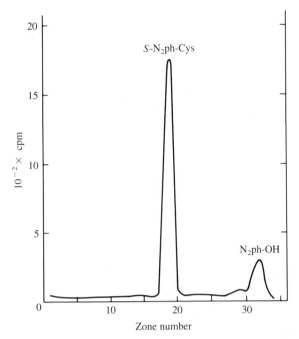

FIGURE 1-3. Graph illustrating the use of a multiplier on the ordinate.

units that eliminate the need for exponential forms should be employed whenever possible. For example, for the sake of clarity, 0.002 M should be expressed as 2 mM rather than 2×10^{-3} M. The use of "2 mM" uses neither zeros nor exponents and expresses the concentration in terms of a whole number. When this is not possible, the following guideline is suggested by the *Journal of Biological Chemistry* in its "Instruction to Authors," found in the first issue of each volume. The quantity expressed in the table or figure is to be preceded by the power of 10 by which its value has been multiplied (see Fig. 1-3). The quantity expressed on the ordinate ranges from 0 to 2000 cpm and has been multiplied by 10^{-2}. The multiplier must apply to the quantity, not to the units. Thus, one should use $10^6 \times concentration$ (M) rather than concentration $(M \times 10^{-6})$ or $10^4 \times velocity$ $(pmol/s)$ rather than $velocity$ $(pmol/s \times 10^{-4})$.

1.3.3 Rules for Making Graphs

Graphs are used to illustrate quantitative data and can quickly indicate the relationship between two or more variables and any special trends that might be present. There are several important rules for making good graphs.

Rule 1: Always label both the *y*-axis (ordinate) and the *x*-axis (abscissa). Place the dependent variable(s) on the ordinate and the independent variable on the abscissa. If possible, enclose the top and right-hand sides of the figure.

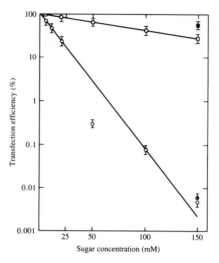

Transfection efficiency of Fl DNA after 8 days of incubation with different sugars and sugar concentrations. Results represent mean $\pm s_{\bar{x}}$ of duplicate experiments: $\square$, glucose; $\bigcirc$, glucose-6-P.

FIGURE 1-4. Graph using error bars.

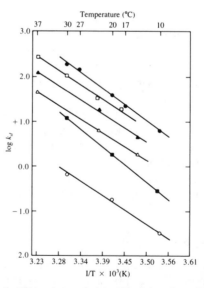

Arrhenius plots for the ATP-dependent proteolysis in rabbit reticulocyte lysates. The different proteins are designated by the following symbols: $\bullet$, ^{125}I–ubiquitin–lysozyme "conjugates"; $\square$, ^{125}I–lysozyme (guanidinated); $\blacktriangle$, ^{125}I–lysozyme after a 1.5-h incubation in 200 μM hemin; $\triangle$, ^{125}I–lysozyme; $\blacksquare$, ^{125}I–albumin (guanidinated); and $\bigcirc$, ^{125}I–albumin.

FIGURE 1-5. Graph using different symbols to indicate different sets of experimental data.

Rule 2: Choose sets of coordinates that are easy to read and have major divisions expressed as integers, multiplied by the appropriate power of 10, if necessary (or a mixture of integers and decimals, if more convenient).

Rule 3: Select scales that have uniform intervals and distribute the data over most of the area of the graph rather than having the data compressed into one region of the graph. Use of computer software designed for making tables and graphs, such as Lotus 1-2-3, simplifies this task.

Rule 4: Clearly indicate experimental points. Circles can be used to indicate the uncertainty (error) of the measurement in both variables. Vertical lines (error bars) can be used to represent uncertainty in the dependent variable (Fig. 1-4).

Rule 5: Keep the number of lines on a single graph to a minimum. Assign distinct symbols (e.g., circles, open and solid, triangles, open and solid) to each set of data. Symbols should be explained in the legend to the graph (Fig. 1-5).

Rule 6: Prepare a title that describes the graph and gives enough experimental detail to permit the material to stand on its own. Readers should be able to understand what the data in the graph mean without reference to the text.

1.3.4 Linearization

Many graphs present data of continuous functions; that is, when the data are plotted, a smooth curve can be drawn connecting the points. This allows interpolation between the data points and perhaps extrapolation beyond the data points when appropriate. Graphing of data and interpolation are most easily done for a straight line. Therefore, whenever possible, data in the form of a straight line should be presented. The classical example in biochemistry is the rearrangement into straight-line form of the Michaelis–Menten equation, which is the equation of a rectangular hyperbola, by taking the reciprocal of both sides of the equation to give a new equation, called the Lineweaver–Burk equation.

$$v_i = \frac{V_m[S]}{K_m + [S]} \qquad \text{Michaelis–Menten equation}$$

$$\frac{1}{v_i} = \frac{K_m}{V_m}\frac{1}{[S]} + \frac{1}{V_m} \qquad \text{Lineweaver–Burk equation}$$

The Lineweaver–Burk equation is the equation of a straight line in which the dependent variable is $1/v_i$, the independent variable is $1/[S]$, K_m/V_m is the slope, and $1/V_m$ is the intercept on the ordinate.

Similarly, the Michaelis–Menten equation can be transformed into another straight-line form called the Eadie–Hofstee equation.

$$v_i = -K_m\left(\frac{v_i}{[S]}\right) + V_m \qquad \text{Eadie–Hofstee equation}$$

In the Eadie–Hofstee equation v_i is the dependent variable, $v_i/[S]$ is the independent variable, $-K_m$ is the slope, and V_m is the intercept on the ordinate. See Chapter 9 (Section 9.6) for the graphical determination of the values of K_m and V_m.

Sometimes, however, more than one straight line can be drawn through the data points, either because of the probable error of the measurements or because of the scatter of the data points. One way to proceed in such cases is to draw two lines on the graph, one with the largest slope consistent with the data and one with the smallest slope consistent with the data. This approach gives the maximum and the minimum values for the slope and the intercepts, which in turn give the range of the calculated values. However, an alternate mathematical method is widely used, the **method of least squares**, or **regression analysis**.

1.3.5 Numerical Curve Fitting: The Method of Least Squares

As indicated above, many types of data collection result in a linear curve when the dependent variable y is plotted against the independent variable x. The equation for a straight line that results when a set of data $(x_1 y_1), (x_2 y_2), \ldots, (x_n y_n)$ is obtained is:

$$y = mx + b \tag{1-5}$$

where m is the slope of the line and b is the y-intercept that results when $x = 0$.

When a set of data points lies almost, but not quite, on a straight line, the experimenter must try to find the line that "best" fits the data points. "Best" can be defined in terms of the so-called **least-squares criterion**, according to which the line must be such that the sum of the squares of the vertical departures of the points from the best line, commonly called the **residuals**, are a minimum.

The residual is defined as $r = y - (mx + b)$; hence the square of the residual is $r^2 = [y - (mx + b)]^2$. When the sum of the squares of the residuals are a minimum, the best line for the data points will result, and the **normal equations** for the linear least-squares line for a set of N data points can be shown to be:

$$\sum y = m\sum x + Nb \tag{1-6}$$

$$\sum xy = m\sum x^2 + b\sum x \tag{1-7}$$

The constants m and b can be obtained by solving for them from the simultaneous equations 1-6 and 1-7 to give:

$$m = \frac{N\sum xy - (\sum x)(\sum y)}{N\sum x^2 - (\sum x)^2} \qquad (1\text{-}8)$$

$$b = \frac{(\sum y)(\sum x^2) - (\sum x)(\sum xy)}{N\sum x^2 - (\sum x)^2} \qquad (1\text{-}9)$$

The line that has a slope of m from Equation 1-8 and a y-intercept of b from Equation 1-9 is then the best fitting line or the linear least-squares line for the data. The normal equations (1-6 and 1-7) can be remembered by observing that Equation 1-6 is obtained by summing on both sides of Equation 1-5, and Equation 1-7 is obtained by multiplying both sides of Equation 1-5 by x and then summing. This is not a derivation of the normal linear least-squares equations, but a means to remember them. For a derivation, see any calculus text.

Thus, to compute the slope m, and the y-intercept b, for a set of N data points using the linear least-squares method, it is necessary to calculate $\sum x, \sum y, \sum x^2$, and $\sum xy$ for the set of data having N variables and to put these values into Equations 1-8 and 1-9.

The manual determination of the values of m and b by the least-squares method is time-consuming, but with the use of computers, the method becomes a practical way for finding the best values for m and b from the set of data, which in turn will give the best-fit linear line with a slope of m and a y-intercept of b for the set of data.

If a given dependent variable and independent variable produce a linear fit, the experimental points will deviate from the best-fit line only by experimental error and the residuals will be a mix of positive and negative values, with no set pattern.

In addition to inspecting the residuals, a quantity called the experimental linear correlation coefficient can be calculated. This coefficient gives information about the overall closeness of a least-squares fit to the data. If the value is zero, there is no correlation. If the data points lie exactly on the best-fitted line, the correlation coefficient will be 1 if the slope is positive and -1 if the slope is negative. In a fairly close fit, the magnitude of the linear correlation coefficient might be 0.99.

Computer programs make it relatively easy to carry out a linear least-squares fit by automatically calculating the slope, the y-intercept, the correlation coefficient, and the residuals for each data point. A wise experimenter, however, will make a plot of the experimental data points and draw the least-squares line on the same graph. Any errors in the results generated by the computer should be evident, as well as any bad data points.

1.3.6 Weighting Factors in Linear Least Squares

One may have a set of data in which the probable errors of the values are not of equal size. An example of such data is the determination of initial velocity of an enzyme-catalyzed reaction as a function of the substrate concentration. At low substrate concentrations, the velocity values are small and subject to larger relative error than at high substrate concentration. In this case, the residuals should be divided by a quantity proportional to the square of the probable error $(1/s^2)$. The effect of weighting factors is to give more importance, or greater weight, to the data points that have a smaller expected error. For a discussion of the use of weighting factors in least-squares analyses, see Reference 3.

1.3.7 Calculating the Area Under a Curve

The area under a peak in a chromatogram from a gas–liquid chromatograph is proportional to the amount of a substance producing the peak. This is also so for many materials that can be chromatographed by different methods. And as discussed in Chapter 3, the area under the absorption peak in a ^{1}H NMR spectrum is proportional to the number of proton nuclei producing the absorption (Section 3.6.1). There are several methods for determining the area of these peaks. One can count the squares on the graph paper, but this can be tedious. Another method is to cut the area out of the graph paper and weigh it, but this approach, too, has limitations. It is also possible to use a planimeter, a mechanical device that traces the peak and registers the area bounded by it.

The area of a symmetrical peak can be approximated by calculating the area by triangulation, as shown in Figure 1-6. Rather than using the width at the base, the width at half-height times the height can be used to obtain the area:

$$\text{area} = \text{width}\left(\text{at } \frac{h}{2}\right) \times h$$

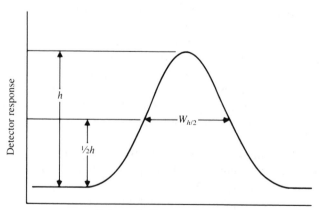

FIGURE 1-6. Approximation of the area of a symmetrical peak by triangulation.

Many instruments are capable of data storage and can be coupled to integrators that measure the area of a peak and present the data in tabular form. If peak area data for corresponding standards are obtained, it is possible to have a direct output of the amount of substance producing the peak.

1.4 CONTROLS AND BLANKS

A blank contains all the elements of an assay solution except the substance being measured, which is usually replaced by an equal volume of buffer, water, or other solvent. For example, in the Bradford assay for protein, 0.10 mL of protein-containing sample is added to 5.0 mL of Bradford reagent, mixed thoroughly, and the absorbance measured at 595 nm. The blank would consist of 0.10 mL of water or buffer added to 5.0 mL of Bradford reagent. The absorbance of the reagent blank can be read against a sample of water or buffer alone, or the blank itself can be used to set the spectrophotometer to zero absorbance. It is safest to read the absorbance of the reagent blank and subtract the absorbance of the blank from the absorbance readings of the samples. Although in theory, the blank should have a minimal absorbance at the wavelength of sample absorbance, it is possible for the reagent blank to have significant absorbance. In such a case one is measuring a small difference between two relatively large absorbance values, and this can lead to serious errors. When the blank has significant absorbance, the method of analysis is not a good one.

In the type of blank prepared for enzyme activity assays, one of the essential components, either enzyme or substrate, is omitted from the blank. The choice of which component to omit is determined by the particular method being used to assay the enzyme activity. In some cases, particularly when a crude enzyme extract is used, both enzyme and substrate can be used in the blank, if the enzyme is first inactivated.

1.5 GENERAL RULES FOR NOMENCLATURE

Ionic charge: Ionic charge is designated as a superscript following the chemical symbol: Ca^{2+}, SO_4^{2-}. The on-line notation, Fe(II) and Fe(III) for the $+2$ and $+3$ charged ions, is also used.

Isotopically labeled compounds: Isotopes are designated by placing the atomic weight as a superscript at the upper left of the element symbol; for most biochemical usage, an isotopically labeled compound is indicated by placing the symbol for the isotope in square brackets directly attached to the name, as in $[^{14}C]$urea, 3-phospho-D-$[^{14}C]$glyceraldehyde. The symbol for the configuration should precede the symbol for the isotope (e.g., D-$[^{14}C]$fructose). However, in common usage, $^{14}CO_2$ and $^{32}P_i$ are acceptable. Likewise, the terms "^{125}I-labeled compound" and "^{14}C steroids" are acceptable, indicating labeling without specifying how the compounds are labeled. For a more specific discussion on nomenclature for labeled compounds, see Chapter 6 (Section 6.12).

Molecular weight and molecular mass: Molecular weight (M_r) is defined as the ratio of the mass of a molecule to one-twelfth the mass of carbon-12. Hence it is a dimensionless number. Molecular mass is the mass of one molecule of a substance and is given in units of daltons (D). The dalton is defined as one-twelfth the mass of carbon-12. For example, we might state: "The molecular mass of lysozyme is 14,000 daltons." Entities that do not have a definable molecular weight can be described in terms of daltons; thus it is correct to say: "The mass of a ribosome is 10^7 daltons." When "molecular weight" is used to refer to the mass of one mole of a substance, however, it is expressed in grams, not daltons. For example, to refer to the mass of one mole of lysozyme, we could say: "The molecular weight of lysozyme is 14,000 grams." However, in stating the molecular weight of macromolecules, "grams" is often omitted, and it is said that "the molecular weight of lysozyme is 14,000," with grams being implied.

Abbreviations: The table in the front of the book gives the abbreviations for the trivial names of commonly referenced biochemical substances that are widely used in the biochemical literature. The abbreviations approved by the *Chemical Abstracts* for biochemical journals, and used in journal article citations, are given in Appendix A. The three-letter and one-letter abbreviations for the 20 commonly occurring amino acids in proteins are given in Appendix B, along with their three-letter DNA code.

1.6 THE LABORATORY OR RESEARCH REPORT

After the design of the experiment, the collection and analysis of the data, and the formation of a conclusion, the writing of a laboratory or research report completes the experiment. The report should be in the general form of a journal article, which usually includes the following parts.

1. *Title*
2. *Summary* or *abstract*
3. *Introduction*, stating the reason for doing the experiment and any relevant background information
4. *Experimental*, including the materials and methods
5. *Results*, including tables and graphs of the data
6. *Discussion*, including the conclusions and interpretations of the data, and any hypotheses that can be formed
7. *References*
8. *Acknowledgments*

Titles should be brief, but specific and rich in informative words. A title should "tell" what is contained in a report in the fewest possible words, and phrases in which more than three words modify another word should be avoided. For example, "*E. coli* phenol-extracted ribosomal-derived proteins" would be a poor title.

The *abstract* or *summary* should state the problem briefly, mention the experiments performed to solve the problem, give the results obtained, and state the conclusions formed.

The *introduction* should state the purpose of the experiments and briefly give the relationship to other work in the field. The *introduction*, however, should not be an exhaustive review of the literature.

The *experimental* section should be brief, but should include enough detail to permit a qualified investigator to repeat the work. Methods published in the literature are referenced, and only new procedures or modified procedures should be described in detail.

Results can be presented in text, tables, graphs, and figures. Tables and graphs should be used whenever possible to give a concise presentation of the results.

The *discussion* should contain the interpretation of the results and the development of hypotheses that explain and are consistent with the data. New results should also be correlated, when possible, with results in the literature. Sometimes, the results and discussion can be combined into one section to give a clearer and logical presentation. Laboratory slang and jargon should be avoided throughout the report.

The *references cited* section lists all the literature cited in the report and should follow a consistent format (see cited references in the *Journal of Biological Chemistry* or other related publications).

The *acknowledgment* recognizes special help received from individuals or groups by way of information, special assistance, use of special equipment, compounds, and so forth. The acknowledgment should not be overdone. That is, usual or expected help from an instructor or major-professor or the expected use of equipment should not be acknowledged. It is the word "special" that should be considered when formulating an acknowledgment.

1.7 LITERATURE CITED

1. W. M. Smart, *Combination of Observations.* Cambridge University Press, New York, 1958.
2. I. M. Kolthoff, E. B. Sandell, E. J. Meehan, and S. Bruckenstein, *Quantitative Chemical Analysis.* Macmillan, Toronto, Ont, 1969.
3. P. R. Bevington, *Data Reduction and Error Analysis for the Physical Sciences.* McGraw-Hill, New York, 1969.

1.8 REFERENCES FOR FURTHER STUDY

1. J. M. Brewer, A. J. Pesce, and R. B. Ashworth, *Experimental Techniques in Biochemistry.* Prentice-Hall, Englewood Cliffs, NJ, 1974, Chapter 2, "Treatment of Gaussian measurement data."
2. R. G. Mortimer, *Mathematics for Physical Chemistry.* Macmillan, New York, 1981, Chapter 10, "The treatment of experimental data."

3. E. R. Tufte, *The Visual Display of Quantitative Information.* Graphics Press, Santa Monica, CA, 1983.

4. W. S. Cleveland, *The Elements of Graphing Data.* Wadsworth, Belmont, CA, 1985.

5. Robert A. Day, *How to Write and Publish a Scientific Paper*, ISI Press, Philadelphia, 1979.

6. *American Chemical Society Style Guide: A Manual for Authors and Editors* (J. Dodd, Ed.). ACS Publications, Washington DC, 1986.

7. *Council of Biology Editors Style Manual*, 5th ed. Council of Biology Editors, Bethesda, MD, 1983.

PREPARATION AND PROPERTIES
OF SOLUTIONS

Preparing and handling solutions is an essential part of experimental biochemistry. When a group of practicing researchers was quizzed about the skills most needed (and sometimes most lacking) for beginning laboratory research, the most frequent response was "competency in preparing reagents and buffers, and in pipetting accurately." This chapter attempts to cover these fundamental topics. Readers desiring more information may consult the references and the bibliography given at the end of the chapter.

2.1 METHODS OF MEASUREMENT

Experimentation in the biochemical sciences frequently requires highly accurate and precise measurements. **Precision** is defined as the agreement among a group (series) of experimental measurements, and **accuracy** is the agreement of the experimental values with the established or "true" value. Good laboratory techniques are necessary in achieving precision in weighing a solid or in measuring the volume of a liquid, and use of the proper equipment can help to ensure accuracy. Section 2.2 describes the equipment to use for accurate measurements.

2.2 VOLUMETRIC GLASSWARE

The common pieces of volumetric glassware are flasks, pipets, and burets. These instruments are capable of measuring liquid volumes with an accuracy of 0.2% or better.

A volumetric flask is constructed of transparent glass with a pear-shaped body and a long, narrow neck (see Figure 2-1). A single line finely etched around the neck of the flask, indicates the level to which the flask should be filled to

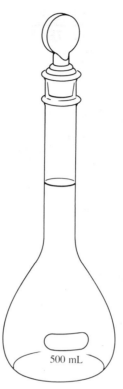

FIGURE 2-1. The volumetric flask. The graduation line is a permanent ground groove in the glass and the stopper is ground glass. The 500 mL flask shown has a tolerance of ±0.20 mL.

contain the stated volume. If the flask is calibrated for 25 mL, it will contain 25.00 ± 0.05 mL, or a 0.2% error limit (Class A, NBS). A 500-mL flask will contain 500.0 ± 0.2 mL or a 0.04% error limit. Thus, as the total volume of the flask increases, the relative error limit decreases.

Volumetric flasks marked *to contain* are used to dilute a solution to a definite volume or to prepare a solution of accurate concentration from a dry solid. If a solution of a dry solid is to be made, the weighed solid is placed in a beaker and, if possible, dissolved in a volume of solvent not more than half the volume of the flask. The solution is then poured into the volumetric flask through a funnel. The beaker is thoroughly rinsed with solvent and the washings poured through the funnel into the flask.

The flask is filled with solvent until the meniscus is a few centimeters below the etched line. After thorough mixing of the contents, the flask is held at eye level and the final dilution made with a pipet or dropper. Solvent is added dropwise until the lowest part of the meniscus just touches the graduation line (see Figure 2-2). For very precise work, the temperature of the solution in the flask is measured and the final dilution is not made until the contents have come close to room temperature.

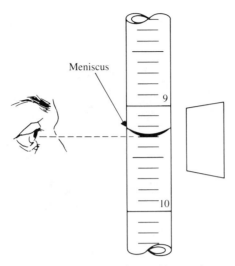

FIGURE 2-2. Reading the meniscus. The meniscus, which is read at the lowest point, is easier to see if a white card is held behind the apparatus. The eye of the viewer must be at the same level as the meniscus.

Pipets are used to measure accurate volumes of liquids. There are many different types of pipets, each with a specific use and precision. Care should be used in selecting pipets, since the precision of experimental results depends on the precision of the pipets used. In general, one selects the pipet that gives the greatest precision for the required conditions of the experiment. Never pipet solutions of toxic, corrosive, and radioactive substances by mouth; special safety precautions for such materials are discussed below.

Pipets can be divided into three general groups, based on their intended use: the volumetric transfer pipet, the measuring pipet, and the air-displacement piston pipet.

Volumetric transfer pipets are calibrated to deliver only one volume (see Figure 2-3A). The tolerance or accuracy of volumetric pipets is set by the manufacturer and is usually indicated on the pipet. A 10-mL pipet may have an accuracy of $\pm 0.2\%$, while a 1-mL or smaller pipet may have an accuracy of $\pm 1.0\%$.

Measuring pipets, which are graduated to deliver varying volumes of liquid, are of two types: serological and Mohr. Serological pipets, commonly called blow-out pipets, are graduated to the tip (Fig. 2-3B). An etched band around the top of the pipet stem indicates that to deliver the stated volume, the liquid left in the tip after draining should be blown out gently. The serological pipet is most accurate when the full volume of the pipet is delivered.

Mohr pipets are also graduated to deliver varying volumes of liquid, but as Figure 2-3C shows, liquid volume is measured only between the calibration marks. This eliminates the calibration errors that occur near the tapered tip of the pipet.

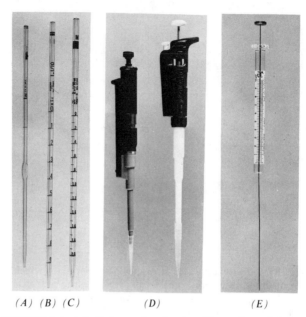

(A) (B) (C)   *(D)*   *(E)*

FIGURE 2-3. (*A*) Volumetric pipets, calibrated to deliver a single volume. (*B*) Serological measuring pipet, calibrated to the tip and meant to be blown out as indicated by the ground band near the top. (*C*) Mohr-type measuring pipet, calibrated to deliver varying volumes; note that the scale ends above the tip. (*D*) Piston pipet, an adjustable-volume device with disposable plastic tip. (*E*) Syringe pipets, each consisting of glass barrel with stainless steel needle and plunger.

Air-displacement piston pipets are popular for multiple measurements. When the plunger is depressed, a calibrated volume of air is displaced from the barrel. A disposable plastic tip is immersed in the sample, and solution is drawn into the tip when the plunger is released (Fig. 2-3*D*). Pipet tip volumes range from a few microliters to several milliliters. These pipets are simple to use and are especially useful in working with radioisotopes or other samples such as toxins or poisons. One need not touch either the sample solution or the disposable plastic tip. There are, however, possibilities for error if the disposable tips do not fit properly and if the amount of liquid remaining in the tip is not constant for each pipetting. The precision limit for this type of pipet is generally 1% or larger. Sample accuracy is stated to be better than 1%.

Syringe pipets of various kinds are used for the precise delivery of liquids in volumes of 1–500 μL (Fig. 2-3*E*). The accuracy is 1% of the syringe volume, with repeatability of dispensed volumes also within 1%. Several sizes are available (e.g., 10, 25, 50, 100, and 500 μL). These pipets consist of a syringe needle and a metal plunger that fits into a carefully bored barrel with volume markings. They are used when small volumes are employed, for example, in the addition of enzymes, substrates, and buffers in enzyme reactions, in the addition of samples

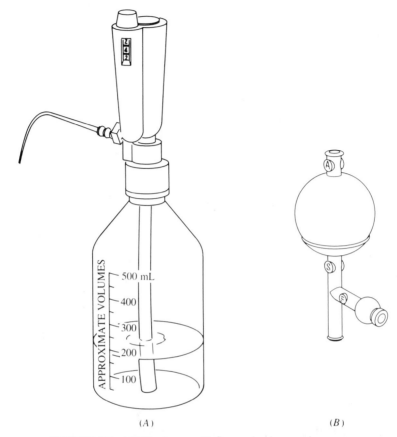

(A)   (B)

FIGURE 2-4. (A) Pipet pump: Teflon and polypropylene pump
for delivering volumes from 0.1 to 50 mL with ±0.1% accuracy.
(B) Propipet (pipet filler): rubber bulb with three stainless steel
ball valves to control the vacuum for filling and delivering.

for thin-layer, gas–liquid, and high performance liquid chromatography, and in
the pipetting of radiochemically labeled compounds.

Toxic or radioactive solutions can be transferred by means of a pipet, but the
solutions should *never* be drawn into a pipet with the mouth. Rather, attach a
device such as a rubber bulb, pipet pump, or syringe to the mouth end of the
pipet and apply suction. These common devices (Figs. 2-3E and 2-4) allow the
safe transfer of hazardous liquids.

Burets are made from heavy-walled glass tubing and are graduated in
subdivisions so that any fraction of the total volume can be measured. Burets are
used primarily for titration—adding a volume of solution necessary to react with
a certain amount of substance. The volumes of titrant are measured accurately
from burets with markings indicating each 0.10 mL and can be estimated to 0.02
mL. Figure 2-5 shows a standard buret with a glass stopcock. Teflon stopcocks
are also available for use with basic solutions or with other solvents that might
cause a glass stopcock to freeze or leak.

FIGURE 2-5. Buret, made from accurate-bore glass tubing, with permanent markings and graduated in tenths of a milliliter.

2.3 USE OF ALL VOLUMETRIC GLASSWARE

2.3.1 Cleaning

Before using a pipet, buret, or volumetric flask, fill the vessel with distilled water and then empty it. If drops of water adhere to the inside surfaces, the glassware should be cleaned with a cleaning solution. Most glassware can be cleaned by soaking in a detergent solution or by applying scouring powder with a brush. However, volumetric glassware is more difficult to clean. Acetone is a good solvent for many organic compounds and is useful for dissolving organic residues, including oily substances. Precipitated protein and other biochemical substances are quite insoluble and require more drastic treatment. Traditional cleaning solutions make use of 7% w/v potassium dichromate in 80% sulfuric acid-water or 10% w/v potassium hydroxide in 85% ethanol-water. Cleaning solution that is drawn or poured into the glassware should be brought into contact with the entire interior surface. After treatment with cleaning solution, glassware should be thoroughly rinsed with distilled water, to ensure removal of chromium ions, which may interfere with some biochemical experiments.

2.3.2 Reading the Meniscus

The sharpness of the meniscus can be increased by holding a white card behind the apparatus. The lowest point of the meniscus is selected as the position of the liquid in the apparatus. The apparatus should be held so that the meniscus is level with the eye (Fig. 2-2).

2.3.3 Calibration

Calibration is the process of determining how much the mark or marks of an apparatus are in error. With volumetric glassware, calibration usually consists in weighing the liquid contained or delivered and then calculating, using the density of the liquid, the true volume of the liquid. Although most volumetric glassware conforms to rigid standards of production, that may not always be the case, and large errors in precision may be introduced if measurements are made with uncalibrated apparatus.[1]

2.4 THE ANALYTICAL BALANCE

Knowledge of the operation of the analytical balance is essential to every experimenter in the biochemical laboratory.

Analytical balances vary in capacity and sensitivity, and in their method of weighing solids and liquids. The sensitivity of a balance is defined as the minimum load that will produce a noticeable change in the scale divisions of the readout. Single-pan balances with manual operation can have a capacity of 200 mg with a precision of ± 0.05 mg. Electronic single-pan balances with electromagnetic force compensation can have a capacity of 160 g with precision of ± 0.1 mg (Fig. 2-6).

The balance must be handled carefully and kept clean at all times. The following list of precautions applies to the use of this equipment.

1. Read the instructions on the use of the analytical balance before making a weighing.
2. Be certain that the balance is properly leveled.
3. Zero the balance and check for proper operation.
4. Place the object to be weighed in the center of the pan; avoid drafts and other air movement.
5. When the balance is stable, record the weight immediately; after the sample has been removed, recheck the zero point of the balance.
6. For experiments requiring high accuracy, have the sample at room temperature and constant humidity.

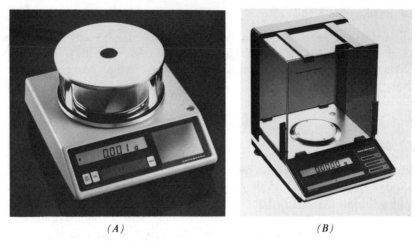

(A)   *(B)*

FIGURE 2-6. Single-pan analytical balance. (*A*) Electronic balance with digital readout and 3-s weighing time. (*B*) Manual balance with glass-enclosed weighing chamber. Courtesy of Brinkmann Instruments.

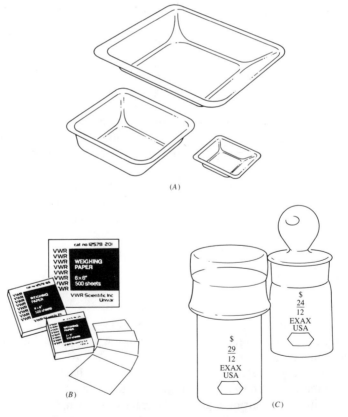

FIGURE 2-7. Weighing containers. (*A*) Plastic boats. (*B*) Papers. (*C*) Bottles.

Samples are seldom weighed directly on the balance pan; a weighing container is generally used. Containers may be aluminum or plastic pans, paper, or glass weighing bottles (Fig. 2-7). Some balances have automatic taring, which will compensate for the weight of the container into which the sample is placed.

Many less accurate balances are available for laboratory use. Electronic single-pan, top-loading balances with microprocessor control, printer, and taring memory can have capacities of 1200 g with a precision of 0.005 g. These balances offer fast operation, high capacity, and acceptable accuracy for many laboratory purposes. Balances of this type will err by ± 0.005 g in a 1-g measurement or 0.5% error, which is frequently satisfactory in laboratory measurements.

2.5 EXPRESSION OF CONCENTRATION AND DILUTION

The three most useful expressions of concentration, *molarity*, *molality*, and *normality*, assume that the molecular weight or the equivalent weight of the solute is known.

Molarity is defined as the number of moles of solute per liter of solution and is denoted by M. This is the most common expression of concentration for solids of known molecular weight. Molarity usually describes solutions of accurate concentration; solutes are weighed on an analytical balance, and solution volumes are measured in volumetric flasks. A more common concentration range for solutions in biochemistry is millimolarity, denoted by mM.

To prepare a liter of 50 mM sodium citrate, one would need to weigh 50 mmol $\times$ 453.6 mg/mmol (or 22.7 g) of sodium citrate and dissolve it in enough water to make 1.00 L of solution.

Molality is defined as the number of moles of solute per 1000 grams of solvent and is denoted by m. Molality can be more precise than molarity, since both solute and solvent are weighed. (A 1.00 millimolal solution contains 1.00 millimole of solute dissolved in 1000 grams of solvent.) However, molality is not widely used in biochemistry.

Normality is defined as the number of gram-equivalent weights of solute per liter of solution, and is denoted by N. This designation is useful for solutions of acids and bases. One gram-equivalent weight of an acid is the quantity of the acid that can donate one mole of hydrogen ions to a base; one gram-equivalent of a base is that amount which can accept one mole of hydrogen ions. For a monoprotic acid such as acetic acid, which yields one mole of hydrogen ions upon complete dissociation, the gram-equivalent weight is equal to the molecular weight.

$$CH_3-COOH \rightleftharpoons CH_3-COO^- + H^+ \qquad \begin{array}{l} \text{1 g-mol} = 60 \text{ g} \\ \text{1 g-equiv} = 60 \text{ g} \end{array}$$

Fumaric acid, a diprotic acid, yields two moles of protons when reacted completely with a base.

$$HOOC-\overset{\overset{\displaystyle H}{|}}{C}=\overset{\overset{\displaystyle H}{|}}{\underset{\underset{\displaystyle H}{|}}{C}}-COOH \rightleftharpoons {}^-OOC-\overset{\overset{\displaystyle H}{|}}{C}=\overset{\overset{\displaystyle H}{|}}{\underset{\underset{\displaystyle H}{|}}{C}}-COO^- + 2H^+$$

1 g-mol = 126 g
1 g-equiv = 63 g

The gram-equivalent weight of fumaric acid, therefore, is half the molecular weight.

To make a 0.50 N solution of fumaric acid, one would weigh 0.50 gram-equivalent of fumaric acid and dissolve in enough water to make 1.00 L of solution. The molecular weight of fumaric acid is 126 g; the equivalent weight is 63 g. For a 0.50 N solution, 0.50 gram-equivalent weight × 63 g/g-equiv = 31.5 g needed to make 1.00 L of 0.50 N solution.

The gram-equivalent weight of an oxidizing or a reducing agent in an oxidation–reduction (redox) reaction is the amount of compound that will transfer one mole of electrons. In the reduction of nicotinamide adenine dinucleotide (NAD^+), two moles of electrons are transferred per mole of NAD^+. Therefore the gram-equivalent weight of NAD^+ is half its molecular weight.

$$NAD^+ + 2e^- + H^+ \rightleftharpoons NADH$$

Expression of concentration as **percent-of-solute** is frequently used for liquids and for solids of undetermined molecular weight. Three types of percent-of-solute designations are used: volume per volume (v/v), weight per volume (w/v), and weight per weight (w/w). For example, a solution with 5.5 mL of liquid solute in 100 mL of solution would be expressed as a 5.5% (v/v) solution, and 5.5 g of solid solute in 100 g of solution would be expressed as a 5.5% (w/w) solution. The most commonly used expression is weight per volume, in which 5.5 g of solid solute diluted to 100 mL with solvent is a 5.5% (w/v) solution.

2.6 MAKING DILUTIONS

Experimental conditions frequently require the dilution of an extract or a standard solution. There are several conventions for indicating the method of dilution. For example, "1-to-5 dilution" can indicate two different cases. First, one part of the original solution could be diluted with four parts of solvent to give a final diluted volume of five parts. This dilution is $\frac{1}{5}$, since the concentration of

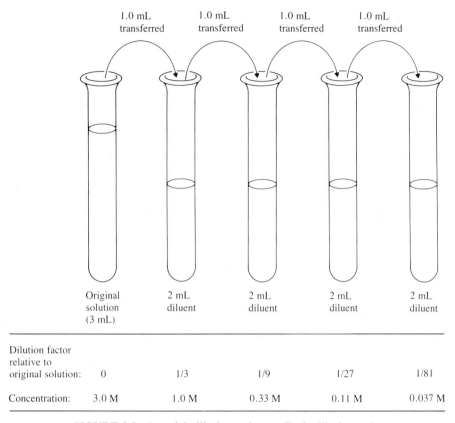

FIGURE 2-8. A serial dilution scheme. Each dilution tube contains an equal volume of diluting solution, and an equal volume is removed from each tube to make the next dilution.

the diluted solution is $\frac{1}{5}$ of the concentration of the original solution. However, "1-to-5 dilution" also could indicate that one part of the original solution is diluted with five parts of solvent to give a solution with $\frac{1}{6}$ of the original concentration. The former convention appears to be more suitable and is more frequently used because the dilution factor is immediately apparent. In this case, it is best to read "dilute 1:5" as "dilute one part solution with solvent to give five parts of total volume."

Serial dilutions involve the systematic dilution of an original solution in fixed steps, such as 1:2, 1:4, 1:8, 1:16 or 1:10, 1:100, 1:1000. This type of dilution scheme, illustrated in Figure 2-8, is frequently used in immunology and microbiology. Serial dilutions require careful attention and accuracy of measurement because errors made in an early part of the process will be carried to all subsequent dilutions.

2.7 STANDARD SOLUTIONS

Many experiments require the use of a standard solution containing a precisely measured amount of a pure substance. For example, in the quantitative determination of amino acids in a protein hydrolyzate, a solution of each pure amino acid is reacted with ninhydrin and the amount of color developed is measured and used for comparison. The standard solutions are prepared by accurately weighing each anhydrous amino acid and adding an accurately measured volume of solvent. Care must be taken to control all variables, such as temperature, state of hydration of the solute, and pH. The accuracy of prepared standards should be at least 1%. Sometimes standard solutions of a single amino acid or a single carbohydrate are required at different concentrations. These can be prepared by making a stock solution from which dilutions are made. For example, if standards containing 1.00, 5.00, 10.0, 50.0, and 100 mg/mL are required, one could prepare a stock solution containing 100 mg/mL. Then a 1:100 dilution would give 1.00 mg/mL, a 1:20 dilution would give 5.00 mg/mL, a 1:10 dilution would give 10.0 mg/mL, and a 1:2 dilution would give 50.0 mg/mL.

2.8 pH AND BUFFERS

2.8.1 pH

The negative logarithm of the hydrogen ion concentration, the pH, is expressed as follows:

$$pH = -\log_{10}[H^+] \tag{2-1}$$

The pH scale is a measure of hydrogen ion concentration that eliminates dealing with large powers of 10 and compresses a large range of concentrations onto a more convenient scale, between 1 and 14 (Fig. 2-9). At a high concentration of H^+ (10^{-1} M), the pH value is low, pH $= 1$, while at low concentration (10^{-12} M), the pH is high, pH $= 12$.

Hydrogen ions in solution arise from the dissociation of acids.

$$HA \rightarrow H^+ + A^- \tag{2-2}$$

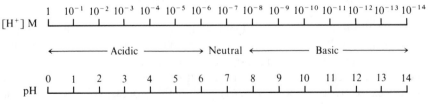

FIGURE 2-9. The pH scale.

Strong acids are considered to be completely dissociated into ions in dilute solution. However, weak acids (or bases) are only partially dissociated in solution, and thus an equilibrium is established between the ions and the undissociated molecules.

$$HA \rightleftharpoons H^+ + A^- \qquad (2\text{-}3)$$

The equilibrium constant K_a', is defined as follows.

$$K_a' = \frac{[H^+][A^-]}{[HA]} \qquad (2\text{-}4)$$

where HA = undissociated acid, H^+ = hydronium ion, and A^- = conjugate base. From this expression, we can derive the **Henderson—Hasselbalch equation**, commonly called the buffer equation, which relates the pH of solution to the pK_a' of the acid and the relative concentrations of the undissociated acid and the conjugate base forms.

Solving first for the hydrogen ion concentration from Equation 2-4:

$$[H^+] = K_a' \frac{[HA]}{[A^-]} \qquad (2\text{-}5)$$

If this equation is now converted to the logarithmic form, we obtain:

$$\log [H^+] = \log K_a' + \log \frac{[HA]}{[A^-]} \qquad (2\text{-}6)$$

Multiplying each term by -1 gives:

$$-\log [H^+] = -\log K_a' - \log \frac{[HA]}{[A^-]} \qquad (2\text{-}7)$$

Since $pH = -\log [H^+]$ and $pK_a' = -\log K_a'$, by substitution:

$$pH = pK_a' - \log \frac{[HA]}{[A^-]} \qquad (2\text{-}8)$$

By inverting the $\log [HA]/[A^-]$ term, we obtain:

$$pH = pK_a' + \log \frac{[A^-]}{[HA]} \qquad (2\text{-}9a)$$

$$pH = pK'_a + \log \frac{[\text{conjugate base}]}{[\text{acid}]} \qquad \text{(2-9b)}$$

This is the **Henderson—Hasselbalch equation**. It is most useful in the preparation of buffers and in understanding how the concentration of the acid and conjugate base forms of a weak acid affect the pH.

2.8.2 Buffers

A buffer, by definition, resists changes in the pH of the solution. A buffer must contain the chemical species for "neutralizing" added amounts of acid or base. Generally, a buffer is a solution of a weak acid and its conjugate base (e.g., acetic acid and sodium acetate) or a weak base and its conjugate acid (e.g., ammonia and ammonium chloride). A buffer is selected on the basis of its pK'_a value and its chemical nature.

The Henderson—Hasselbalch equation (Equation 2-9a) gives the relationship between pH, pK'_a, and the ratio of the concentration of the salt and acid forms of the buffer. As shown by this equation, when the concentrations of the conjugate base and the undissociated acid are equal, [conjugate base] = [acid], the pH of the solution equals the pK of the buffer.

$$pH = pK'_a + \log 1$$
$$\log 1 = 0$$
$$pH = pK'_a$$

When [conjugate base] = $10 \times$ [acid], then:

$$\log \frac{[\text{conjugate base}]}{[\text{acid}]} = \log 10 = 1$$

and

$$pH = pK'_a + 1$$

When [conjugate base] = $\frac{1}{10}$ [acid], then:

$$\log \frac{[\text{conjugate base}]}{[\text{acid}]} = \log \tfrac{1}{10} = -1$$

and

$$pH = pK'_a - 1$$

Thus, buffers are most effective in the range pH = $pK'_a \pm 1$. Outside that range, the concentration of either the acid or the conjugate base is too small to effectively resist the effect of added hydrogen or hydroxide ion.

Once the desired pH range for an experiment has been decided, one can select a buffer on the basis of pK value. From the list of common biological buffers given in Table 2-1, eight buffers could serve to control the pH in the range of pH $= 7$. Since temperature affects the dissociation of some weak acids and bases and thus the pK value, buffers should always be pH-adjusted for the temperature at which they will be used. Table 2-1 shows the effect of temperature on some important buffers.

For example, from Table 2-1, the widely used buffer Tris [tris(hydroxymethyl)aminomethane] shows a change in pK_a' of -0.031 for each degree change in temperature from the reference temperature of 20°C (Δp$K_a'/$°C). Therefore at 10°C the change in p$K_a' = (10$°C $- 20$°C$)(-0.031) = 0.31$; the pK_a' at 10°C is $8.3 + 0.31 = 8.6$. Likewise, at 30°C, the change in p$K_a' = (30$°C $- 20$°C$)(-0.031) = -0.31$, and the pK_a' at 30°C is 8.0.

2.8.3 Buffer Capacity and Ionic Strength

The molar strength of the two components of a buffer should be chosen to give adequate buffering capacity. The total ionic strength of the buffer may also be an important consideration–for example, in enzyme studies. Buffer strength is expressed in terms of the total concentration of conjugate base or conjugate acid. Again let us use Tris as an example.

$$CH_2OH \qquad\qquad CH_2OH$$
$$\text{HOCH}_2\text{—C—NH}_3^+ \rightleftharpoons \text{HOCH}_2\text{—C—NH}_2 + H^+ \qquad pK_a' = 8.3$$
$$CH_2OH \qquad\qquad CH_2OH$$
$$\textbf{(I)} \qquad\qquad\qquad \textbf{(II)}$$

The pK_a' for Tris is 8.3 at 20°C; therefore this substance can effectively serve as a buffer between pH 7.3 and 9.3. To make a 50-mM Tris buffer with pH $= 7.8$, start with either the free base (**II**) or the hydrochloride salt, Tris–HCl. A 50-mM solution contains 50 mmol of solute per liter of solution. Tris, the free base form, has a molecular weight of 121.1 g. The amount of solid Tris needed to make one liter of 50 mM solution is:

$$50 \text{ mmol} \times 121.1 \text{ mg/mmol} = 6055 \text{ mg Tris}$$

However, the pH of the buffer would have to be adjusted to 7.8 by adding acid; hydrochloric acid would be the best choice because Tris–HCl is soluble and chloride ion is compatible with most biological systems.

Since the desired pH is 0.5 unit below the pK_a', the Henderson—Hasselbalch equation (2-9) can be used to determine the amount of HCl needed to bring the pH down to 7.8.

TABLE 2-1. Properties of some common biological buffers

Name	Structure	pK'_a at 20°C	$\Delta pK'_a/°C$
Phosphoric acid	H_3PO_4	2.12	—
Citric acid	$HOOCCH_2COH(COOH)CH_2COOH$	3.06	—
Formic acid	$HCOOH$	3.75	—
Succinic acid	$HOOCCH_2CH_2COOH$	4.19	—
Sodium dihydrogen citrate	$HOOCCH_2COH(COONa)CH_2COOH$	4.74	—
Acetic acid	CH_3COOH	4.75	—
Disodium hydrogen citrate	$HOOCCH_2CHOH(COONa)CH_2COONa$	5.40	—
Sodium hydrogen succinate	$HOOCCH_2CH_2COONa$	5.48	—
MES [2-(N-morpholino)ethanesulfonic acid][a]	$O\!\!\diamond\!\!\overset{+}{N}HCH_2CH_2SO_3^-$	6.15	−0.011
ADA [N-(2-acetamido)iminodiacetic acid][a]	$H_2NCOCH_2\overset{+}{N}H(CH_2COO^-)CH_2COONa$	6.6	−0.011
PIPES [piperazine-N,N'-bis(2-ethanesulfonic acid)][a]	$NaO_3SCH_2CH_2\overset{+}{N}\!\!\diamond\!\!\overset{+}{N}HCH_2CH_2SO_3^-$	6.8	−0.0085
Imidazole hydrochloride	$\begin{array}{c}CHNHCl\\ \parallel \quad CH\\ CHNH\end{array}$	7.00	—
BES [N,N'-bis(2-hydroxyethyl)-2-aminoethanesulfonic acid][a]	$(HOCH_2CH_2)_2\overset{+}{N}HCH_2CH_2SO_3^-$	7.1	−0.027
MOPS [3-(N-morpholino)propanesulfonic acid][a]	$O\!\!\diamond\!\!\overset{+}{N}HCH_2CH_2CH_2SO_3^-$	7.20	—
Sodium dihydrogen phosphate	NaH_2PO_4	7.21	—
HEPES [N-2-hydroxyethylpiperazine-N'-2-ethanesulfonic acid][a]	$HOCH_2CH_2\overset{+}{N}H\!\!\diamond\!\!NCH_2CH_2SO_3^-$	7.55	−0.014
Tricine [N-tris(hydroxymethyl)methylglycine][a]	$(HOCH_2)_3C\overset{+}{N}H_2CH_2CH_2COO^-$	8.15	−0.021
Glycine amide hydrochloride	$NH_2COCH_2NH_3Cl$	8.2	−0.029
Tris [tris(hydroxymethyl)aminomethane] hydrochloride	$(HOCH_2)_3C\overset{+}{N}H_3Cl$	8.3	−0.031
Bicine [N,N'-bis(hydroxyethyl)glycine][a]	$(HOCH_2CH_2)_2\overset{+}{N}HCH_2COO^-$	8.35	−0.018
Glycylglycine	$H_3\overset{+}{N}CH_2CONHCH_2COO^-$	8.4	−0.028
Boric acid	H_3BO_3	9.24	—
Disodium hydrogen phosphate	Na_2HPO_4	12.32	—

[a] Good's buffer.

$$\log \frac{[\text{conjugate base}]}{[\text{acid}]} = \text{pH} - pK_a' = 7.8 - 8.3$$

$$\log \frac{[\text{conjugate base}]}{[\text{acid}]} = -0.5$$

or

$$\log \frac{[\text{acid}]}{[\text{conjugate base}]} = 0.5 \quad \text{and} \quad \frac{[\text{acid}]}{[\text{conjugate base}]} = 3.16$$

Since we are starting with 50 mmol of the conjugate base form of Tris, the final ratio will be:

$$\frac{[\text{acid}]}{50 \text{ mM} - [\text{acid}]} = 3.16$$

$$[\text{acid}] = 158 \text{ mM} - 3.16 \, [\text{acid}]$$

$$4.16 \, [\text{acid}] = 158 \text{ mM}$$

$$[\text{acid}] = 37.98 \text{ mM} = 38 \text{ mM}$$

Thus, 50 mmol of Tris plus 38 mmol of hydrochloric acid diluted to one liter gives 50 mM buffer, pH 7.8, at 20°C. The final pH should be verified by use of a pH meter.

The chemical nature of a buffer is especially important in biological investigations. Solubility, stability, defined interactions with other ions in the system, light absorption, and potential inhibition or stimulation of the function of the biological system are all criteria to be considered in buffer selection. Good and co-workers[2] have designed a group of buffers especially for use with biological systems, with pK_a values between 6 and 9, and well-defined interactions with mineral cations. These zwitterionic buffers do not pass through biological membranes and do not complex cations such as Ca^{2+} and Mg^{2+}. All these buffers have a tertiary amine function that dissociates in the pH range of 6–9 (see Table 2-1).

2.9 pH MEASUREMENT

Since many biochemical reactions are pH dependent, accurate measurement and control of pH is important. Measurement of pH with a pH meter is a simple process, but careful control is necessary to avoid errors.

2.9.1 pH Meters

A pH meter is a potentiometer used to measure the H^+ concentration in solution (Fig. 2-10). An electric potential is measured that depends on the voltage difference between a reference electrode, usually calomel, and a glass electrode that is sensitive to H^+ concentration. The glass membrane acts as if it is selectively permeable to H^+ while other cations and anions are excluded. This permeability results in a potential across the membrane that is a linear function of the pH.

$$E = \text{constant} + \frac{2.303 \, RT}{F} \times \text{pH} \qquad (2\text{-}10)$$

The magnitude of the potential difference is measured with a voltmeter. The constant in Equation 2-10 depends on a number of factors and varies from electrode to electrode. Therefore, in the measurement of pH one must standardize the electrode and meter with solutions of known H^+ concentration. The calibration circuit of the pH meter allows the operator to adjust the pH reading of the meter to the pH of the standard buffer solution, thereby eliminating the contribution of the constant from the pH reading.[3] The measured potential is

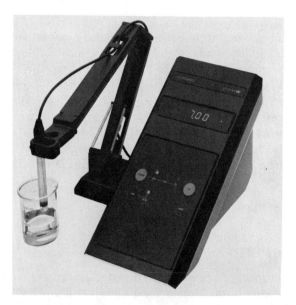

FIGURE 2-10. Digital pH meter with glass combination electrode. Courtesy of Corning Glass Works.

also a function of temperature as indicated in Equation 2-10. For accurate pH measurement, the temperature compensation control on the meter must be adjusted to the temperature of the solution being measured.

2.9.2 Making the pH Measurement

Before a pH meter is used, the electrodes must be rinsed with distilled water and allowed to drain. They can be blotted with a tissue. The meter should be standardized with two buffers, one with pH below and one with pH above the values to be measured, such as pH 4 and 7, or pH 7 and 10. The temperature of the standard buffers should be measured and the temperature compensation control adjusted to the temperature of the solution. When the meter has been standardized and readings are stable, the pH of the test solution can be measured. Meters with two separate electrodes, the glass and calomel electrodes, are used if the volume of solution is adequate. A combination electrode, in which both the reference electrode and the glass electrode are incorporated into one slim tube is useful for very small volumes. Errors in pH measurement may arise in the presence of high Na^+ concentration, at high ionic strength such as with ammonium sulfate solutions, and with electrodes fouled by proteins or other biopolymers. After use, electrodes should be rinsed with water and stored according to the manufacturer's directions.

2.10 LITERATURE CITED

1. S. E. Manahan, *Quantitative Chemical Analysis*. Brooks/Cole, Monterey, CA, 1986, Chapter 2.
2. N. E. Good et al., *Biochemistry*, *5*: 467 (1966).
3. D. C. Wharton and R. E. McCarty, *Experiments and Methods in Biochemistry*. Macmillan, New York, 1972, Chapter 6.

2.11 REFERENCES FOR FURTHER STUDY

1. S. E. Manahan, *Quantitative Chemical Analysis*. Brooks/Cole, Monterery, CA, 1986.
2. D. Pietrzyk and C. Frank, *Analytical Chemistry*, 2nd ed. Academic Press, New York, 1979.
3. R. Dilts, *Analytical Chemistry*. Van Nostrand, New York, 1974.

CHAPTER 3

SPECTROSCOPIC METHODS

Spectrophotometry and colorimetry are analytical methods of measuring the amount of light absorbed by a substance in solution. They are commonly used techniques for quantitatively determining substances encountered in biochemistry. All substances in solution absorb light of one wavelength and transmit light of other wavelengths. Absorbance is a characteristic of a substance just like melting point, boiling point, density, and solubility. But because it can be related to the amount of the substance in solution, absorbance can be used to quantitatively determine the amount of the substance in solution.

3.1 ELECTROMAGNETIC RADIATION AND SPECTRA

Light or electromagnetic radiation is composed of photons moving in a wave that oscillates along the path of motion. The **wavelength** of light is defined as the distance between adjacent peaks in the wave and can be further defined by the equation:

$$\lambda = \frac{c}{v} \tag{3-1}$$

where λ is the wavelength, c is the speed of light, and v is the frequency or number of waves passing a certain point per unit of time. Photons of different wavelengths have different energies that are given by:

$$E = h\frac{c}{\lambda} = hv \tag{3-2}$$

where h is Planck's constant. Thus, the shorter the wavelength, the greater the energy.

Electromagnetic radiation can be divided into various regions according to wavelength (Fig. 3-1): the ultraviolet region has wavelengths of 200–400 nm,

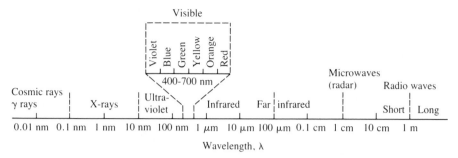

FIGURE 3-1. Spectral regions of electromagnetic radiation and their wavelengths.

the visible region has wavelengths of 400–700 nm, the infrared region has wavelengths of 700 nm to 500 μm, and the radiowave region has wavelengths of 1–5 m.

In the visible region, light of different wavelengths has different colors: violet and blue in the low wavelength region and orange and red in the high wavelength region. When a substance in solution appears blue, it means that the substance is absorbing red light and transmitting blue light. A substance that appears red is absorbing blue light and transmitting red light. A substance is said to have an absorption spectrum in the region in which it is absorbing light. Figure 3-2 shows absorption spectra for a blue substance and a red substance.

Molecules possess both kinetic energy and the energy associated with their bonding electrons. The absorption of light does not directly affect the kinetic energy, but it can affect the energy of the bonding electrons. The energy of the bonding electrons can be further divided into three types: electronic, vibrational, and rotational. Absorption of electromagnetic radiation involves a change in one or more of these bonding energies. Each of the electronic, vibrational, and rotational energy components can have only certain definite values or energy levels. A molecule whose electronic, vibrational, and rotational energies are all at their lowest values is said to be in the **ground state**. When a molecule is irradiated by photons whose energies just correspond to the difference in energy between the ground state and some higher or excited state of the molecule, the photons are absorbed and the molecule is raised to the higher energy level (Fig. 3-3).

Spectra arise when molecules absorb photons of specific energy. Transitions between different electronic levels give rise to spectra in the ultraviolet or visible regions; transitions between vibrational levels within the same electronic state give rise to spectra in the near-infrared region (700 nm to 200 μm); and transitions between rotational levels belonging to the same vibrational state give rise to spectra in the far-infrared region (greater than 200 μm).

Bonding electrons do not absorb electromagnetic radiation of very long wavelengths (radiowaves), but nuclei of specific kinds of atoms do and can give rise to nuclear magnetic resonance (NMR) spectra (see Section 3.6).

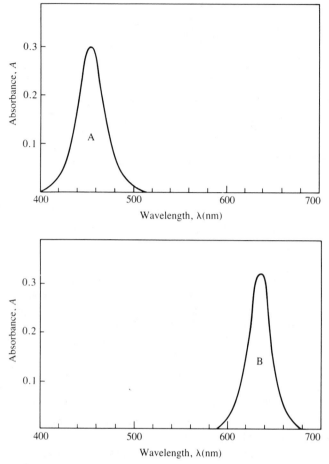

FIGURE 3-2. Absorption spectra for (*A*) a red-colored substance and (*B*) a blue-colored substance.

Theoretically, since only discrete packets of energy (specific wavelengths) are absorbed by the bonding electrons, the spectrum should consist of sharp lines. The existence of many possible vibrational levels of each electronic energy level, however, increases the number of possible transitions and gives rise to several distinct lines that together make up the broad peaks in the spectrum as shown in Figures 3-2 and 3-6.

Whether any electronic transitions in a molecule can be brought about by the energy of a photon in the visible or ultraviolet region depends on the nature of the bonding electrons in the ground state. In general, the absorption of visible and ultraviolet light by organic compounds can occur only when there is some unsaturation in the molecule. A specific grouping of atoms having unsaturation and absorbing light is called a **chromophore** or **chromophoric group**. The common

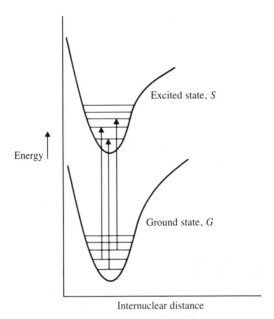

FIGURE 3-3. Energy-level diagram showing three transitions from the ground state G to the first excited electronic state S. The horizontal lines represent the permitted vibrational levels. In the transitions indicated, the bond lengths in the excited state are the same as in the ground state. The vertical, electronic transitions, however, can meet the upper, excited curve at different vibrational levels, giving a whole series of transitions, hence absorption lines that will result in the spectral absorption peak.

chromophoric groups include carbon–carbon double and triple bonds and the carbonyl, carboxyl, amido, azo, nitrile, nitroso, nitro, imidazole, indole, purine, and pyrimidine groups. Any molecule containing one or more such groups will have an absorption band somewhere in the visible or ultraviolet regions. Conjugation of the unsaturated bonds will also contribute to specific absorption bands. The overlapping molecular orbitals containing the electrons in a conjugated system give rise to delocalization of the electrons. The result is a decrease in the energy required for a transition and a spectral peak at higher wavelengths.

Once arrived at, an excited electronic, vibrational, or rotational state does not continue forever because the energy originally gained from photon absorption is lost by collisions with other molecules, such as solvent molecules. Eventually, the energy is transformed into kinetic or thermal energy and the molecule is re-turned to the ground state with the liberation of heat. Special arrangements of chromophoric groups can also result in the release of radiation from the energy of the absorbed photons, giving rise to the process known as **fluorescence** (see Section 3.4).

3.2 QUANTITATIVE ASPECTS OF LIGHT ABSORPTION

3.2.1 The Lambert–Beer Law

The amount of light passing through a substance is called **transmittance**, T, or **percent transmittance**, $\%T$, and is defined by the following equations:

$$T = \frac{I}{I_0} \tag{3-3}$$

$$\%T = \frac{I}{I_0} \times 100 \tag{3-4}$$

where I_0 is the intensity of the incident light and I is the intensity of the transmitted light. The amount of light of a specified wavelength absorbed by the substance depends on the length of the light path through the substance. The negative logarithm of the transmittance, the **absorbance** A, is directly proportional to the amount of light absorbed and to the length of the light path and is described by the **Lambert law**, which is expressed in Equation 3-5.

$$-\log T = -\log \frac{I}{I_0} = A = k_1 b \tag{3-5}$$

Here b is the length of the medium, usually a solution in a cell, and k_1 is a constant. A comparison of the scales for percent transmittance and absorbance (Fig. 3-4) may be used to convert percent transmittance into absorbance.

The negative logarithm of the transmittance is also directly proportional to the concentration of the absorbing substance c and is described by **Beer's law**, which is expressed in Equation 3-6.

$$-\log \frac{I}{I_0} = -\log T = A = k_2 c \tag{3-6}$$

Combining the two laws as the **Lambert–Beer law** gives Equation 3-7:

$$-\log \frac{I}{I_0} = -\log T = A = abc \tag{3-7}$$

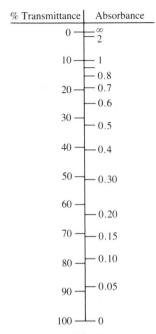

FIGURE 3-4. A comparison of the percent transmittance and absorbance scales.

where a is a constant called the **extinction coefficient** incorporating k_1 and k_2. The extinction coefficient is dependent on the wavelength of the light passing through the substance and on the chemical nature of the substance; b is the path length (cm); and c is the concentration of the substance.

3.2.2 Definition of Extinction Coefficients

The units of the extinction coefficient depend on the units of c and vice versa. Frequently a is expressed as the **molar extinction coefficient** (a_M), and c is then expressed in moles per liter; a may also be expressed as the millimolar (a_{mM}) or micromolar ($a_{\mu M}$) extinction coefficient, and the concentration is then expressed as millimolar or micromolar, respectively. The units for a_M are liters per mole per centimeter (L mol^{-1} cm^{-1}). When the molecular weight of a substance is not known, as frequently occurs in biochemical measurements of macromolecules such as proteins and nucleic acids, the extinction coefficient may be based on a weight/volume concentration such as a 1% solution (i.e., 10 mg/mL) and would then be designated $a_{1\%}$.

A substance with a molar extinction coefficient of 3782 L mol^{-1} cm^{-1} would give an absorbance of 3782 when its concentration was one mole per liter and the path length was one centimeter. Although an absorbance of 3782 could never be measured, the absorbance of a dilution of 1:10,000 or 1:20,000 of the 1 M solution could be accurately measured. Likewise, a substance with an $a_{1\%}$ of 0.555 would give an absorbance of 0.555 when its concentration was 1% or 10 mg/mL and the path length was 1 cm.

3.2.3 Determining an Ultraviolet or Visible Spectrum of a Substance and the Wavelength of Maximum Absorption

To use the Lambert–Beer law to determine the concentration of a substance, light of a specified wavelength must be chosen. A spectrum of the pure substance—that is, the absorbance of the substance as a function of the wavelength of the incident light—is necessary. This is most easily obtained using a recording spectrophotometer with a double beam, which automatically changes the wavelength of the incident light and records the absorbance. The double beam permits a correction to be made for the solvent blank or base line as a function of the wavelength. The solvent blank is placed into one of the beams (the reference beam), and the absorbing sample is placed into the other beam (the sample-beam).

An absorbance spectrum can also be obtained with a single beam instrument, but the process is tedious because the baseline blank must be determined for each wavelength and then the sample inserted and the absorbance determined for the sample at each wavelength. Furthermore, it is impractical to make these measurements for every wavelength, so the spectrum will be discontinuous.

The wavelength that gives the maximum absorption of light is used to obtain the most accurate measurements of the changes in absorbance with changes in concentration.

3.2.4 Determining an Extinction Coefficient and Concentration

If the extinction coefficient for a substance at the maximum absorbance is known and the path length is fixed (most frequently at 1.00 cm), the concentration of the substance can be determined. The extinction coefficient may be obtained from the literature or determined by measuring the absorbance at different concentrations of the substance. A plot of the absorbance versus concentration should give a linear curve whose slope is the extinction coefficient when the cell length is 1.00 cm (Fig. 3-5).

The measurement of the amount of light absorbed may be either as percent transmittance (%T) or as absorbance (A). Absorbance is used more often than percent transmittance because this variable is linear with the concentration of the absorbing substance, whereas percent transmittance is exponential. The side-by-side absorbance and percent transmittance scales (Fig. 3-4) show that when the amount of light absorbed is greater than 50% (>0.30 absorbance), errors become magnified. The measurement of concentration, therefore, is best achieved between 0.05 and 0.30 absorbance or between 90 and 50% transmittance. The errors in measuring absorbance values of 1 or 2 could be very large.

When the extinction coefficient is known and a fixed path length established, the concentration of an unknown amount of the substance can be determined by

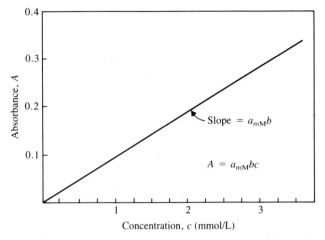

FIGURE 3-5. Plot of absorbance versus concentration.

measuring the absorbance of the substance and applying the Lambert–Beer law:

$$c = \frac{A}{ab} \qquad\qquad (3\text{-}8)$$

The limit of sensitivity of a spectrophotometric analysis is determined by the value of the extinction coefficient. The higher the value of the extinction coefficient, the lower is the concentration that may be measured. If the molar extinction coefficient is $10{,}000$ L mol^{-1} cm^{-1} and the minimum detectable absorbance is 0.01, then for a cell with a 1.00-cm path length, the minimum molar concentration that can be measured is 1.00×10^{-6} M.

$$a_M = 10{,}000 \text{ L mol}^{-1} \text{ cm}^{-1}$$
$$A = 0.01$$
$$b = 1 \text{ cm}$$
$$c = \frac{A}{ab} = \frac{0.01}{(10^4 \text{ L mol}^{-1} \text{ cm}^{-1})(1 \text{ cm})} = 10^{-6} \text{ mol/L}$$

If the molar extinction coefficient, however, were 10 times greater (i.e., $100{,}000$ L mol^{-1} cm^{-1}), the minimum molar concentration that could be measured would be 10 times lower, or 1.00×10^{-7} M.

Substances that have very high extinction coefficients give high absorbance values, usually a desirable characteristic, as indicated above. To obtain reliable absorbance values between 0.10 and 0.30, however, the experimental parameters must be modified. Either the sample must be diluted or, if it is not desirable to dilute the sample, the path length must be decreased. This may be conveniently done by placing a **spacer** into the 1.00-cm cell containing the substance. The spacer is a piece of glass or quartz whose width is precisely known (e.g., 9.90 mm or 9.98 mm). The path length then becomes 1.00 cm minus the width of the spacer, 0.010 cm or 0.002 cm respectively. If the concentration of a substance is too low, or if its extinction coefficient is low, a cell larger than 1.00 cm can be used (e.g., 5.0 cm or 10.0 cm) to obtain an accurate absorbance. Cells for the visible region may be made either of glass or quartz, but cells for the ultraviolet region must be of quartz because glass absorbs in the ultraviolet region.

A colorimeter permits the selection of the wavelength of the incident light by the use of a colored filter. The color of the filter employed also depends on the absorption maximum of the substance or the color of the substance. A blue filter would be used for a red substance, a red filter for a blue substance, and a blue or green filter for a yellow substance.

A colorimeter is used in the visible spectral region for routine analyses when a more expensive spectrophotometer is not required. A standard curve of absorbance or response versus concentration of the substance being determined is prepared, and the concentration of the unknown is determined from this curve. Such analyses are used in the clinical laboratory, especially with the automated clinical analyzers.

Many organic substances are not colored, hence do not absorb in the visible region of the electromagnetic spectrum. They may, however, absorb in the

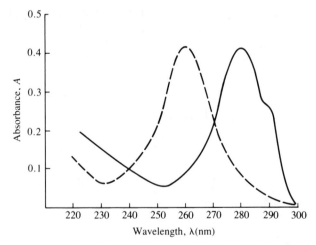

FIGURE 3-6. Ultraviolet spectra of protein (solid curve) and nucleic acid (dashed curve). The protein is serum albumin in water at pH 6 and the nucleic acid is bacterial DNA in water at pH 7.

ultraviolet region. Proteins, for example, have an absorption maximum around 280 nm due to their tyrosine and tryptophan amino acids, and nucleic acids absorb around 255–260 nm due to their purine and pyrimidine constituents (Fig. 3-6; see also Section 7.2.3.2). The coenzyme NADH absorbs at 340 nm.

Chemical reactions are sometimes run to generate a substance that has an absorption maximum in the visible or ultraviolet region so that the substance can be quantified. This may involve the addition of an indicator, an oxidation or reduction, a dehydration and condensation, or the addition of a complexing or chelating agent. The type of reaction is determined by the particular chemistry of the substance being measured (see Chapter 7).

3.2.5 Determining the Concentration of Two Substances in a Mixture

Spectrophotometric methods can be used to determine the concentrations of two or more absorbing substances in a mixture if the spectrum of the mixture is simply the sum of the spectra of the individual components. This results if the absorbing substances do not interact with each other when they are mixed. In most instances, the absorption curves will overlap to some extent, so the total absorbance, measured at the wavelength of maximum absorption of one component, will include the absorption of the other component. This is illustrated for a two-component mixture in Figure 3-7. At each wavelength of maximum absorption of the two components, the extinction coefficients of the two components should be sufficiently different from each other that one is high and the other low at λ_1 and vice versa at λ_2. The resulting absorbance at the two wavelengths is the additive absorbance of the two components. Hence the Lambert–Beer law may be applied, using the known extinction coefficients of

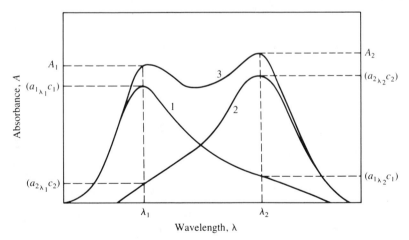

FIGURE 3-7. Absorption spectrum (3) of a mixture of two components; 1 and 2 are the individual spectra of the two components; $a_{1_{\lambda_1}}$ and $a_{2_{\lambda_1}}$ are the extinction coefficients of components 1 and 2 at λ_1, and $a_{1_{\lambda_2}}$ and $a_{2_{\lambda_2}}$ are the extinction coefficients of 1 and 2 at λ_2.

the two substances at the two wavelengths, to give two simultaneous equations 3-9 and 3-10:

$$A_1 = (a_{1_{\lambda_1}})(c_1) + (a_{2_{\lambda_1}})(c_2) \qquad (3\text{-}9)$$

$$A_2 = (a_{1_{\lambda_2}})(c_1) + (a_{2_{\lambda_2}})(c_2) \qquad (3\text{-}10)$$

where A_1 and A_2 are the absorbance values at wavelengths λ_1 and λ_2, respectively, for the mixture, $a_{1_{\lambda_1}}$ and $a_{2_{\lambda_1}}$ are the extinction coefficients for components 1 and 2 at λ_1, and $a_{1_{\lambda_2}}$ and $a_{2_{\lambda_2}}$ are the extinction coefficients for components 1 and 2 at λ_2. The unknown concentrations, c_1 and c_2, can be obtained by solution of the simultaneous equations. For a comparison of the spectra of the two separate components and in a mixture, see Figure 3-7.

The method is most accurate when component 1 is responsible for most of the absorption at λ_1 and component 2 is responsible for most of the absorption at λ_2. Furthermore, the concentrations of the two components in the mixture must be relatively equal. When, for example, the concentration of component 2 is low with respect to component 1, the differences in the absorbances of the two components at λ_2 will be small and the error in determining component 2 will be large (Fig. 3-8). A similar method for the simultaneous determination of two radioactive isotopes in a mixture is given in Chapter 6 (Section 6.9.3).

3.2.6 Difference Spectroscopy

In difference spectroscopy, the absorption spectra of two samples of slightly different composition or physical state are compared. Common features in the

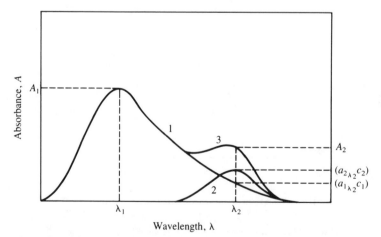

FIGURE 3-8. Spectra of two components in a mixture, with component 2 at a much lower concentration than component 1. Symbols are the same as in Figure 3-7.

spectra cancel, and the absorption bands that are recorded can be interpreted in terms of the known differences between the samples. Biochemists use difference spectroscopy to study the conformation of proteins and nucleic acids in solution. A protein or nucleic acid is examined under different solvent conditions of pH, temperature, concentration, organic solvent, and so on. The study of a spectrum of a protein or nucleic acid by difference spectroscopy under two different solvent conditions is called **solvent perturbation**. For example, helix–coil transitions of proteins can be studied as a function of temperature, pH, salt, and so on by measuring the difference spectra under different conditions of these parameters. In addition, the fraction of bases in nucleic acids that are not base-paired can be determined by measuring the difference spectra in H_2O and D_2O.

The techniques of difference spectroscopy consist of measuring the absorption of a test sample or model compound against a reference solution of the sample. The difference spectra are usually obtained with a recording double-beam spectrophotometer. Double compartment cells are also used to subtract the solvent contributions to the difference spectra. Because the difference spectra may often constitute only a few percent of their direct spectra, meaningful results can be obtained only when great care is taken with volumetric manipulations of the solutions and the instrumental measurements. Examples of the use of difference spectroscopy in the study of solvent perturbations, effects of pH, temperature, and denaturation of proteins are given in Reference 1.

3.3 ULTRAVIOLET AND VISIBLE SPECTROMETRY

3.3.1 Instrumentation

The amount of light that is absorbed by a substance may be measured by spectrophotometers and colorimeters. These instruments have several parts, which include a light source, a monochromator or colored filter to give a selected wavelength, a variable slit, a sample holder, a photodetector, and a meter. Different types of light sources are required for the different spectral regions. Tungsten and deuterium lamps are used for the visible (400–700 nm) and the ultraviolet (200–400 nm) regions, respectively. The wavelengths of light in these regions are selected by a monochromator in spectrophotometers and by colored filters in colorimeters. A monochromator is composed of prisms or diffraction gratings and gives light of a narrow band of wavelengths in both the visible and the ultraviolet regions. Colored filters give a relatively broad band of wavelengths exclusively in the visible region. The light then passes through an adjustable slit, which controls the intensity, and into the sample, from which it is detected and quantitated by a phototube or photomultiplier tube and measured by a galvanometer and/or recorder. A schematic diagram representing the parts of a spectrophotometer and a colorimeter is shown in Figure 3-9. Recording spectrophotometers are automated to change the wavelength continuously and record the amount of light transmitted at each wavelength.

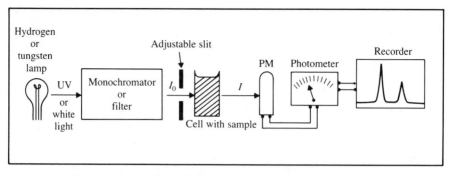

FIGURE 3-9. Schematic diagram of the parts of a spectrophotometer and colorimeter: I_0 is the incident light of wavelength selected by a monochromator (spectrophotometer) or
by a filter (colorimeter), and I is the transmitted light; PM is a
photomultiplier tube or phototube.

3.3.2 Applications of Ultraviolet and Visible Spectroscopy

The most common and obvious application of visible and ultraviolet spectroscopy is the determination of the concentration of a substance in solution if
the extinction coefficient is known and the Lambert–Beer law is obeyed. This
may be extended to the measurement of reactions if one of the reactants or
products of the reaction has an absorbance. The progress of the reaction can then
be determined by measuring the amount of loss of the reactant or the yield of
the product. This approach has been applied especially to enzyme-catalyzed
reactions to assay the effect of the enzyme (Section 9.5 discusses the assay of
enzymes).

An absorbance spectrum may be used, at least presumptively, to identify
substances in a sample. For example, proteins and nucleic acids have characteristic absorbances at 280 and 260 nm, respectively. A measurement of the ratio
of absorbance at 260 and 280 nm would give the relative amounts of nucleic acid
and protein in a sample (Section 7.2.3.2). Difference spectra can give the effects
of solvent changes, helix–coil transitions, protein–protein association and
dissociation, the number of hydrogen bonds, and the number of chromophoric
("colored") groups on the surface or buried in a protein or other macromolecule.

3.4 FLUORESCENCE PHOTOMETRY

3.4.1 Fluorescence

A specialized adaptation of spectrophotometry is fluorimetry, in which the
special property of fluorescence of some organic compounds is used. Many
organic substances will emit light of a longer wavelength after absorbing light of
a shorter wavelength. If there is no measurable time delay, the process is called

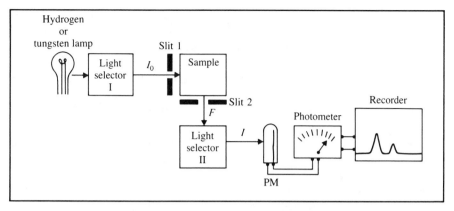

FIGURE 3-10. Schematic diagram of the parts of a fluorometer: light selectors I and II are monochromators or filters, I_0 is the selected wavelength of exciting light, F is the emitted fluorescent light, I is the selected wavelength of fluorescent light, and PM is a photomultiplier tube.

fluorescence. In a susceptible molecule, the absorption of light (usually ultraviolet or visible) of appropriate wavelength raises the molecule from its ground state to an excited state. Fluorescence results when the molecule returns to the ground state and energy is released as heat and as light of lower energy, hence of longer wavelength than the incident or exciting light. At room temperature, most fluorescing compounds emit light having a relatively wide band of wavelengths. Thus two kinds of light are involved, the **exciting light** and the **emitted light**, the former with lower wavelength than the latter.

3.4.2 Instrumentation

Instruments for measuring fluorescence range from simple fluorometers to more complicated spectrofluorometers.[2] Both contain four main parts: a light source and wavelength selector (filter or monochromator), a sample cell, a wavelength selector for the emitted light (filter or monochromator), and a photomultiplier tube to measure the amount of the emitted fluorescent light (Fig. 3-10). The incident light is selected for a specific exciting wavelength to pass through the sample, which produces the fluorescent light that passes out of the sample through a secondary wavelength selector, which may be set to select wavelengths of the maximum fluorescent emission. This fluorescent emission is detected by a photomultiplier tube. The tube transforms the light energy into electrical energy, which is measured by a meter.

The cells are the same as used in ultraviolet and visible spectrophotometry. Glass may be used if the exciting wavelengths are above 320 nm; fused silica or quartz must be used if the exciting wavelength is below 320 nm.

Fluorescence measurements are made by measuring the fluorescence of a series of various concentrations of the test substance to give a standard curve. To establish the true zero concentration, the experimenter determines the fluorescence of both the solvent and the cell.

The advantage of fluorescence photometry is that substances may be determined at very low concentrations because fluorescence is 10^3 to 10^4 times more sensitive than ultraviolet and visible spectrophotometric methods. Furthermore, the determination of two compounds in the same sample may be facilitated by the specificity that may be shown in the fluorescent process: for example, (a) the two compounds may absorb at identical wavelengths, but only one may fluoresce; (b) the compounds may absorb at the same wavelength but fluoresce at different wavelengths; or (c) the compounds may absorb and emit at the same wavelength, but the fluorescence of one may be selectively quenched so that the fluorescence of the other can be measured.

Fluorescence intensity and wavelength often vary with solvent, pH, presence of quenching agents, temperature, and intensity of the incident light. These variations can offer advantages and disadvantages. For example, the addition of quenching agents or changes in pH can eliminate or reduce the fluorescence, which means that one substance often can be selectively affected, hence preferentially determined. The disadvantage is that both the standards and the unknown material must be measured under the same conditions of solvent,

TABLE 3-1. Direct fluorimetry of some compounds of biological interest

Compound	Solvent	pH	λ_e	λ_f
Antimycin A	Water	8	350	420
Azoguanine	Water	7	285	405
Benzo(e)pyrene	Pentane	—	329	389
Brucine	Water	7	305	500
Codeine	Water	7	285	350
Epinephrine	Water	7	285	325
Estradiol	H_2SO_4–ethanol	—	285	325
Hydroxyamphetamine	Water	1	275	300
Indole acetic acid	Water	8	295	345
Lysergic acid diethylamide	Acid	—	325	445
Morphine	Water	7	285	350
Norepinephrine	Water	7	285	325
Phenobarbital	Water	13	265	440
Pyridoxal (vitamin B_6)	Water	12	310	365
Quinine	Water	1	250	450
Streptomycin	Water	13	366	445
Testosterone	H_2SO_4–ethanol	—	475	530
Vitamin A	1-Butanol	—	340	490
Warfarin	Methanol	—	290	325

Header spanning columns λ_e and λ_f: Wavelengths (nm)[a]

Source: Adapted from reference (1).
[a] λ_e = wavelength of excitation radiation; λ_f = wavelength of maximum fluorescence.

TABLE 3-2. Indirect fluorimetric methods for some compounds of biological interest

Compound	Reagent and reaction	Solvent or pH of aqueous solution	Wavelength (nm)[a]	
			λ_e	λ_f
Acetylsalicylic acid	Hydrolysis to salicylic acid	10	313	442
Alloxan	Derivatization with o-phenylenediamine	7	405	520
Amino acids	Reaction with dansyl chloride	7	335	578
Digitoxin	Heat with strong acid	Acid	350	490
Epinephrine	Oxidation to trihydroxyindole	3.5	436	540
Formaldehyde	Reaction with acetylacetone and ammonia	6	410	510
Glutathione	Condensation with o-phthaldehyde	8	343	425
Hydrocortisone	Heat with ethanol and concentrated H_2SO_4	Conc. acid	470	525
Malic acid	Reaction with resorcinol	8.5	490	530
Mescaline	Reaction with ammonia and formaldehyde	Acid	375	575
Morphine	Heat with concentrated H_2SO_4	10	365	420
Oxalic acid	Reduction followed by reaction with resorcinol	12	490	530
Sulfonamides	Reaction with 4,5-methylene-dioxyphthaldehyde	Acid	320	375
Tetracycline	Complex with Ca^{2+} and barbituric acid	8	405	530
Vitamin C	Reaction with o-phenylenediamine	9	350	430
Vitamin D	Reaction with acetic anhydride–H_2SO_4	Acid	390	470

Source: Adapted from Reference 1.
[a] λ_e = wavelength of excitation radiation; λ_f = wavelength of maximum fluorescence.

temperature, pH, and presence of quenching agents. For biochemical substances extracted from cells, these variables must be carefully considered and controlled to obtain an accurate determination of the concentration.

Fluorescence will usually occur with organic compounds containing aromatic rings or highly conjugated alkene systems such as occur in carotenes. For organic compounds that are nonfluorescent or weakly fluorescent, indirect methods may be used by converting the compound into a fluorescent derivative. Table 3-1 lists the solvent, pH, exciting wavelength, and emission wavelength for 19 fluorescent compounds of biochemical interest. Table 3-2 gives 16 fluorescent derivatives for determining substances by fluorescence.

3.5 INFRARED SPECTROSCOPY

3.5.1 Infrared Spectrophotometry

Infrared spectrophotometry is similar in many respects to ultraviolet and visible spectrophotometry. The differences involve the spectral region, which includes radiation at wavelengths between 700 nm and 500 μm. The spectral range of greatest use is the mid-infrared region, 2.5–50 μm. Infrared absorption spectra result from the absorption of infrared light by atoms in a molecule due to their twisting, bending, rotating, and vibrating motions within the molecule. These complex motions produce an absorption spectrum that is characteristic of the functional groups (e.g., methyl, methylene, carbonyl, amide) comprising the molecule and of the overall molecular configuration as well. Infrared spectrophotometry, thus, is most widely used for the qualitative identification of the functional groups present in a molecule. Furthermore, because the infrared spectrum of an organic molecule is unique, it may be used in the identification of the molecule itself.

Infrared spectra are conventionally plotted differently from UV–visible spectra in that the abscissa is in terms of wavenumbers instead of wavelengths. A wavenumber is a reciprocal wavelength and is expressed in reciprocal centimeters (cm^{-1}). In the relation:

$$\text{Wavenumber} = \frac{1}{\lambda} = \frac{v}{c} \qquad (3\text{-}11)$$

λ is the wavelength, v is the frequency (in s^{-1}), and c is the speed of light in a vacuum (3.0×10^{10} cm s^{-1}). In addition, the ordinate is most frequently plotted as percent transmittance rather than absorbance (Fig. 3-11).

The infrared spectrum consists of the absorption bands of the specific functional groups in the molecule. Because of the variety of different functional groups in organic compounds, infrared spectra, in contrast to UV–visible spectra, usually have a number of absorption bands (Fig. 3-11A). These absorption bands are modified by the interactions of the individual groups with the surrounding atoms in the molecule (compare Fig. 3-11A, B). This gives a spectrum that is unique for each particular compound. Thus, by a study of an infrared spectrum, it is possible to determine that certain functional groups are present or absent in a compound and how they are arranged.[3] Table 3-3 gives characteristic infrared absorption bands for some functional groups. For relatively simple organic compounds, such as glycine and glycylglycylglycine, the absorption bands are sharp, as indicated in Figure 3-11 (A, B). For larger, more complicated molecules (e.g., macromolecules), each bond type exists in large numbers and in many more configurations; thus each absorption band is shifted to an extent that depends on the location of the bonds in the molecule. As a result,

G620-1 Glycine

$H_2NCH_2CO_2H$ MW 75.07 mp 245°C (dec.)

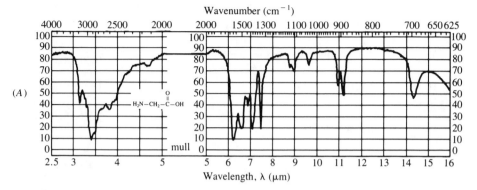

(A)

G820-4 Glycylglycylglycine

$H_2NCH_2CONHCH_2CONHCH_2CO_2H$ MW 189.17
mp 245°C (dec.)

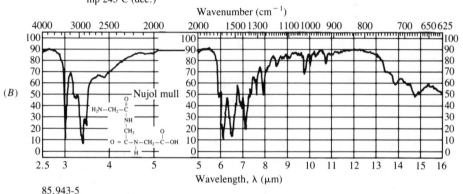

(B)

85.943-5
Polyglycine
mp >300°C Beil. 4(2),771

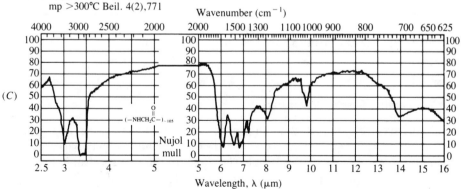

(C)

FIGURE 3-11. Infrared spectra of glycine and glycine peptides.
(A) Glycine. (B) Glycylglycylglycine (C) Polyglycine with an
average molecular weight of 6000. (Reprinted by permission
from *The Aldrich Library of Infrared Spectra*, 3rd ed., 1981.)

TABLE 3-3.   Infrared absorption bands for some common functional groups

Group	Compound	Wavenumber (cm^{-1})
C—H	Alkanes	3000–2850
		1580–1450
C—H	Alkenes	3100–2900
		1000–900
		700–525
C—H	Alkynes	3300–3200
		2200–2100
		700–600
C—H	Aromatic rings	3150–3000
		1630–1475
		775–675
C=C	Alkenes	1680–1640
C≡C	Alkynes	2260–2100
C—C	Aromatic rings	1600, 1500
—C—O	Alcohols, ethers, esters	1300–1100
C=O	Aldehydes, ketones, carboxylic acids, esters	1770–1690
O—H	Alcohols, phenols	3640–3610
N—H	Amines	3500–3300
C—N	Amines, amides	1360–1180
N—H	Amides	3500–3025
P=O	Phosphates	1130–1040
S—H	Sulfhydryls	2600–2400

the absorption bands overlap, and the spectrum appears to contain a few relatively broad absorption regions. Compare Figure 3-11A, the spectrum of glycine, and Figure 3-11C, the spectrum of polyglycine, a polymer of glycine. An extensive listing of absorption bands of functional groups was published by Colthup[4] in 1950 and has been widely reproduced.[5-7] Furthermore, an atlas of the infrared spectra of 12,000 organic compounds has been published.[8] By comparing the spectrum of an unknown with published spectra, presumptive evidence for the nature of the unknown often may be obtained.

3.5.2 Infrared Spectrophotometers

Infrared spectrophotometers are in principle the same as ultraviolet and visible spectrophotometers, although the source, the detector, and the materials used for the optical components are different. The standard infrared spectrophotometer is a filter–grating or prism–grating system with a range of 4000–650 cm^{-1} (2.5–15.4 μm). The instruments give high resolution that permits separation of close absorption bands. The source of the radiation is an object heated to 1200–1750°C, and a thermocouple (incandescent wires, silicon carbide rod, or rare earth oxides), instead of a photocell, is used as a detector. The instruments have a double beam.

3.5.3 Sample Handling for Infrared Spectroscopy

Aqueous solutions cannot be used in infrared spectroscopy because of the strong absorption by water. Also, glass and quartz, which are used for cells in UV and visible spectroscopy, are not transparent in the infrared region, hence are commonly replaced by sodium chloride windows. The light path is 0.1 to 1 mm, and spacers are used to vary the path length. If the sample is a liquid (other than an aqueous solution), a spectrum may be obtained directly by placing a drop of the liquid between two salt plates to give a layer 0.01 mm thick or less. Solid samples can be dissolved in carbon tetrachloride, which is useful between 4000 and 1330 cm^{-1}, or carbon disulfide, which is useful between 1300 and 625 cm^{-1}. Other solvents that can be used in selected regions are acetonitrile, benzene, chloroform, deuterium oxide, and dimethylformamide.

Solids that cannot be dissolved in one of these solvents can be suspended in a suitable medium to form a two-phase mixture called a **mull**. The sample is ground in a mortar and the mulling agent, a heavy hydrocarbon oil (Nujol), is added to give a thick paste. If the hydrocarbon bands interfere, a halogenated polymer, perfluorokerosene or a chlorofluorocarbon grease, is used. The spectrum is obtained from a thin film between salt plates.

In another type of procedure the finely ground sample is mixed with potassium bromide powder to form a transparent pellet, from which a spectrum is obtained. For polymers and natural product macromolecules, the samples may be dissolved in a volatile solvent and evaporated on a salt plate to give a thin film from which a spectrum can be obtained.

3.5.4 Applications of Infrared Spectroscopy

A major use of infrared spectroscopy (IRS) is the identification of functional groups in organic molecules, especially in the determination of the structure of biomolecules that have been isolated from living cells and in the confirmation of the structure of synthetic molecules prepared for the study of their interaction with living cells, enzymes, membranes, and so on. For example, IRS has been used to determine the structure of carbohydrates,[9] and phospholipids,[10] and amino acids, polypeptides, and proteins.[11] Differences in the structures of related polysaccharides have been determined using difference IRS.[12,13] This technique also has been used to determine the mode of binding of molecular oxygen to the iron in oxyhemoglobin,[14] as well as to measure the hydrogen bonding between the peptide carbonyl and the proton on the peptide nitrogen. Hydrogen–deuterium exchange studies on proteins by IRS can give information about the probability of exposure of peptide bonds to the solvent. Information on changes in conformation due to changes in random coil, α-helical, and β-pleated sheet structures can be obtained by hydrogen–deuterium exchange IRS studies. The ionization of carboxyl groups in proteins can be monitored using IRS. The ionized residues have a band (C=O stretching band) around 1570 cm^{-1} that changes to around 1710 cm^{-1} when protonated.

3.6 NUCLEAR MAGNETIC RESONANCE SPECTROMETRY

Nuclear magnetic resonance (NMR) spectrometry is another form of absorption spectrometry, similar to ultraviolet, visible, or infrared spectrometry. In NMR studies, a molecule is placed in a strong magnetic field and nuclei of certain types absorb electromagnetic radiation in the radio frequency region. Absorption of the radio waves is a function of the particular frequency of the radio waves, the strength of the magnetic field, the environment of the nucleus, and therefore the structure of the molecule.

All nuclei carry a positive charge, and for nuclei of some types, this charge "spins" on the nuclear axis. The angular momentum of the spinning charge can be described in terms of quantum spin numbers, I, which can have values of $0, \frac{1}{2}$, $1, \frac{3}{2}$, and so on. Each proton and neutron has its own spin corresponding to $\pm\frac{1}{2}$, and I is a resultant of these spins. If the sum of protons and neutrons in the nucleus is even, I is zero or integral $(1, 2, \ldots)$. When $I = 0$, the nucleus has no net spin and will not give an NMR signal. If the sum of protons and neutrons is odd, I is half-integral $(\frac{1}{2}, \frac{3}{2}, \frac{5}{2}, \ldots)$; the nucleus has a net spin and will give an NMR signal. Thus, ^{12}C and ^{18}O, both of which have a quantum spin number of $I = 0$, have no net spin and do not give NMR spectra; ^{1}H, ^{13}C, ^{19}F, and ^{31}P, however, all have the spin number of $I = \frac{1}{2}$ and a net nuclear spin, and will give NMR signals. We will be discussing the latter isotopes, which are the ones most frequently studied by biochemists using NMR spectrometry.

The spinning, charged nuclei generate a magnetic moment along the axis of spin, so that they act like tiny bar magnets. If these nuclear magnets are placed in an external magnetic field, their magnetic moments can be aligned either with or against the external magnetic field. The most stable alignment is *with* the external field. Energy must be absorbed to "flip" the nuclear magnet over into the higher energy alignment *against* the field (Fig. 3-12). It has been found that radio frequencies have sufficient energies to produce this "flip" of the nuclei.

When a substance containing $I = \frac{1}{2}$ nuclei is placed into a magnetic field of constant strength, some of the nuclei become aligned with the magnetic field. A spectrum can be obtained by passing radiation of steadily changing radio frequency through the substance and observing the frequency at which radiation is absorbed. This is called continuous wave operation. Other continuous wave instruments are designed with the radio frequency constant and the strength of the magnetic field variable. At some value of the field strength, the energy required to flip the nucleus matches the energy of the radiation; then some of the nuclei absorb the energy, and a signal is observed.

In instruments with Fourier transform capabilities, the magnetic field strength is fixed and the sample is subjected to a pulse of 5–50 microseconds duration with a wide range of radio frequencies. Some of the nuclei with net spin absorb the radio frequency energy necessary to flip to the higher energy state against the external magnetic field. After the pulse has passed, some of the nuclei flip back, returning to the lower energy state, and reemit the absorbed energy, which is detected by a radio receiver and recorded. The nuclei are said to resonate, and the result of detecting the emitted radiation is an NMR spectrum.

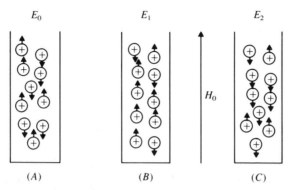

FIGURE 3-12. Representation of the alignment of nuclei of quantum spin number $I = \frac{1}{2}$ in a magnetic field H_0. (*A*) A sample of random nuclei alignment in the absence of an external magnetic field. (*B*) Alignment of some of the nuclei with the external magnetic field, giving an energy of E_1. (*C*) Alignment of some of the nuclei against the external magnetic field after absorption of the energy of the radio waves, giving an energy of E_2, in which $E_1 < E_2$. The energy necessary to "flip" the nuclei from alignment with the field to alignment against the field is $E_2 - E_1$.

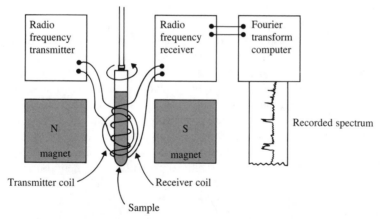

FIGURE 3-13. Schematic diagram of an NMR spectrometer in which a spinning sample is placed in a fixed magnetic field with a radio frequency transmitter to pulse with radio frequency waves and a receiver to detect the radio frequencies emitted by the resonating nuclei. The various radio frequencies that are absorbed and reemitted by the nuclei are stored and compiled by a computer to output the final spectrum.

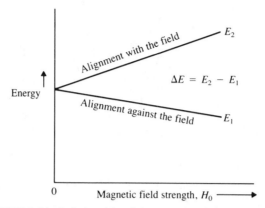

FIGURE 3-14. Relationship of the magnetic field strength, H_0, to the energy ($\Delta E = E_2 - E_1$) necessary to "flip" nuclei, with quantum spin number $I = \frac{1}{2}$, from alignment with the magnetic field to alignment against the field.

In Fourier transform nuclear magnetic resonance spectrometry (FT-NMR), all the resonant frequencies are monitored simultaneously, recorded, and stored in a computer. The pulse of radio frequencies is repeated many times and the emitted signals that are identical are added together, resulting in an enhanced spectrum with a high signal-to-noise ratio.

The fixed magnetic field NMR instrument shown schematically in Figure 3-13 has a strong magnet with a homogeneous field, a radio frequency transmitter and receiver, and a computer to store, compile, and integrate the signals. A sample holder positions the sample relative to the magnetic field, the transmitter coil, and the receiver coil, and spins the sample to increase the probability of the sample being continuously exposed to a homogeneous magnetic field. The sample holder may also have a variable temperature control. The amount of energy required to cause a particular nucleus to resonate depends on factors such as the external field strength (Fig. 3-14), the electronic configuration around the particular nucleus, the type of molecule, and the intra- and intermolecular interactions. NMR spectra, thus, can answer many questions about what particular nucleus is absorbing the energy, where it is located in the molecule, how many there are, and the location and type of neighboring atoms.

NMR instruments are built with different magnetic field strengths. They are listed according to the radio frequencies required for the proton to resonate. For example, a 60-megahertz (MHz) instrument has a magnetic field strength such that a proton will resonate when it absorbs a radio frequency of 60 MHz. A 300-MHz instrument has a stronger magnetic field: a proton must absorb a radio frequency of 300 MHz to resonate.

NMR instruments can have magnetic fields that require protons to absorb 60–600 MHz to resonate. Table 3-4 lists the characteristics of NMR instruments with different field strengths. Instruments with higher magnetic field strengths give better resolution of the peaks in the spectrum. At higher field strengths, the ratio of the chemical shifts to the coupling constant (see below) is increased,

TABLE 3-4. Field strengths and radio frequency resonances for NMR instruments in the 60–600 MHz range

NMR instrument designation[a]	Magnetic field strength (T)[b]	Resonance (MHz)		
		^{13}C	^{19}F	^{31}P
60 MHz	1.40	15.09	56.45	24.29
90 MHz	2.10	22.63	84.67	36.43
200 MHz	4.70	50.29	188.15	80.96
300 MHz	7.05	75.44	282.23	121.44
400 MHz	9.40	100.59	376.31	161.92
500 MHz	11.75	125.73	470.38	202.40
600 MHz	14.10	150.88	564.46	242.88

[a] NMR instruments are designated according to the radio frequency of proton resonance.
[b] A tesla (T) is the unit of magnetic field strength and is equal to one newton per coulomb-meter per second ($1 \text{ T} = 1 \text{ N C·m}^{-1} \text{s}^{-1}$).

spreading the chemical shifts in the NMR spectrum over a wider range. The higher field strengths also reduce complicating second-order spectrum effects and produce a gain in sensitivity.

3.6.1 Proton NMR Spectrometry

The theory and practice of nuclear magnetic resonance was developed from NMR studies of protons. Protons of different kinds, that is, protons attached to a molecule at different atoms and in different molecular environments give different signals. The relative positions of the signals in the spectrum tell something about the electronic environment of the protons in the molecule. The intensities of the signals tell how many protons of each kind there are. The splitting of a signal into several signals (peaks) gives information about the environment of the proton and nearby protons.

In any given molecule, protons in the same environment give identical signals, and protons in different environments give different signals. Identical protons are structurally equivalent. If the NMR spectra of methane and ethane are obtained, we find that although the signals come at different positions, each compound gives only one.

$$CH_4 \qquad CH_3—CH_3$$

Methane   Ethane

Methane has four equivalent protons and ethane has six equivalent protons. If we obtain the NMR spectrum of propane, we find two signals of two types: one from the six methyl protons (a) and one from the two methylene protons (b). The ratio of the intensities of the two signals, a and b, is 3:1.

$$\overset{a}{CH_3}—\overset{b}{CH_2}—\overset{a}{CH_3}$$

Propane

Replacing one of the protons of propane with a substituent produces further changes. For example, 1-chloropropane gives three signals with intensities in the ratio of 3:2:2 because it has three types of protons (*a*, *b*, and *c*), and 2-chloropropane gives two signals in the ratio of 6:1 because it has two different protons (*a* and *b*) in this ratio.

$$
\overset{a}{C}H_3-\overset{b}{C}H_2-\overset{c}{C}H_2-Cl \qquad\qquad \overset{a}{C}H_3-\underset{b}{\overset{Cl}{C}H}-\overset{a}{C}H_3
$$

1-Chloropropane   2-Chloropropane

To be chemically equivalent, the protons must also be sterically equivalent, as illustrated by the following examples in which the sterically different protons are indicated by different letters.

$$
\begin{array}{cccc}
\overset{a}{CH_3}\ \ \overset{b}{H} & \overset{a}{CH_3}\ \ \overset{d}{H} & \overset{a}{CH_3}\ \ \overset{a}{CH_3} & \overset{a}{H}\ \ \overset{b}{H} \\
\ \diagdown\ \diagup\ & \ \diagdown\ \diagup\ & \ \diagdown\ \diagup\ & \ \diagdown\ \diagup\ \\
C{=}C & C{=}C & C{=}C & C{=}C \\
\diagup\ \ \diagdown & \diagup\ \ \diagdown & \diagup\ \ \diagdown & \diagup\ \ \diagdown \\
\underset{a}{CH_3}\ \ \underset{b}{H} & \underset{b}{CH_3}\ \ \underset{c}{CH_3} & \underset{b}{H}\ \ \underset{b}{H} & \underset{d}{CH_3}\ \ \underset{c}{H}
\end{array}
$$

2 NMR signals   4 NMR signals   2 NMR signals   4 NMR signals

The position of the signals in the spectrum indicates the types of proton (primary, secondary, tertiary, aromatic, benzylic, acetylenic, vinylic, etc.) in the molecule. These protons of different kinds have different electronic environments, which determine the number and location of the signals generated.

The inductive effect plays an important role in the positions of absorption. A more **shielded proton** is one surrounded by a relatively greater electron density than another proton and absorbs radio frequencies of higher energy or **upfield** from that proton. A less shielded, or **deshielded** proton has relatively less electron density around it than another proton and absorbs radio frequencies of lower energies or **downfield** from that proton. For example, the proton attached to the carbonyl carbon of acetaldehyde (CH_3—C$\underline{H}$O) is more deshielded, due to the electron-withdrawing properties of the carbonyl oxygen, than the protons of the methyl group of an alkane (R—C$\underline{H}_3$), which are surrounded by a higher electron density since there is no electron-withdrawing group. The carbonyl proton absorbs downfield from the methyl protons. These shifts in the NMR signals are termed **chemical shifts** and are expressed in parts per million (ppm) of the total magnetic field. When the magnetic field is 60 MHz, 1.0 ppm equals 60 Hz. For practical purposes, the reference point from which chemical shifts are measured is the signal obtained from the 12 protons of tetramethylsilane (TMS), $(CH_3)_4Si$, which is given the value of 0.0 ppm. This is called the delta (δ) scale and is the scale most commonly used to measure chemical shifts. Because of the low electronegativity of silicon, the shielding of the protons of TMS is greater than in most organic molecules, and therefore the TMS is upfield from the signals of organic protons. Most proton chemical shifts have δ values between 0 and 15 ppm. The δ scale is used instead of frequency units because δ values are relative and thus independent of the magnetic field strength.

The NMR signal for a proton is split if the proton has neighboring protons on adjacent carbons that are nonequivalent to it. For example, in ethane, the six protons are all equivalent and do not split each other. But, with propane there are two types of protons that split the signals of each other—the methyl protons split the methylene protons and the methylene protons split the methyl protons. The splitting of signals is called **spin–spin splitting**, and protons splitting the signals are called **coupled protons**. The number of spin–spin splitting peaks can be predicted by applying the **$n + 1$ rule**: that is, by counting the number n of nonequivalent adjacent protons and adding 1.

$$CH_3—CH_3 \qquad\qquad CH_3—CH_2—CH_3$$

Ethane: equivalent protons   Propane: nonequivalent protons
No splitting   Splitting

Thus the signal from the methyl protons of propane is split into three peaks because there are two methylene protons $(2 + 1 = 3)$ and the signal from the methylene protons is split into seven peaks because there are six methyl protons $(6 + 1 = 7)$. If a proton has no adjacent protons, it gives a single peak; if there is one nonequivalent, adjacent proton, the signal is split into two peaks (a doublet); and if there are two nonequivalent, adjacent protons, the signal is split into three peaks (a triplet), and so on.

The separation between any two peaks of a split signal, called the **coupling constant** (J), varies with the environment of the protons and their geometrical relationship. The existence of coupling constants provides a method of determining which protons are coupled together in a complex spectrum. If the J values are the same, the protons may be coupled. If they are different, the protons in question are not coupled but are being split by some other proton(s). The coupling constants for H_a split by H_b and for H_b split by H_a are both indicated by J_{ab} because the J value is the same for each of the two doublets.

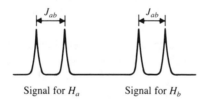

Signal for H_a Signal for H_b

For propane, the distance between the two adjacent peaks in the triplet for the methyl protons is J_{ab} and the distance between each of the peaks in the methylene septet also is J_{ab}. The total width of the triplet is $2J_{ab}$ and the total width of the septet is $6J_{ab}$.

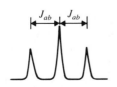 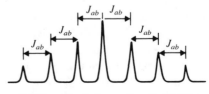

Splitting of methyl protons in propane   Splitting of methylene protons in propane

The proton chemical shifts for groups that can form hydrogen bonds such as —OH or —NH$_2$ are temperature dependent, unlike the chemical shifts for CH protons. At higher temperatures, where hydrogen bonding is low, the chemical shifts are upfield, and at lower temperatures, where hydrogen bonds are favored, the proton absorptions are shifted downfield as much as 4–5 ppm. Chemical shifts of the hydrogen bonding groups can also be affected by the solvent, depending on whether the solvent is hydrogen bonding or non-hydrogen bonding. Concentration effects can also be important. Intermolecular hydrogen bonding is favored by high concentrations, giving downfield chemical shifts; intramolecular hydrogen bonding is concentration independent, but will still display downfield chemical shifts.

A proton NMR spectrum is obtained on a solution of the sample in a suitable solvent, which usually contains no protons (e.g., carbon disulfide or deuterated chloroform); for water-soluble compounds, D$_2$O may be used. A small amount of tetramethylsilane is added as an internal standard, except when D$_2$O is used, in which TMS is insoluble. The usual NMR tubes hold about 0.5 mL of solution. The concentration of the sample is usually between 10 and 30% w/v.

In summary, the number of absorption signals in a proton NMR spectrum indicates the number of distinct types of protons in the sample; the position of the signals in the spectrum indicates the type of proton; the area under the signal peak indicates the number of protons of that type; and the splitting of the signal indicates the nature of the neighboring protons.

3.6.2 Carbon-13 NMR Spectrometry

Direct observations of the carbon skeleton of a molecule by ^{13}C NMR spectrometry became possible in the early 1970s. The ^{13}C nucleus, like the ^{1}H nucleus, has a quantum spin number of $\frac{1}{2}$. The natural abundance of ^{13}C, however, is only 1.1% that of the common isotope, ^{12}C, giving an NMR sensitivity much less than that of ^{1}H. The development of Fourier transform instrumentation has greatly facilitated the obtaining of a ^{13}C NMR spectrum. The study of ^{13}C NMR spectra can be further aided by the use of ^{13}C-enriched compounds, which are becoming increasingly available for biochemical studies.

If each of the carbons in a molecule is distinct, there will appear a ^{13}C NMR signal for each carbon. These signals will be split by neighboring protons, giving multiple peaks. This complication can be eliminated by decoupling the protons by irradiating them at their resonating frequency. For example, sucrose, which has 12 distinct carbons, gives 12 distinct ^{13}C NMR peaks in a proton-decoupled spectrum (Fig. 3-15). The presence of equivalent carbon atoms in a molecule will give a signal at the same position, and there will be a reduction in the number of peaks compared with the actual number of carbon atoms. Dioxane, a molecule with four equivalent carbon atoms gives a single ^{13}C NMR peak. The ^{13}C NMR spectrum of t-butyl alcohol has two peaks, one for the methyl carbons and another for the tertiary carbon. The methyl carbon peak is much larger than the tertiary carbon peak, but not three times as large, as would be the case if the area of the peak were proportional to the number of carbons contributing to the signal.

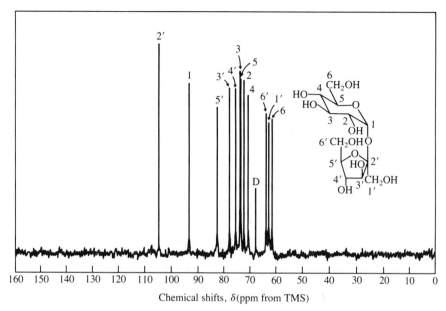

FIGURE 3-15. Identification of 12 peaks from the 12 carbon atoms of sucrose in the ^{13}C proton-decoupled NMR spectrum: D = a ^{13}C dioxane peak. The chemical shifts are given in ppm from tetramethylsilane.

A number of complications prevent the resulting peak areas in a ^{13}C NMR spectrum from integrating to give the correct number of carbon atoms contributing to each peak as is observed for ^{1}H NMR signals. Nevertheless, a comparison of peak areas or peak heights for the same type of carbon in two different samples of identical concentration can give semiquantitative results (Fig. 3-16). The use of carbon isotopes has greatly increased the scope of nuclear magnetic resonance techniques in the study of molecular structure. This is because ^{13}C has a wide range of chemical shifts, is very sensitive to structural changes, and gives a relatively simple spectrum. The simple spectrum results from the low abundance of ^{13}C in the sample, giving molecules that only rarely have adjacent ^{13}C atoms and thus do not display ^{13}C–^{13}C coupling.

3.6.3 NMR Spectra of Other Nuclei

The NMR spectra of other nuclei (e.g., ^{19}F and ^{31}P) can also be studied. Even though both these isotopes have a natural abundance of 100%, they are usually present in limited numbers in biochemical compounds, and the use of Fourier transform instrumentation is necessary to obtain their NMR spectra. Because the number of ^{19}F or ^{31}P nuclei in a sample is limited, these isotopes give relatively simple spectra compared with ^{1}H and ^{13}C NMR spectra. Figure 3-17 shows the ^{19}F NMR spectrum for 6,6'-dideoxy-6,6'-difluorosucrose. Because of the unique

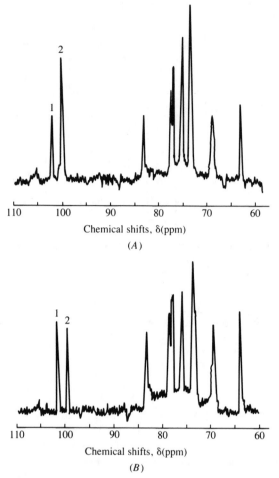

FIGURE 3-16. Comparison of ^{13}C proton-decoupled NMR spectra of two different α-glucans with different degrees of α-1→3 branch linkages. Peak 1 is due to C_1 linked to C_3 (the branch linkage) and peak 2 is due to C_1 linked to C_6 (the main chain linkage). Spectrum *A* is of a glucan sample with a lower degree of α-1→3 branch linkages than the sample in spectrum *B*. The height and area of peak 1 in spectrum *B* is greater than in spectrum *A*. [From G. L. Côté and J. F. Robyt, *Carbohydr. Res. 127*: 95 (1984).]

types of peaks obtained, this ^{19}F NMR spectrum confirms the structure of 6,6′-dideoxy-6,6′-difluorosucrose. The ^{31}P spectrum of ATP (Fig. 3-18) has three signals for the α-, β-, and γ-phosphorus atoms with different chemical shifts and splittings.

As previously indicated, in a 300-MHz instrument, protons will resonate when they absorb radio frequencies of 300 MHz; other nuclei, however, will resonate when they absorb radio frequencies of different energies. For example,

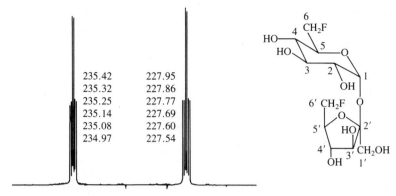

235.42	227.95
235.32	227.86
235.25	227.77
235.14	227.69
235.08	227.60
234.97	227.54

Chemical shifts, δ(ppm from $CFCl_3$)

FIGURE 3-17. The ^{19}F NMR spectrum of 6,6'-dideoxy-6,6'-difluorosucrose has two distinct sextets centered at 235 and 227 ppm due to signals from a fluorine atom on C_6 coupled to the two hydrogen atoms on C_6 giving a triplet, each of which is further split by H_5, resulting in the sextet. Similarly, the signals arising from fluorine on $C_{6'}$ is a triplet that is split by $H_{5'}$, giving a sextet. If fluorine substitution had occurred at $C_{1'}$, a unique triplet would have resulted, and because the requisite hydrogen on $C_{2'}$ is not present, it would not have been split into a sextet. [From J. N. Zikopoulous, S. H. Eklund, and J. F. Robyt, *Carbohydr. Res. 104*: 245 (1982).]

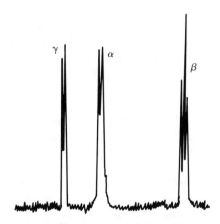

FIGURE 3-18. The ^{31}P spectrum of ATP. Three signals result from the α-, β-, and γ-phosphorus atoms with different chemical shifts and splittings characteristic of their respective positions in the structure.

in a 300-MHz instrument, ^{13}C resonates at 75 MHz, ^{19}F resonates at 282 MHz, and ^{31}P resonates at 121 MHz. Table 3-4 lists the resonant radio frequencies of the different isotopes in NMR instruments of different field strengths. Because of the differences in resonance frequencies for the various isotopes, different NMR instrument probes are required for each type of isotope. The probe consists of sample holder, radio frequency transmitter, and receiver appropriate for the specific isotope being studied. Each of the types of nucleus we have considered has chemical shifts in specific regions of the NMR spectrum: ^1H has chemical shifts between 0 and 20 ppm, ^{13}C has chemical shifts between 50 and 225 ppm, ^{19}F has chemical shifts between 0 and -200 ppm, and ^{31}P has chemical shifts between -20 and $+20$ ppm.

3.6.4 Uses of NMR in Biochemistry

A common use of NMR techniques has been the determination and confirmation of the structures of molecules isolated from living cells and their environment, and the determination of the structures of synthetic molecules that have been developed by biochemists to be used as enzyme–substrate analogues, inhibitors, hormones, drugs, and so on. NMR spectrometry also may be used in the measurement of a wide variety of ligands binding to enzymes, including metal ions, anions, coenzymes, and small organic molecules.[15] Highly reactive and unstable, covalent enzyme–substrate intermediates have been detected by a combination of cryogenic and NMR techniques.[16] Conformational effects due to the substitution of fluorouracil into tRNA have been studied by ^{19}F NMR.[17]

NMR of a variety of isotopes (^1H, ^2H, ^3H, ^{13}C, ^{15}N, and ^{31}P) has been studied directly in cells and organelles to identify and measure the *in vivo* concentrations of compounds involved in metabolic processes. This includes the measurement of intracellular hydrogen ion concentration (pH) by ^{19}F and ^{31}P NMR. These applications of NMR techniques are finding uses in the study of the development, growth, and metabolism of organs, tissues, and cancerous tumors. Metabolic pathways in live animals or humans can be studied by NMR by using surface-coil probes to detect compounds in localized regions near the surface of the animal. For example, the surface coil can be placed on the head of an anesthetized rat to obtain the ^{31}P spectrum of phosphorus-containing metabolites in the rat's brain.

Kinetics of enzyme-catalyzed reactions *in vitro* and *in vivo* can be measured by following the changes in the concentrations of substrate or products. Changes in hydrogen bonding and hydrophobic interactions of macromolecules can be determined, as well as the dynamics of molecular movement in membranes.[18]

A modified technique has emerged in which the specimen is deliberately placed in a nonuniform magnetic field, instead of a homogeneous magnetic field in accordance with the usual practice. This nonuniform magnetic field is produced by small magnetic field gradients along the x-, y-, and z-axes. Different parts of the specimen are thus localized by different NMR frequencies, and an NMR "image" of the specimen can be obtained. So far, the majority of the images are from ^1H frequencies, which essentially measure the distribution of mobile water in the specimen. From a series of these projections, the object can be reconstructed by a computer process known as **image reconstruction**. NMR imaging is harmless and painless and results in a "picture" of soft tissues that other methods cannot give.

New techniques and new instruments have allowed the development of two-dimensional NMR and solid-state NMR, which can be used in studying the three-dimensional structures of biopolymers such as proteins, polysaccharides, and nucleic acids. These approaches offer particular advantages for studying biopolymers that are not or cannot be crystallized. Two-dimensional NMR is especially useful in studying biopolymer structure in solution or in a lipid environment, and both two-dimensional and solid state NMR can be used to study biopolymer motions.

3.7 LITERATURE CITED

1. T. T. Herskovits, "Difference spectroscopy," in *Methods in Enzymology*, Vol. 11, (C.H.W. Hirs, Ed.) Academic Press, New York, (1967), pp. 748–775.

2. S. G. Schulman, *Fluorescence and Phosphorescence Spectroscopy: Physicochemical Principles and Practice*. Pergamon Press, New York, (1977).

3. J. E. Stewart, in *Interpretive Spectroscopy*, S. K. Freeman, Ed. Van Nostrand Reinhold: New York, (1965).

4. N. B. Colthup, "Spectra–structure correlations in the infra-red region," *J. Opt. Soc. Am. 40*: 397 (1950).

5. H. H. Willard, L. L. Merritt, Jr., and John A. Dean, *Instrumental Methods of Analysis*, 5th ed. Van Nostrand Reinhold, New York (1974), pp. 171–172.

6. R. P. Bauman, in L. Meites and H. C. Thomas, Eds., *Advanced Analytical Chemistry*. McGraw-Hill: New York, (1958), pp. 322–323.

7. R. L. Shriner, R. C. Fuson, D. Y. Curtin, and T. C. Morrill, *The Systematic Identification of Organic Compounds*, 6th ed. New York, (1980), p. 122.

8. C. J. Pouchert, *The Aldrich Library of Infrared Spectra*, 3rd ed. Aldrich Chemical Co., Milwaukee (1981).

9. H. Spedding, "Infrared spectroscopy and carbohydrate chemistry," *Adv. Carbohydr. Chem. 19*: 23 (1964).

10. A. Wright, M. Dankert, P. Fennessey, and P. W. Robbins, "Characterization of a polyisoprenoid compound, functional in O-antigen biosynthesis," *Proc. Natl. Acad. Sci. USA. 57*: 1798 (1967).

11. G.B.B.M. Sutherland, "Infrared analysis of the structure of amino acids, polypeptides, and proteins," *Adv. Protein Chem. 7*: 291 (1952).

12. F. R. Seymour and R. L. Julian, "Fourier-transform infrared difference-spectrometry for structural analysis of dextrans," *Carbohydr. Res. 74*: 63 (1979).

13. F. R. Seymour, R. L. Julian, A. Jeanes, and B. L. Lamberts, "Structural analysis of insoluble D-glucans by Fourier-transform infrared difference-spectrometry: Correlation between structures of dextrans from strains of *Leuconostoc mesenteroides* and of D-glucans from strains of *Streptococcus mutans*," *Carbohydr. Res. 86*: 227 (1980).

14. C. M. Barlow, J. C. Maxwell, W. J. Wallace, and W. S. Caughey, "Elucidation of the mode of binding of oxygen to iron in oxyhemoglobin by infrared spectroscopy," *Biochem. Biophys. Res. Commun. 55*: 91 (1973).

15. T. J. Williams and R. W. Henkens, "Dynamic ^{13}C NMR investigations of substrate interaction and catalysis by cobalt(II) in human carbonic anhydrase," *Biochemistry, 24*: 2459 (1985).

16. N. E. Mackenzie, J.P.G. Malthouse, and A. I. Scott, "Studying enzyme mechanisms by ^{13}C NMR," *Science, 225*: 883 (1984).

17. J. Horowitz, M. L. Cotton, C. C. Hardin, and P. Gollnick, "Characterization of the fluorodihydrouracil substituent in 5-fluorouracil-containing *Escherichia coli* transfer RNA," *Biochim. Biophys. Acta, 741*: 70 (1983).
18. P. M. Macdonald, B. D. Sykes, and R. N. McElhaney, "^{19}F-NMR studies of lipid fatty acyl chain order and dynamics in membranes," *Biochemistry, 24*: 2412 (1985).

3.8 REFERENCES FOR FURTHER STUDY

1. H. H. Willard, L. L. Merritt, Jr., J. A. Dean, and F. A. Settle, Jr., *Instrumental Methods of Analysis*, 6th ed. Wadsworth, Belmont, CA, (1981), Chapters 1, 2, 3, 4, and 7.
2. D. Freifelder, *Physical Biochemistry — Applications to Biochemistry and Molecular Biology*. Freeman, New York, (1976), Chapters 14 and 15.
3. I. D. Campbell and R. A. Dwek, *Biological Spectroscopy*. Benjamin/Cummings, Menlo Park, CA, (1984).
4. Aksel A., Bothner-By, J. D. Glickson, and B. D. Sykes, Eds. *Biochemical Structure Determination by NMR*. Dekker, New York, (1982).
5. R. K. Harris, *Nuclear Magnetic Resonance Spectroscopy*. Pitman, Marshfield, MA, (1983).
6. R. G. Shulman, Ed. *Biological Applications of Magnetic Resonance*. Academic Press, New York, (1979).

CHAPTER 4

CHROMATOGRAPHIC TECHNIQUES

PART I
TYPES OF CHROMATOGRAPHY AND CHROMATOGRAPHIC TECHNIQUES

Chromatography is the separation of chemical substances by partitioning them between two media. One medium, or phase, is stationary, and the other is moving; the first may be a solid or liquid, and the second may be a liquid or gas. The substances to be separated are distributed between the stationary phase and the moving phase. Different substances are distributed to different degrees and are thereby separated from one another. The process, which may be either analytical or preparative, has had and continues to have great impact on biochemical research. Chromatography provides a means for separating compounds that are very similar, hence difficult or impossible to analyze by other methods. In some cases, chemically similar isomers, such as D-glucose and D-galactose, may be separated.

The Russian botanist Michael Tswett (Mikhail Semenovich Tsvett, 1872–1919) is usually credited with the invention of chromatography. In 1906 Tswett published a paper in *Berichte der deutschen botanischen Gesellschaft* in which he described the separation of plant pigments on calcium carbonate packed into a narrow glass tube. The various colored pigments were separated on the carbonate by elution with petroleum ether.

A much earlier description of a chromatographic process, however, was made by Pliny the Elder (23–79 AD) in *Historia Naturalis*, in which he describes an Egyptian process (*ca.* 500 BC) whereby pigments in a dye were separated by spotting them in solution onto papyrus. This could be considered to be a very early paper chromatographic separation. In 1855 the German chemist F. F. Runge actually described paper chromatography and included illustrations of chromatograms in his book, *Der Bildungstrieb der Stoffe*.

These methods received little attention until the announcement in 1931 by Richard Kuhn and Edgar Lederer that crystalline plant carotene, long regarded as a single chemical compound, could be separated into two components by adsorption chromatography. The real explosion in the use of chromatography, however, started in 1941 with the publication of the theory of partition chromatography by Martin and Synge,[1] followed in 1944 by the description of paper chromatography by Consden, Gordon, and Martin.[2] Paper chromatography (PC), which separates compounds by passing solvent over a sheet of filter paper to which the sample has been applied, developed rapidly in the 1950s along with gas–liquid chromatography (GLC), which separates gaseous or volatile compounds on a narrow-bore column of adsorbent. Thin-layer chromatography (TLC), which separates compounds by passing a solvent through a thin layer (0.25 mm) of adsorbent spread onto a support such as a glass plate, was developed in the 1960s. High performance liquid chromatography (HPLC), which separates compounds by forcing sample and eluting solvent under pressure through a narrow-bore column of adsorbent, was developed in the 1970s. Paper chromatography and thin-layer chromatography are still widely used, and frequently are preferred to gas–liquid chromatography and high performance liquid chromatography because of their power in effecting separations with very simple and inexpensive equipment.

There are essentially three types of chromatography: adsorption, partition, and gel permeation. The many varieties of chromatography, such as ion-exchange, paper, thin-layer, gas–liquid, high performance liquid, affinity, and gel filtration, are special adaptations of one or more of the three basic types.

The principal application of chromatography is in the separation of mixtures of compounds for analytical or preparative purposes. The process also can serve, however, in the identification, isolation, purification, and quantitation of individual compounds. These applications, in turn, may be used in the determination of the homogeneity of chemical substances, the determination of molecular structure, the determination of reaction products and enzyme–substrate specificities, and the monitoring of chemical and biochemical products and processes.

4.1 ADSORPTION CHROMATOGRAPHY

In adsorption chromatography the compounds to be separated are adsorbed onto the surface of a solid material. The compounds are desorbed from the solid adsorbent by eluting solvent. Separation of the compounds depends on the relative balance between the affinity of the compounds for the adsorbent and their solubility in the solvent. The relative affinities of the compounds to be separated from the adsorbent are functions of the chemical nature of the substances, the nature of the solvent, and the nature of the adsorbent. Solid adsorbents commonly used are alumina, silica gel, charcoal, cellulose, starch, calcium phosphate gels, calcium hydroxylapatite, and sucrose. Solvents commonly used are hexane, benzene, petroleum ether, diethyl ether, chloroform, methylene chloride, various alcohols (ethyl, propyl, n-butyl and t-butyl alcohols), and various aqueous buffers and salts, some in combination with organic solvents.

Adsorption chromatography is frequently performed in a column that is packed with the adsorbent. A slurry of the adsorbent is prepared and poured into the column with an inert support at the bottom. Suitable supports include sintered plastic discs, or sheets of woven nylon or Teflon fabrics of 400 mesh or less.

The adsorbent bed must be homogeneous and free of bubbles, cracks, or spaces between the adsorbent and the walls of the column. The choice of the eluting solvent, although very important, depends on the nature of the substances to be separated and the adsorbent, and hence affords considerable latitude. The process of eluting the sample components from the adsorbent by the solvent is termed development. As illustrated in Figure 4-1, the compounds in the mixture that are more soluble in the solvent and have less affinity for the adsorbent move more quickly down the column.

If the substances are colored, as they were in Tswett's experiment, they are readily visible as they separate. However, many substances are not colored, and in these instances, as the development proceeds, fractions are collected at the bottom of the column, and the different fractions are analyzed for compounds of the types that are being separated. For example, if proteins are being separated,

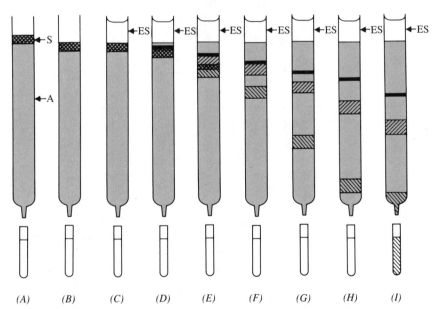

| (A) | (B) | (C) | (D) | (E) | (F) | (G) | (H) | (I) |

FIGURE 4-1. Adsorption chromatography: A = adsorbent, S = sample, ES = eluting solvent. (*A*) Application of sample to the top of the column. (*B*) Adsorption of sample onto adsorbent. (*C*) Addition of elution solvent. (*D*) and (*E*) Partial fractionation of sample components 2 and 3. (*F*) Complete fractionation of sample. (*G*) and (*H*) Separation of all three components at various stages on the adsorbent. (*I*) Elution of the first component from the column.

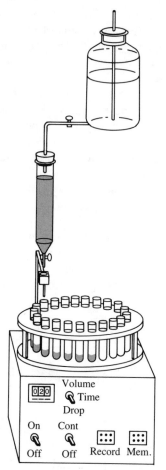

FIGURE 4-2. Collection of fractions from a column by an automatic fraction collector.

the fractions would be analyzed for protein by measurement of the UV absorbance at 280 nm or by other chemical means, as described in Chapter 7 (Section 7.2.3). Likewise, if carbohydrates or nucleic acids are being separated, analytical measurements for carbohydrates or nucleic acids would be made. Figure 4-2 illustrates the collection of fractions by an automatic fraction collector, a device that accumulates from an elution column the same predetermined volume in each of a series of tubes that automatically change position when the proper volume has been collected. This may be accomplished in various ways—for example, with a volumetric siphon of set volume, with a timer, or by counting drops with a drop counter. The latter is frequently used and is usually the most reliable and flexible. The fraction collector may be equipped with a detection cell that automatically measures some parameter of the solution going into the tubes and may be correlated with fraction number and automatically recorded. The detection cell is frequently a small spectrophotometer that can measure absorbances at a fixed wavelength or at variable wavelengths. Other detecting cells use index of refraction, optical rotation, and other properties.

PACKING THE COLUMN AND APPLYING THE SAMPLE

The column is packed by filling it approximately 25% full of solvent or eluant. The column outlet is opened and approximately 5% of the solvent is allowed to flow out to remove air from the space beneath the column bed support and the outlet tubing. The outlet is closed and the slurry of adsorbent or gel is added to the column from a reservoir (Fig. A). All the adsorbent should be poured in a single operation. The adsorbent is allowed to settle by gravity with the column outlet closed. After the bed has formed, the outlet is opened and the excess liquid on the top of the column is allowed to enter the bed, until approximately 1 cm remains above the bed surface. (If the resultant bed length is insufficient, resuspend the top 2–3 cm of the bed by stirring, and pour in additional slurry.) As a general rule, the bed length should be 10–20 times the column diameter. The resolution of two separated zones increases as the square root of the column length.

FIGURE A. Packing the column with adsorbent.

When the packing is completed, the reservoir is removed, and the solvent is allowed to flow just to the top of the bed. The solution of the substances to be separated, in a volume of about 1–5% of the bed volume, is slowly added to the top, taking care to not disturb the bed. The sample is allowed to flow into the bed, and the top of the column is carefully washed with a volume of solvent of 1–5% of the bed volume. An additional amount of solvent is added to the top, and constant elution is started with an appropriate solvent. *At no time during the column packing or sample application should the adsorbent be allowed to go "dry."*

The substances adsorbed on the column support can be eluted in three ways: (*a*) in the simplest method, a single solvent continuously flows through the column until the desired compounds have been separated and eluted from the column; (*b*) stepwise elution, in which two or more different solvents of fixed

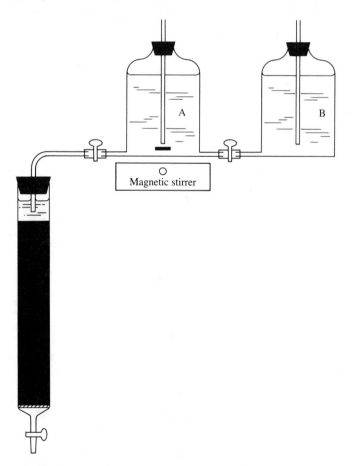

FIGURE 4-3. Gradient elution. Flow of solvent B into solvent A with mixing, continuously changing the composition of solvent A as it flows into the column.

volume are added in sequence to elute the desired compounds; and (*c*) gradient elution, in which the composition of the solvent is continuously changing. The latter method is used to effect separations that are difficult because of a tendency of components to be eluted in broad, trailing bands when a single solvent is used. Gradient elution frequently provides a means of sharpening the bands. A simple linear gradient has two solvents, A and B, in which A is the starting solvent and B is the final solvent. Solvent B is allowed to flow into solvent A as solvent A flows into the column. The composition of solvent A is, thus, constantly changing as solvent B is constantly being added to A (Fig. 4-3). Gradients other than linear gradients (*e.g.*, exponential, concave, or convex) may be obtained by introducing a third vessel and varying the composition of the solvents in the vessels.[2] These eluting methods are also used with other column chromatographic methods, described in subsequent sections.

4.2 ION-EXCHANGE CHROMATOGRAPHY

Ion-exchange chromatography is a variation of adsorption chromatography in which the solid adsorbent has charged groups chemically linked to an inert solid. Ions are electrostatically bound to the charged groups; these ions may be exchanged for ions in an aqueous solution. Ion exchangers are most frequently used in columns to separate molecules according to charge. Because charged molecules bind to ion exchangers reversibly, molecules can be bound or eluted by changing the ionic strength or pH of the eluting solvent.

Two types of ion exchanger are available: those with chemically bound negative charges are called **cation exchangers**, and those with chemically bound positive charges are called **anion exchangers**. The charges on the exchangers are balanced by counterions such as chloride ions for the anion exchangers and metal ions for the cation exchangers. Sometimes buffer ions are the counterions. The molecules in solution, which are to be adsorbed on the exchangers, also have net charges, which are balanced by counterions. As an example of an ion-exchange process, let us say that the molecules to be adsorbed from solution have a negative charge (X^-), which is counterbalanced by sodium ions (Na^+). Such negatively charged molecules can be chromatographed on an anion exchanger (A^+), which has chloride ions as the counterion to give A^+Cl^-. When (Na^+X^-) molecules in solution interact with the ion exchanger, the X^- displaces the chloride ion from the exchanger and becomes electrostatically bound to give A^+X^-, simultaneously releasing sodium ions. This process of ion exchange is illustrated in Figure 4-4. A similar, but opposite, process would take place for positively charged molecules (Y^+Cl^-), which would be chromatographed on cation exchangers (C^-Na^+). Thus, the cation exchangers will bind positively charged molecules from solution, and the anion exchangers will bind negatively charged molecules from solution.

One of the inert materials used in this type of chromatography is cross-linked polystyrene, to which the charged groups are chemically bound. In the separation of biologically important macromolecules, such as enzymes and proteins,

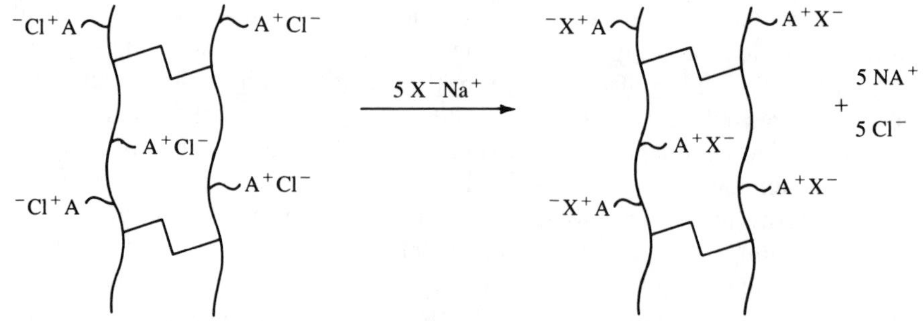

FIGURE 4-4. The process of anion-exchange chromatography.

cellulose and cross-linked dextran (Sephadex, see Section 4.6) are used as the solid supports and charged groups such as diethylaminoethyl (DEAE) or carboxymethyl (CM) are chemically linked to them to give anion and cation exchangers, respectively. The preparation and commercial availability of these materials beginning in the 1960s provided the biochemist with powerful tools for the separation of proteins and nucleic acids.[3,4] Figure 4-5 presents partial structures of DEAE–cellulose and CM–cellulose.

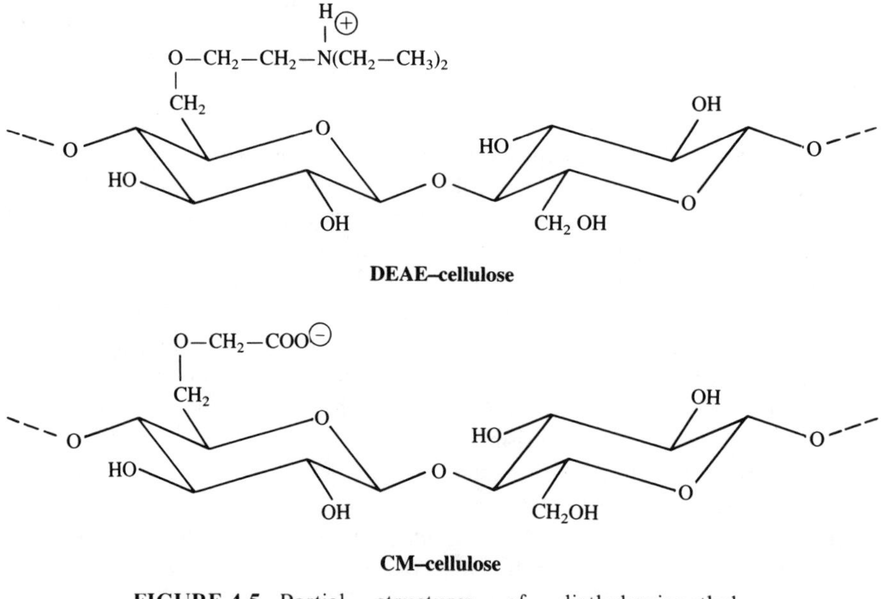

DEAE–cellulose

CM–cellulose

FIGURE 4-5. Partial structures of diethylaminoethyl–cellulose and carboxymethyl–cellulose. The DEAE and CM groups are shown attached to the C_6–hydroxyl group of glucose. The DEAE and CM groups are also found attached to the hydroxyl groups of C_2 and C_3. The total degree of substitution of the DEAE and CM groups must be less than one group per five glucose residues to maintain a water-insoluble product.

TABLE 4-1.   Pretreatment steps for DEAE–cellulose and CM–cellulose ion exchangers

Cellulose	First treatment	Intermediate pH	Second treatment
DEAE	0.5 M HCl	4	0.5 M NaOH
CM	0.5 M NaOH	8	0.5 M HCl

The dry ion-exchange celluloses are pretreated with acid and base to swell the exchangers so that they become fully accessible to the charged macromolecules in solution. The weighed exchanger is suspended in 15 volumes (w/v) of the "first treatment," acid or alkali depending on the exchanger (Table 4-1), and is allowed to stand at least 30 minutes but not more than 2 hours. The supernatant is decanted and the exchanger is washed until the effluent is at the "intermediate pH." The exchanger is stirred into 15 volumes of the "second treatment" and allowed to stand for an additional 30 minutes. The second treatment is repeated and the exchanger is washed with distilled water until the effluent is close to neutral pH. The treated exchanger is placed into the acid component of the buffer (the pH should be less than 4.5) and degassed under vacuum (<10 cm Hg pressure) with stirring, until bubbling stops. The exchanger is then titrated with the basic component of the buffer to the desired pH, filtered, and suspended in fresh buffer to complete the pretreatment. The exchanger is allowed to settle and the "fines" (fragments < 10 μm in diameter) above the settled exchanger are removed by decantation. Buffer is added to the exchanger so that the final volume of the slurry is 150% of the settled wet volume of the exchanger. The column is then packed with the slurry of the exchanger, the sample is applied, and elution is performed as described for adsorption chromatography in Section 4.1.

Three general methods are used for eluting molecules from the exchanger: (a) changing the pH of the buffer to a value at which binding is weakened (i.e., the pH is lowered for an anion exchanger and raised for a cation exchanger), (b) increasing the ionic strength by increasing the concentration of salt in the elution solvent, thereby weakening the electrostatic interactions between the adsorbed molecule and the exchanger, and (c) performing affinity elution. In affinity elution the adsorbed molecule is usually a macromolecule that is desorbed from the affinity ligand by adding a molecule that is charged and of opposite sign to the net charge on the macromolecule and has a specific affinity for the macromolecule. Thus, the reduction of the net charge on the macromolecule weakens its electrostatic interaction with the exchanger sufficiently to permit the elution of the macromolecule from the affinity ligand.

4.3 LIQUID–LIQUID PARTITION CHROMATOGRAPHY

In partition chromatography two liquid phases are used. The phase that is bound to the "inert" support is called the **stationary phase**. The other liquid, the **mobile phase**, passes over the stationary phase, distributing the substances to be separated between the two liquids. The substances are separated by a differential solubility in the two liquid phases. The process, however, is usually complicated by adsorption of the substances and the mobile phase to the "inert" support.

Partition chromatography can be performed on columns of cellulose powder, but the most common application is the use of paper sheets (paper chromatography) or a thin layer of silica gel or alumina on a glass plate (thin-layer chromatography). Cellulose, being a carbohydrate, can bind a relatively large amount of water. Partitioning thus occurs between the bound water (the stationary phase) and the developing solvent (the mobile phase). Frequently one of the components of the developing solvent is water. When water and the organic component are immiscible, a third solvent component is added that is capable of dissolving both water and the water-immiscible, organic solvent to give a one-phase solvent—for example, in a solvent composed of water, ethanol, and nitromethane, the ethanol is capable of dissolving both water and nitromethane.

4.4 PAPER CHROMATOGRAPHY

Paper chromatography is a type of liquid–liquid partition chromatography that may be performed by ascending or descending solvent flow. Each mode has its advantages and disadvantages. Ascending chromatography involves relatively simple and inexpensive equipment compared with descending chromatography and usually gives more uniform migration with less diffusion of the sample "spots." Descending chromatography, on the other hand, is usually faster because gravity aids the solvent flow, and with substances of relatively low mobility, the solvent can run off the paper, giving a longer path for migration. To resolve compounds with low mobility, ascending chromatography may be performed more than once utilizing a multiple-ascent technique.

For descending chromatography, papers 22 cm wide and 56 cm long can be used. To facilitate the flow of solvent from the paper, the bottom of the paper is serrated with a pair of pinking shears. Three pencil lines are drawn 25 mm apart at the top of the sheet, and small aliquot of the sample (10–50 μL) is placed at a marked spot on the third line. The spot is kept as small as possible by adding the aliquot in small increments, with drying in between. This may be expedited with a hair dryer. Several samples, including standards, are placed 15–25 mm apart. The paper is then folded along the other two lines and placed in the solvent trough of the descending tank (Fig. 4-6), which has been equilibrated with solvent beforehand to ensure a saturated atmosphere. The paper is irrigated with solvent until the solvent reaches the bottom or for a longer period, allowing the solvent to flow off the end of the paper, if necessary. The chromatogram is then removed, dried, and developed to reveal the locations of the compounds. (Part II gives methods of locating carbohydrates, amino acids, proteins, nucleotides and nucleic acids, and lipids.)

In ascending chromatography, a paper approximately 25 cm × 25 cm is used. A pencil line 20–25 mm from the bottom is drawn across the paper and

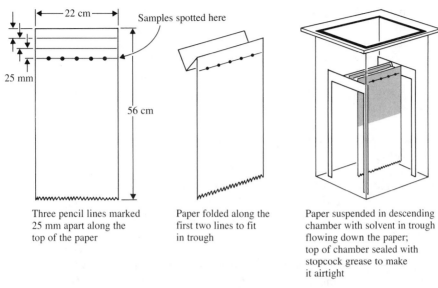

FIGURE 4-6. Steps in descending paper chromatography.

aliquots (10–50 μL) of the samples and standards are spotted approximately 15–25 mm apart along the line. The spots are dried and the paper is rolled into a cylinder and stapled so that the ends of the paper are not touching (Fig. 4-7). Solvent is poured into the bottom of a chromatographic chamber, and the cylinder is placed inside. The chamber is closed and solvent is allowed to flow up

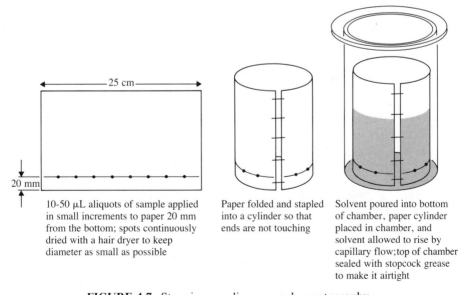

FIGURE 4-7. Steps in ascending paper chromatography.

the paper by capillary action. The chamber may be a simple wide-mouth, screw-top, gallon jar or a cylinder with a ground-glass edge and a glass plate top. As with descending chromatography, the chamber should be equilibrated with solvent beforehand. Contrary to a popular misconception, if the chamber has been sealed and is airtight, the paper does not have to be removed as soon as the solvent reaches the top. When multiple ascents are performed, the paper is removed, thoroughly dried, and returned to the chamber for another ascent of solvent.

The resolved compounds on a paper chromatogram may be detected by their color if they are colored, by their fluorescence if they are fluorescent, by a color that is produced from a chemical reaction on the paper after spraying or dipping the chromatogram with various reagents, or by autoradiography if the compounds are radioactive. Specific methods for various classes of compounds are discussed in Part II. Autoradiography, a method for detecting radioactive compounds, is discussed in Chapter 6 (Section 6.11).

Identification of compounds on a chromatogram is usually based on a comparison with authentic compounds (standards). A quantitative comparison may be made by measuring the R_f, which is the ratio of the distance the compound migrates to the distance the solvent migrates. A better comparison is the ratio of the distance a particular compound migrates to the distance a particular standard migrates. For example, in the separation of carbohydrates, the standard might be glucose and the ratio would be R_{Glc} or for amino acids, the standard might be glycine and the ratio would be R_{Gly}.

A useful modification is two-dimensional paper chromatography, in which the sample is spotted in the lower left-hand corner and irrigated in the first dimension with solvent A. The chromatogram is removed from the solvent, dried, turned 90°, and irrigated in the second dimension with solvent B, giving a two-

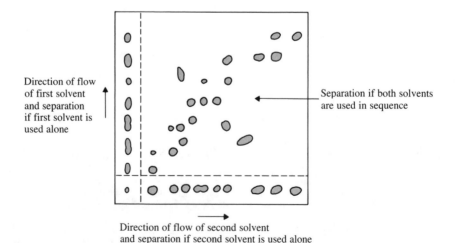

Direction of flow of first solvent and separation if first solvent is used alone

Separation if both solvents are used in sequence

Direction of flow of second solvent and separation if second solvent is used alone

FIGURE 4-8. Two-dimensional paper or thin-layer chromatography.

dimensional separation (Fig. 4-8). An application of this procedure has been developed for the study of enzyme specificity in which a solution of the enzyme is sprayed onto the chromatogram between the first irrigation and the second to see what products are formed by the action of the enzyme on the compounds separated in the first dimension.[5]

Paper chromatography has been used to establish the structural homology of a series of oligomers obtained by enzymic synthesis, by acid or enzymic hydrolysis, or by isolation from a natural source. The R_f of each separated homologue is determined and a French–Wild plot is made by plotting $\log [R_f/(1 - R_f)]$ against the number of monomers in the oligomer. If the isolated compounds fall on a straight line of this plot, they belong to a homologous series, differing from each other by one monomer residue (Fig. 4-9).[6]

Compounds separated by paper chromatography may be quantitatively determined. Aliquots ($50-200 \, \mu L$) of the solution containing the substances to be separated and quantitatively determined are streaked along the separation line. Aliquots of the solution ($5-10 \, \mu L$) are also spotted on the two outside edges of the streak and are used as location standards. The chromatogram is irrigated in the usual way, and vertical sections of the location standards are cut out and developed to reveal the positions of the compounds. After drying, these standards are placed alongside the streaked sections and the undeveloped compounds are located; horizontal strips containing the individual compounds are cut out and

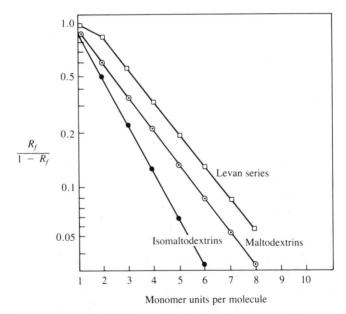

FIGURE 4-9. French–Wild plots $(\log R_f/1 - R_f$ versus number of monomer units per molecule) correlating paper chromatographic mobility with the number of homologous monomer residues in oligosaccharide molecules.

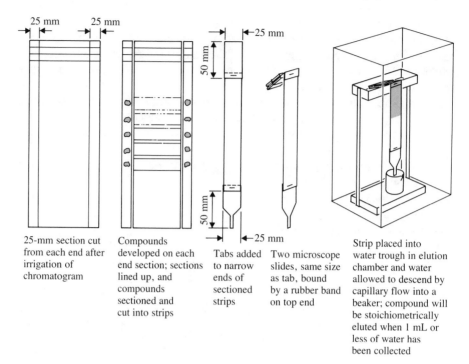

25-mm section cut from each end after irrigation of chromatogram

Compounds developed on each end section; sections lined up, and compounds sectioned and cut into strips

Tabs added to narrow ends of sectioned strips

Two microscope slides, same size as tab, bound by a rubber band on top end

Strip placed into water trough in elution chamber and water allowed to descend by capillary flow into a beaker; compound will be stoichiometrically eluted when 1 mL or less of water has been collected

FIGURE 4-10. Elution of compounds from paper chromatograms for preparative chromatography or quantitative determination.

eluted with water. To accomplish the elution, tabs of chromatographic paper are stapled to the narrow ends of each strip. As shown in Figure 4-10, one end is fitted with two pieces of glass (cut microscope slides), which are held together with rubber bands, and the bottom end is cut tapered, like a pipet tip. This assembly is placed so that one end lies in a chromatographic trough containing water, and the elution of the strip occurs by capillary flow of the water down the paper strip into a beaker. Usually less than 1 mL of water is sufficient to effect quantitative elution. The samples are quantitatively diluted to a specific volume, and a chemical analysis is performed for the specific compound separated. This technique also may be used as a preparative procedure to obtain small quantities of pure compound from a mixture of compounds.

In an alternate quantitative procedure, the compounds in the sample are radioactively labeled and separated in the usual way, and an autoradiogram is prepared. The labeled compounds are located on the chromatogram by comparing their positions on the autoradiogram (Section 6.11). The radioactive compounds are cut out and placed into a liquid scintillation cocktail, and the radioactivity is determined by heterogeneous liquid scintillation counting (Section 6.9.4).

4.5 THIN-LAYER CHROMATOGRAPHY

Thin-layer chromatography (TLC) was developed to separate classes of compounds, initially in the lipid field, where paper chromatography had not been successful. As TLC developed, it was found to have wide applicability for many classes of compounds and to have many advantages over paper chromatography. The method involves the use of finely powdered adsorbents such as silica gel, Celite, alumina, cellulose, and cellulose derivatives. The powdered solid adsorbent, usually mixed with a binder (e.g., 10% calcium sulfate), is spread as a thin (0.2–0.5 mm) uniform coat over the surface of a glass plate or a plastic sheet. Fortunately, the availability of highly reproducible commercial plates for the most commonly used adsorbents has decreased the need for investigators to prepare their own TLC plates. The glass-backed plates are the most popular and come in various sizes (20 cm × 20 cm, 20 cm × 10 cm, and 20 cm × 5 cm). For rapid, exploratory tests, small TLC plates of microscope slide size (2.5 cm × 7.5 cm) are also available. Plastic-backed TLC sheets, which may be cut to any size, are particularly useful in quantitative determinations, especially for radioactive compounds. After TLC separation of the radioactive compounds, an autoradiogram is prepared. The radioactive compounds are cut from the plastic TLC sheet and the pieces are placed into a liquid scintillation cocktail for measurement of radioactivity by heterogeneous, liquid scintillation spectrometry (Section 6.9.4).

The advantages of TLC over paper chromatography are greater resolving power, greater speed of separation, and relatively wide choice of adsorbents. The high resolving power and speed of separation of this technique are due to the very fine particle size (<0.1 mm) of the adsorbent, with the consequently high surface-to-volume ratio, giving a large active surface area for a given amount of adsorbent. It would be impractical to use such small particles in columns because the flow rate would be extremely slow.

The fibrous nature of cellulose in paper produces capillary diffusion, which increases spot size. The inorganic adsorbents used in TLC, however, do not have a fibrous structure, so capillary diffusion is eliminated, resulting in smaller spots and decreased running time due to higher resolution. Furthermore, the very high surface area of the adsorbent also increases the flow rate of the solvent. Because of the relatively inert nature of the inorganic adsorbents, corrosive detecting reagents, such as sulfuric acid, can be sprayed onto the plates, giving sensitive and simplified detecting systems. Because the spots are small, the material is more concentrated in the spot, thus giving a high level of detection sensitivity. The lower limit of the amount of material that may be detected on a thin-layer chromatogram is 0.1–0.05 of that detected on a paper chromatogram (i.e., 2–4 μg vs. 20–40 μg).

Thin-layer chromatograms are prepared much like ascending paper chromatograms. The sample (1–10 μL) containing 1–10 μg is spotted along a line 15–20 mm from the bottom. Spot diameters are kept as small as possible by adding very small amounts of sample with drying in between. After the samples

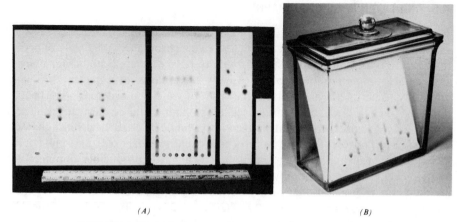

(A)   *(B)*

FIGURE 4-11. Thin-layer chromatography. (*A*) Thin-layer plates of various sizes. (*B*) Chromatographic development of a thin-layer plate.

have been added and dried, the TLC plate is placed into a chromatographic chamber with solvent in the bottom. Solvent is allowed to move up the plate by capillary flow (Fig. 4-11). When the solvent reaches the top, the plate is removed and dried. If the compounds are not colored, the plate may be developed by spraying an appropriate reagent (see Part II) or, if necessary, the plate may be returned to the chamber and a second ascent made. Like paper chromatography, TLC can be conducted in two dimensions, (as in Figure 4-8), or multiple ascents can be made.

4.6 GEL PERMEATION CHROMATOGRAPHY

Gel chromatography, sometimes also called gel filtration, was introduced in 1959,[7] with the use of cross-linked dextran as a material for desalting macromolecules and effecting group separations based on molecular weights. The technique has provided a very mild method for purifying enzymes, polysaccharides, nucleic acids, and proteins. The method separates molecules according to their molecular size.

Gel particles form the stationary phase of this type of chromatography; the mobile phase is the solution of molecules to be separated and the eluting solvent, which most frequently is water or a dilute buffer. The sample is applied to the gel; if the molecules are too large for the pores, they never enter the gel and move outside the gel bed with the eluting solvent. Thus, the very large molecules in a mixture move the fastest through the gel bed and the smaller molecules, which can enter the gel pores, are retarded and move more slowly through the gel bed. In gel chromatography, molecules are, therefore, eluted in order of decreasing molecular size (Fig. 4-12).

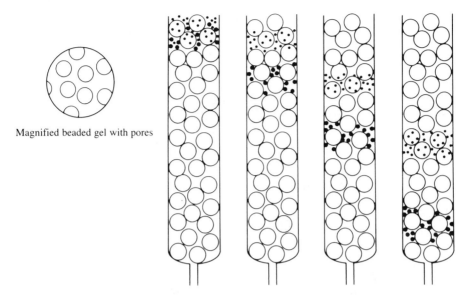

Magnified beaded gel with pores

FIGURE 4-12. Gel permeation chromatography. Open circles represent porous gel molecules; large solid circles represent molecules too large to enter the gel through the pores, and smaller solid circles represent molecules capable of entering the gel pores.

The gels are relatively inert, cross-linked polymers, and the amount of cross-linking determines the average pore size of a gel. Different degrees of cross-linking, thus, give different pore sizes and different molecular sieving or separation ranges. Three types of polymers are principally used—dextran, polyacrylamide, and agarose.

Dextran is a polysaccharide composed of $(\alpha\text{-}1\rightarrow6)$-linked glucose residues with α-1,3 branch linkages. It is synthesized from sucrose by an enzyme produced by the bacterium *Leuconostoc mesenteroides* B-512F. The dextran is cross-linked to various extents by reaction with epichlorohydrin to give gel beads with different pore sizes (Fig. 4-13). Cross-linked dextrans are commercially produced by Pharmacia Fine Chemicals, Inc., (Uppsala, Sweden), and sold under the trade name *Sephadex*. Sephadex gels in the so-called G-series, where the G-numbers refer to the amount of water gained when the beads are swelled in water (Table 4-2), have different degrees of cross-linking, hence different pore sizes. This gives gels that have capabilities of separating different ranges of molecular weights and have different molecular exclusion limits. The exclusion limit is the molecular weight of the smallest peptide or globular protein that will not enter the gel pore. Sephadex G-10, the highest cross-linked dextran, has a water regain of about 1 mL/g of dry gel and Sephadex G-200, the lowest cross-linked dextran, has a water regain of about 20 mL/g of dry gel. In the swelling process, the gels become filled with water.

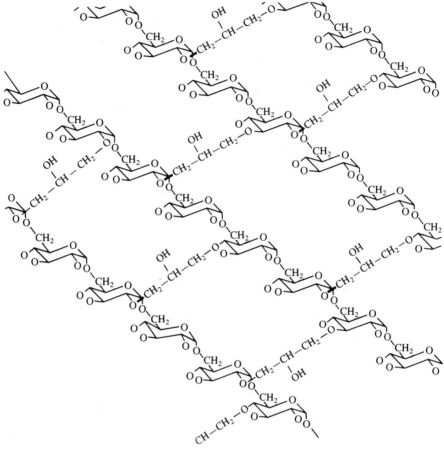

FIGURE 4-13. Structure of epichlorohydrin cross-linked dextran.

Polyacrylamide gels are long polymers of acrylamide cross-linked with N,N'-methylene-bisacrylamide (Fig. 4.14). The gels are commercially produced by Bio-Rad Laboratories, Richmond, California, as the Bio-Gel P series. Like the Sephadex G series, the Bio-Gels differ in degree of cross-linking and in pore size; the Bio-Gels, however, have a wider range of pore sizes than is available in the Sephadex G series. See Table 4-2 for the exclusion limits and properties of the different Bio-Gels.

Agarose is a gel material with pore sizes larger than cross-linked dextran or polyacrylamide. Agarose is the neutral polysaccharide fraction of agar. It is composed of a linear polymer of D-galactopyranose linked β-1→4 to 3,6-anhydro-L-galactopyranose, which is linked α-1→3 (Fig. 4-15). When the polysaccharide is dissolved in boiling water and cooled, it forms a gel by forming inter- and intramolecular hydrogen bonds. The pore sizes are controlled by the concentration of the agarose. High molecular weight materials such as protein

TABLE 4-2. Properties of gels used in gel-permeation chromatography[a]

Gel	Water regain (mL/g)	Exclusion limit[b]	Maximum hydrostatic pressure (cm H_2O)[c]	Maximum flow rate (mL/min)[c]
Sephadex G-10	1.0	700	200	100
Sephadex G-15	1.5	1,500	200	100
Sephadex G-25	2.5	5,000	200	50
Sephadex G-50	5.0	30,000	200	25
Sephadex G-75	7.5	70,000	160	6.4
Sephadex G-100	10.0	150,000	96	4.2
Sephadex G-150	15.0	300,000	36	1.9
Sephadex G-200	20.0	600,000	16	1.0
Sepharose 6B	NA	4×10^6	200	1.2
Sepharose CL 6B	NA	4×10^6	>200	2.5
Sepharose 4B	NA	20×10^6	80	0.96
Sepharose CL 4B	NA	20×10^6	120	2.17
Sepharose 2B	NA	40×10^6	40	0.83
Sepharose CL 2B	NA	40×10^6	50	1.25

Gel	Water regain (mL/g)	Exclusion limit[b]	Maximum hydrostatic pressure (cm H_2O)[c]	Maximum flow rate (mL/min)[c]
Bio-Gel P-2	1.5	1,800	>100	110
Bio-Gel P-4	2.4	4,000	>100	95
Bio-Gel P-6	3.7	6,000	>100	75
Bio-Gel P-10	4.5	20,000	>100	75
Bio-Gel P-30	5.7	40,000	>100	65
Bio-Gel P-60	7.2	60,000	100	30
Bio-Gel P-100	7.5	100,000	100	30
Bio-Gel P-150	9.2	150,000	100	25
Bio-Gel P-200	14.7	200,000	75	11
Bio-Gel P-300	18.0	400,000	60	6
Bio-Gel A-0.5 m	NA	500,000	>100	3.0
Bio-Gel A-1.5 m	NA	1.5×10^6	>100	2.5
Bio-Gel A-5 m	NA	5×10^6	>100	1.5
Bio-Gel A-15 m	NA	15×10^6	90	1.5
Bio-Gel A-50 m	NA	50×10^6	50	1.0
Bio-Gel A-150 m	NA	150×10^6	30	0.5

Source: Data taken from the manufacturer's technical information.
[a] Bio-Gel is a trade name of Bio-Rad Laboratories, and Sephadex and Sepharose are trade names of Pharmacia Fine Chemicals.
[b] Given for peptides and globular proteins. The exclusion limit is the molecular weight of the smallest peptide or globular protein that will not enter the gel pore.
[c] Maximum hydrostatic pressure and maximum flow rates are given for 100–200 mesh particle size. For larger and smaller sizes the pressure and flow rates will be higher and lower, respectively.

```
                              HN₂
                               |
                              C = O
                               |
  − − −CH₂−CH−CH₂−CH−CH₂−CH− − −
              |                    |
             C = O                C = O
              |                    |
             HN                   NH₂
              |
             CH₂
              |
             HN
              |
             C = O
              |
  − − −CH₂−CH−CH₂−CH−CH₂−CH− − −
                  |        |
                 C = O    C = O
                  |        |
                 HN       NH₂
                  |
                 CH₂
                  |
        NH₂       NH
         |        |
   O = C     O = C
         |        |
  − − −CH₂−CH−CH₂−CH−CH₂−CH− − −
                          |
                         C = O
                          |
                         NH₂
```

FIGURE 4-14. Structure of cross-linked polyacrylamide.

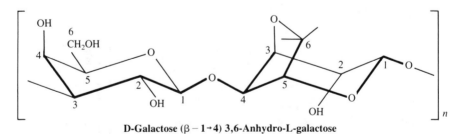

D-Galactose (β − 1→4) 3,6-Anhydro-L-galactose

FIGURE 4-15. Structure of the repeating unit of agarose, D-galactopyranose linked β-1→4 to 3,6-anhydro-L-galacto-pyranose, which is linked α-1→3 to the next D-galacto-pyranose residue.

aggregates, chromosomal DNA, ribosomes, viruses, and cells have been fractionated on agarose gels. Bio-Rad markets the agarose Bio-Gel A series with different molecular exclusion limits, and Pharmacia markets agarose as Sepharose and Sepharose CL. The latter is Sepharose cross-linked by reacting with alkaline 2,3-dibromopropanol to give an agarose gel with increased thermal and chemical stability. Table 4-2 gives the properties of the different Sephadex, Bio-Gel, and Sepharose gels.

The separations that may be achieved by gel permeation chromatography are based on differences in the molecular sizes of the molecules. The method is used for both preparative and analytical purposes. The latter has been especially useful in determining the molecular weights of proteins.[8-10] The proteins are chromatographed on a gel column and the elution volume of the protein determined. Proteins with known molecular weights are also chromatographed and the elution volumes determined. Then, from a plot of log molecular weight versus elution volume, the molecular weight of an unknown protein may be determined (Fig. 4-16).

Gel chromatography provides a rapid and mild method of removing salts and other small molecules from high molecular weight biomolecules. The sample containing the biomolecules and the salt is passed over a gel column whose exclusion limit is below the molecular weight of the biomolecules. The

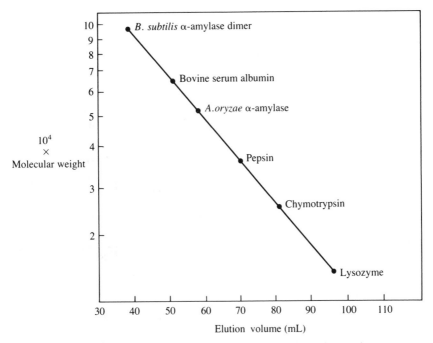

FIGURE 4-16. Molecular weight determination of proteins by gel permeation chromatography using Sephadex G-100 as the gel bed: log molecular weight is plotted versus elution volume.

biomolecules, which do not enter the gel, emerge in the void volume of the column, while the salts enter the gel and are retarded, and therefore are removed from the biomolecules.

4.6.1 Preparation of Gels and Columns

Sephadex and Bio-Gel P are supplied as dry powders that must be swollen in excess solvent (usually water or buffer). During swelling, excessive stirring should be avoided. The gel should be stirred occasionally with a glass rod, but never with a magnetic stirrer. Fine particles (fines) that are suspended above the gel are

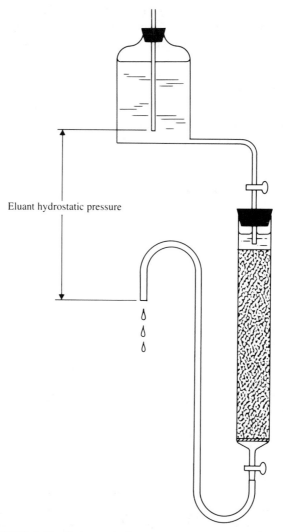

Eluant hydrostatic pressure

FIGURE 4-17. Column elution setup for descending gel permeation chromatography.

removed by decanting. Swelling time depends on the degree of cross-linking. Sephadex G-10 to G-50 and Bio-Gel P2 to P30 are allowed to swell for 3 hours at 20°C or 1 hour at 90°C; Sephadex G-75 to G-200 and Bio-Gel P60 to P300 are allowed to swell for 72 hours at 20°C or 5 hours at 90°C. Sepharose and Bio-Gel A gels are supplied preswollen.

A fairly thick slurry of the gel is degassed under reduced pressure. (Note, however, that the mixture should not be so thick that it doesn't pour easily or that it retains air bubbles.) The procedures for column packing and sample application are similar to that for adsorption chromatography. A gel column elution setup is shown in Figure 4.17. Maximum flow rates and hydrostatic pressures are given in Table 4-2. Too much hydrostatic pressure should be avoided, especially with the softer gels (i.e., the Bio-Gel A series and those in the Bio-Gel P series with the higher numbers, Sephadex gels with the higher numbers, and Sepharose), since they are easily compressed and deformed, and the flow rates become extremely slow and impractical. The gel columns may be stored and used several times. The columns should be stored with an antimicrobial agent such as 0.02% sodium azide or 0.002% chlorhexidine.

4.6.2 Definitions Used in Gel Permeation Chromatography

Total bed volume V_t is the total volume occupied by the packed gel bed. This is obtained most easily by a water calibration of the column before packing. **Void volume** (V_0) is the volume of the space surrounding and outside the particles of gel. It is usually determined by measuring the volume necessary to elute a solute that is excluded from the pores of the gel. A commonly used solute is blue dextran, a high molecular weight polysaccharide-dye complex.

Elution volume V_e is the volume of solvent necessary to elute a solute from the time the solute enters the gel bed to the time it begins to emerge at the bottom of the column. The V_e is measured as the volume of solvent that has flowed through the column when the leading side of the solute peak is extrapolated to the base line of an elution profile. A **distribution coefficient** K_{avg}, which is characteristic of the solute and is analogous to the R_f value used in paper chromatography, is obtained by the following relation:

$$K_{avg} = \frac{V_e - V_0}{V_t - V_0} \qquad (4\text{-}1)$$

4.7 AFFINITY CHROMATOGRAPHY

Affinity chromatography is a specialized type of adsorption chromatography in which a specific type of molecule is covalently linked to an inert solid support. This specific molecule, called a **ligand**, has a high binding affinity for one of the compounds in a mixture of substances. The process uses the unique biological property of the substance to bind to the ligand specifically and reversibly, and provides a high degree of selectivity in the isolation and purification of biological molecules.

A solution containing the substance to be purified, usually a macromolecule such as a protein (enzyme, antibody, hormone, etc.), polysaccharide, or nucleic acid, is passed through a column containing an insoluble, inert polymer to which the ligand has been covalently attached. The ligands may be specific competitive inhibitors, substrate analogues, product analogues, coenzymes and so on. Molecules in the mixture not having affinity for the ligand pass through the column, while molecules that have specific affinity for the ligand are bound and retained on the column. The specifically adsorbed molecule(s) can be eluted by changing the ionic strength, the pH, or by the addition of a competing ligand. In one chromatographic step, the method is capable of isolating a single substance in a pure form. It has, thus, become a powerful tool in the isolation and purification of enzymes, antibodies, antigens, nucleic acids, polysaccharides, coenzyme or vitamin binding proteins, repressor proteins, transport proteins, drug or hormone receptor structures, and other biochemical materials.

4.7.1 The Inert Support and the Ligand

The inert solid supports are the same materials discussed in the preceding sections: cross-linked dextran, cross-linked polyacrylamide, agarose, and cellulose. The macromolecules to be separated should not be retarded by a gel filtration process but should be retarded only by the specific interaction with the ligand. The ligand must be a molecule that displays special and unique affinity for the macromolecule to be purified. It also must have a chemical group that can be modified for covalent linkage to the solid support without destroying or seriously decreasing its interaction with the macromolecule to be purified. Also for successful affinity chromatography, the chemical groups of the ligand that are critical for the binding of the macromolecule to be purified must be sufficiently distant from the solid support to minimize steric interference with the binding process. This steric problem has been solved by adding a long, hydrocarbon chain, spacer arm to the solid support and coupling the ligand to the end of the arm. Alternatively the hydrocarbon arm may be attached to the ligand and the arm attached to the solid support.

4.7.2 Attachment of the Ligand to the Solid Support

The polysaccharide solid supports—cross-linked dextran, agarose, and cellulose—can be activated by reaction with alkaline cyanogen bromide.[11,12] The products that are formed upon coupling of the activated polysaccharides with amino compounds are derivatives of amino carbonic acid.[13] The reactions are the following:

$$\text{(a)}$$

CH$_2$OH ... —O, OH, OH $\xrightarrow[\text{OH}^-]{\text{CNBr}}$ CH$_2$—O—C(=NH)—Br ... —O, OH, OH

$\xrightarrow[\text{H}_2\text{O} \quad \text{NH}_3 \quad \text{HBr}]{\text{H}_2\text{NR}}$ CH$_2$—O—C(=O)—NHR ... —O, OH, OH   (a)

If the ligand contains an amino group, it can be coupled directly to the activated polysaccharide. A spacer arm can be introduced by sequential reaction with a diaminoalkane and glutaraldehyde. The amino group on the ligand can then be coupled to the free aldehyde group.

CH$_2$—O—C(=NH)—Br $\xrightarrow[\text{H}_2\text{O} \quad \text{NH}_3 \quad \text{HBr}]{\text{H}_2\text{N—(CH}_2)_n\text{—NH}_2}$ CH$_2$—O—C(=O)—NH—(CH$_2$)$_n$—NH$_2$

$\xrightarrow[\text{H—C(=O)—(CH}_2)_3\text{—C(=O)—H} \quad \text{H}_2\text{O}]{}$ CH$_2$—O—C(=O)—NH—(CH$_2$)$_n$—N=CH—(CH$_2$)$_3$—CHO   (b)

If the ligand contains an aldehyde group instead of an amino group, it can be coupled directly to the free amino group of the diaminoalkane.

Ligands may be coupled to polyacrylamide by displacing the amide group of the polyacrylamide by heating with a diaminoalkane (c),[14] followed by reaction with glutaraldehyde (d).

—C(=O)—NH$_2$ + H$_2$N—(CH$_2$)$_n$—NH$_2$ $\xrightarrow[\text{NH}_3]{90°}$ —C(=O)—NH—(CH$_2$)$_n$—NH$_2$ (c)

$$\overset{O}{\underset{\|}{\text{C}}}\text{—NH—(CH}_2)_n\text{—NH}_2 + \text{H—}\overset{O}{\underset{\|}{\text{C}}}\text{—(CH}_2)_3\text{—}\overset{O}{\underset{\|}{\text{C}}}\text{—H} \xrightarrow[\text{H}_2\text{O}]{}$$

$$\overset{O}{\underset{\|}{\text{C}}}\text{—NH—(CH}_2)_n\text{—N}{=}\text{CH—(CH}_2)_3\text{—CHO} \quad \text{(d)}$$

The Schiff base that results from the reaction of glutaraldehyde with an amino group may be stabilized by reduction with sodium cyanoborohydride[15] without affecting the aldehyde group. The ligand can then be coupled to the aldehyde group.

$$\overset{O}{\underset{\|}{\text{C}}}\text{—NH—(CH}_2)_n\text{—N}{=}\text{CH—(CH}_2)_3\text{—CHO} \xrightarrow{\text{NaCNBH}_3}$$

$$\overset{O}{\underset{\|}{\text{C}}}\text{—NH—(CH}_2)_n\text{—NH—CH}_2\text{—(CH}_2)_3\text{—CHO} \quad \text{(e)}$$

Another method of activating polyacrylamide is to form the hydrazide derivative by reaction with hydrazine hydrate. When an amino, aldehyde, or hydrazide group is incorporated onto the solid support, the support becomes activated so that ligands may be attached through amino, carboxyl, phenolic, or imidazole groups.[16] Methods for attaching 160 different ligands are given in Reference 17.

Many proteins have been purified by affinity chromatography. Examples of a few of the purified proteins are staphylococcal nuclease, α-chymotrypsin, 3-deoxy-D-arabinoheptulosonate 7-phosphate synthetase, avidin, thyroxine binding protein, and ribonuclease.[17,18]

4.7.3 Hydrophobic Chromatography

A special adaptation of affinity chromatography is **hydrophobic chromatography**, in which alkyl chains are added to the solid support giving sites for interaction by hydrophobic bonding. The preparation of the hydrophobic-affinity supports is obtained by carrying out the reactions described above. Carbohydrate supports are activated by alkaline cyanogen bromide to which aminoalkanes or diaminoalkanes of different chain lengths are added. Polyacrylamides are reacted directly by heating with the alkylamines. Aminoalkanes and diaminoalkanes containing 1 to 12 carbon atoms have been used to give a homologous series of hydrophobic chromatographic adsorbents. Each member in the series offers hydrophobic chains of a different length, which interact with accessible hydrophobic sites on various proteins, binding only some proteins out of a mixture. Resolution of the bound proteins is achieved by gradually increasing the hydrophobicity of the eluting solvent.[19] Another type of hydrophobic chromatographic material is O-(phenoxyacetyl)cellulose, which is prepared by derivatizing either fibrous or microcrystalline cellulose with phenoxyacetyl chloride.[20]

4.8 GAS–LIQUID CHROMATOGRAPHY

Gas–liquid chromatography is a type of partition chromatography; the stationary phase is a high-boiling liquid and the mobile phase is an inert gas. Substances to be separated are partitioned between the stationary and mobile phases because of the differential solubility in the two phases. For the separation of volatile samples, an inert carrier gas sweeps through a long narrow tube packed with a liquid phase. The sample is introduced at one end of the column and is carried through by the inert gas. The column is usually placed in an oven and the separation carried out at one preferred temperature (isothermal) or at variable temperatures. The sample components leave the other end of the column and pass through a detector. The detector response is recorded, and both the position (time) and the size of the response are used to identify and quantify substances being separated (Fig. 4-18). The **retention time** (position of the peak) is used for identification by comparison with the retention time of known standards; the area under the peak is used to estimate the amount of substance present (Fig. 4-19).

Three types of columns are commonly used: packed, wall-coated empty capillary tubes, and support-coated open tubes. In packed columns the support material is usually an inert solid such as diatomaceous earth or fire brick. The

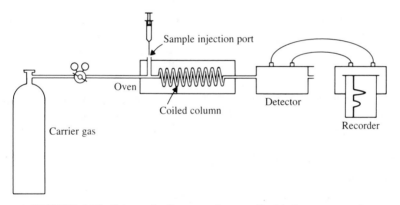

FIGURE 4-18. Schematic diagram of a gas–liquid chromatograph.

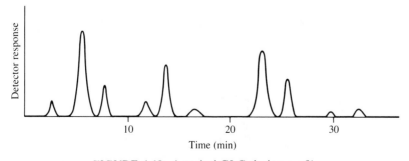

FIGURE 4-19. A typical GLC elution profile.

stationary phase, a high-boiling, thermally stable liquid, is coated onto the inert support. The actual stationary phase liquid is chosen on the basis of the type of substances to be separated (see Part II). Different compounds interact with the stationary phase to different extents and therefore are separated as they are carried through the column. Nonpolar compounds interact with a nonpolar stationary phase, such as siloxanes, and relatively more polar compounds interact with more polar phases such as polyethylene glycols.

Wall-coated, open tubular columns (WCOT) are usually made of glass and have the stationary liquid phase deposited directly onto the glass surface. Support-coated, open tubular columns (SCOT) generally have the liquid phase deposited on a surface of the tube covered with some type of support material. Both these column types are capable of high resolution of very small size samples (and can be used at lower temperatures). Columns can be prepared by the individual investigator, but most researchers use prepacked or custom-packed columns.

The carrier gas should be chemically inert, like helium or nitrogen. The role of the carrier gas is a passive one—merely to provide an inert medium to sweep the sample molecules through the column. The carrier gas should not dissolve in the stationary phase or be adsorbed in it.

There are a number of types of detectors available, but only two are commonly used: flame ionization and thermal conductivity. The flame ionization detector (FID) consists of a hydrogen–air flame burning at the end of a capillary tube. When organic compounds are introduced into the flame, ions are formed and the ions are collected by applying a voltage across the flame. The current resulting from the ion collection is amplified and recorded. The FID responds with high sensitivity to almost all organic compounds but does not respond to water, carbon monoxide, carbon dioxide, or the inert gases. Therefore, the flame ionization detector produces a stable base line, is minimally affected by changes in temperature, and has a relatively high sensitivity for detection of organic compounds (Fig. 4-20).

The thermal conductivity detector consists of a hollow tube with an electrically heated filament situated in the central axis. The filament reaches a steady temperature when the heat gained from electrical energy equals the heat lost to the carrier gas, and the electrical resistance of the wire is constant. When a

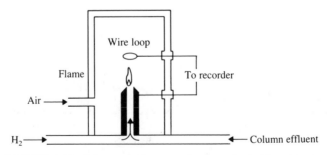

FIGURE 4-20. Schematic diagram of a flame ionization detector.

sample component flows past the filament, it conducts heat away at a rate different from the rate of the carrier gas and the electrical resistance of the filament changes. This change in resistance is registered by a recorder. The detector responds to all substances with a thermal conductivity different from that of the carrier gas, and it is sensitive to changes in temperature and flow rates of the carrier gas, as well. It is less sensitive than the flame ionization detector, but it is nondestructive and allows complete sample recovery.

4.9 HIGH PERFORMANCE LIQUID CHROMATOGRAPHY

High performance liquid chromatography (HPLC) is a newcomer in the field of liquid chromatography. Operating at pressures up to 5000 psi, these instruments give analysis times comparable to GLC. The key to the use of such high pressures was the development of support materials capable of withstanding these high pressures, hence decreasing the distance solute particles had to travel. HPLC was introduced by Horvath who coated glass beads with a layer of ion-exchange resin,[21] and Kirkland who coated glass beads with a thin layer of silica gel.[22] Further increases in chromatographic efficiency and resolution are obtained with smaller particles (5-μm diameter) and other chemically bonded phases on the particle surface. HPLC offers analysis times of 5–30 minutes and is suitable for thermally labile compounds, strongly polar compounds, and biopolymers.

Because of the high pressures used in HPLC, the types of stationary phase supports are limited, and separation is achieved by variation in the composition of the mobile phase. With small column volumes and relatively high flow rates, changes in solvent can be made easily. In addition, multicomponent mobile phases and gradients can be used.

The essential components of a high performance liquid chromatography system (Fig. 4-21) include the following: a high pressure pumping system; a

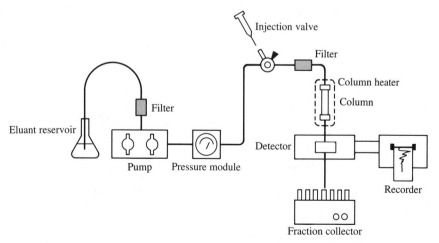

FIGURE 4-21. Schematic diagram of a high performance liquid chromatograph.

narrow-bore, short column with one of a variety of stationary phases bonded to small particles; an on-line, sensitive detection system.

The pump is a critical part of the system. It must be capable of handling pressures up to 5000 psi with a highly reproducible flow rate with minimal pulsation. Constant flow of the mobile phase is essential for achieving sharp separations and reproducible retention times. Constant flow is especially critical if peak area is used to quantitate the analysis. Constant flow can be assured by pulse dampers, multiple-piston pumps, or piston-driven syringe pumps. To generate solvent gradients, two pumps can be used to deliver programmed amounts of two or more solvents. Since samples are applied to the column while under high pressure, special injectors are required. The high pressure sample injection valve (see Fig. 4-22) is the most popular method of sample introduction. After a sample has been loaded with a syringe into the sample loop at standard pressure, the loop is connected to the column. Sample loops can be interchanged to give a volume range of $10 - 2000 \, \mu L$.

The heart of the chromatograph is the column–injector combination. Because of the high pressures present in the column, the separation media must be specially prepared to withstand the extremes of pressure. This is accomplished by using rigid core particles with a shallow pore layer (pellicular) where separation takes place, or fully porous particles of very small diameter ($\leq 5 \, \mu$m).

All the major classes of chromatographic separations are possible with these packings, including adsorption, liquid–liquid partition, ion exchange, exclusion, and affinity. Special mention is made of two types of separation methods developed for HPLC; reserved phase and ion pair partition.

Normal phase chromatography involves the use of polar solid supports and an organic, relatively nonpolar mobile phase solvent. Because highly polar molecules are strongly attracted to the stationary phase, problems of long retention times and peak trailing are associated with this type of system. Reversed-phase HPLC involves the use of a nonpolar stationary phase and a polar mobile phase. The polar molecules now have a higher affinity for the mobile phase and

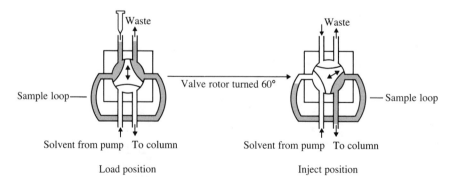

Load position   Inject position

FIGURE 4-22. Schematic diagram of a valve-type sample injection system.

tend to elute relatively quickly. The reversed-phase method is especially useful for polar biomolecules.

Paired-ion chromatography involves the pairing of charged polar substances with their counterion, thus creating a less polar species:

$$R_3NH^+ + ClO_4^- \rightarrow R_3NHClO_4$$

Ion pairs partition into the nonpolar phase, while the free ion R_3NH^+ is attracted to the polar phase.

The three principal types of stationary phases are:

1. *Adsorbents.* Microparticulate silica and alumina are most common.
2. *Size exclusion gels.* Silica and rigid styrene-divinyl benzene copolymers that can withstand high pressure and a variety of aqueous and organic solvents are used.
3. *Chemically bonded stationary phases.* A wide variety of substituents are chemically bonded to the silanol (Si—OH) groups on silica. These include long-chain hydrocarbon groups (e.g., $C_{18}H_{37}$) and such ion-exchange groups

as $-\overset{|}{\underset{|}{Si}}-(CH_2)_n-\overset{+}{N}R_3$, $-\overset{|}{\underset{|}{Si}}-(CH_2)_n-\langle\text{C}_6\text{H}_4\rangle-SO_3^-$, and other polar

groups containing cyano and nitro groups.

The ideal detector for HPLC has high sensitivity and a linear response range over several orders of magnitude of concentration, and can be used for many different compounds. At present two types of detectors are in common use: ultraviolet–visible light absorption, and differential refractive index. The photometric detector can be of fixed wavelength, such as 254-nm radiation from a low-pressure mercury lamp, or of variable wavelength, covering the range from 190 to 750 nm using a deuterium tungsten lamp combination. The volume of the detector cell is very small ($1–20\,\mu L$) with a light path of 1 cm, allowing for high sensitivity of detection. Since the light absorption is measuring a property of the solute rather than a bulk property of the solute–solvent combination, it is ideal for use with gradient elution in which the properties of the solvent are changing. The absorbance method is nondestructive but is useful only with compounds that have a strong UV or visible light absorption.

The differential refractometer detector continuously measures the difference in refractive index between the pure solvent and the solute–solvent phase as it elutes from the chromatography column. This type of detector is of universal application; the only requirement is a difference in refractive index between the solute(s) and the solvent. However, the differential refractometer cannot be used for gradient elution schemes because the base line continuously varies as the solvent composition changes. Other types of detectors are fluorometric, electro-chemical, electron capture, and postcolumn reaction. HPLC systems can be used for preparative separations as well as analytical separations. HPLC has become an important chromatographic technique because it can separate a wide variety of complex mixtures in a very short time.

PART II
SPECIFIC APPLICATIONS OF CHROMATOGRAPHY

4.10 CARBOHYDRATES

4.10.1 Adsorption Chromatography of Carbohydrates

Charcoal has wide use as an adsorbent in the preparative separation of carbohydrates, especially in the separation of mixtures of homologous oligosaccharides.[23] Several types of homologous oligosaccharides have been separated by charcoal chromatography—for example, maltodextrins,[24] isomaltodextrins,[25] cellodextrins,[26] and xylodextrins.[27] The charcoal (Darco G60, Atlas Chemical Industries, Wilmington, DE) is mixed with Celite 560 (Johns Manville) in a dry-weight ratio of 2:1. Elution may be stepwise with increasing concentration of aqueous ethanol[23] or gradient elution (see Section 4.1) with n-butyl alcohol or t-butyl alcohol.[24] Fractions containing oligosaccharides with 2–10 monomer residues can be separated (see Fig. 4-23 for an elution profile of the separation of maltodextrins).

Dextrin derivatives (e.g., acetates) may be separated on silica gel using benzene–ethylacetate (2:1, v/v) eluant with a flow rate of 400 mL/h.[28]

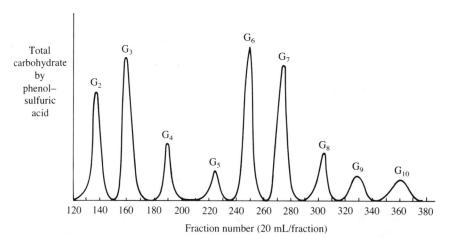

FIGURE 4-23. Elution profile of a charcoal column separation of maltodextrins produced by *Bacillus subtilis* α-amylase digestion of starch: G_n refers to the number of glucose residues in the molecule (e.g., G_4 a tetrasaccharide containing four glucose residues). Maltodextrins (1 g) were added to a column of 100 g of charcoal and 50 g of Celite 560. Separation was achieved by gradient elution with t-butyl alcohol at a flow rate of 3 mL/min.

4.10.2 Ion-Exchange Chromatography of Carbohydrates

Two types of ion-exchange chromatography have been used for the separation of carbohydrates. The first is an anion-exchange resin in the borate form, which separates carbohydrates by forming borate complexes with *cis*-hydroxyl groups of the sugars and sugar alcohols. The second is based on the weak anion exchanger DEAE, which is coupled to dextran, polyacrylamide, or cellulose. It has been used to separate charged carbohydrates containing carboxylate or sulfate groups, found as sugar sulfates, sugar pyruvates, uronic acids, or sialic acids.

The borate exchange method was first reported by Khym and Zill,[29,30] who separated mixtures of carbohydrates on Dowex 1 anion exchanger in the borate form by stepwise elution with borate buffers of pH between 8 and 9. A modification and improvement of the method[31] uses buffers with a pH of 7. For example, a column of Aminex A-14 (Bio-Rad) in the borate form gives a separation of 18 carbohydrates with stepwise elution using two pH 7 buffers and elevated temperatures of 40 and 60°C; see Figure 4-24 for an elution profile.

In the second type of ion-exchange chromatography, the products of partial acid hydrolysis of plant gums can be separated on DEAE–Sephadex G-25 in the formate form. The neutral sugars without borate are eluted with water, and the acidic components, which are bound to the ion-exchange groups, are separated by stepwise elution using increasing concentrations (0.05–0.5 M) of formic

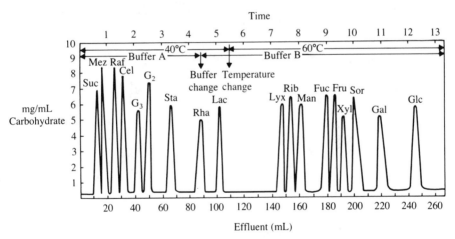

FIGURE 4-24. Elution profile of carbohydrates on Aminex A-14 in the borate form using two pH-7 buffers: buffer A, 0.15 M boric acid, 0.5 M 2,3-butanediol; buffer B, 0.8 M boric acid, 1.0 M 2,3-butanediol. Carbohydrates: Suc = sucrose, Mez = melezitose, Raf = raffinose, Cel = cellobiose, G_3 = maltotriose, G_2 = maltose, Sta = stachyose, Rha = rhamnose, Lac = lactose, Lyx = lyxose, Rib = ribose, Man = mannose, Fuc = fucose, Fru = fructose, Xyl = xylose, Sor = sorbitol, Gal = galactose, Glc = glucose.

acid.[32] Oligogalacturonic acids from the enzymatic hydrolysis of pectic acid can be separated on DEAE–Sephadex G-50 using 0.05–0.275 M solutions of sodium chloride as eluant.[33] In the phosphate form, DEAE–cellulose can be used to fractionate acidic polysaccharides[34] and plant gum polysaccharides[35] by elution with 0.1–0.5 M phosphate buffers, and microcrystalline DEAE–cellulose (Whatman DE-52) can be used to separate a mixture of milk sialyl oligosaccharides, by using pH 5.4 pyridinium–acetate buffer with stepwise elution between 0.012 and 0.06 M.[36]

4.10.3 Paper Chromatography of Carbohydrates

Both analytical and preparative chromatography can be done using Whatman 3MM paper. This paper is heavier than Whatman 1 and is easier to handle when wet with solvent and developing reagents.

The solvent systems recommended for paper chromatography of carbohydrates are frequently ternary mixtures containing water, an organic solvent that is completely miscible with water (e.g., ethanol or pyridine), and an organic solvent that is insoluble or only slightly miscible with water (e.g., nitromethane, 1-butanol, or ethylacetate).

TABLE 4-3. **Descending paper chromatographic mobilities relative to that of D-glucose (R_{Glc}) for several mono-, di-, and trisaccharides and alditols.**

	R_{Glc}	
Sugar	Solvent I[a]	Solvent II[b]
D-Glucose	1.0	1.0
D-Glucitol	0.9	2.4
D-Mannose	1.0	1.2
D-Mannitol	0.96	1.9
D-Fructose	1.2	1.7
D-Galactose	0.8	1.5
D-Xylose	1.3	2.5
Xylitol	—	3.2
D-Arabinose	—	2.3
D-Arabinitol	—	3.0
D-Ribose	1.5	4.4
Ribitol	—	2.9
2-Deoxy-D-ribose	—	5.0
2-Deoxy-D-ribitol	—	4.0
L-Rhamnose	1.6	3.4
L-Rhamnitol	—	4.0
Sucrose	0.72	—
Maltose	0.65	—
Maltotriose	0.38	—
Raffinose	0.27	—

[a] Ethyl acetate–pyridine–water (10:4:3) at 25°C.
[b] Nitromethane–acetic acid–ethanol–water saturated with boric acid (8:1:1:1) run at 37°C.

Descending paper chromatography using ethylacetate–pyridine–water (10:4:3, v/v/v) will separate several mono-, di-, and trisaccharides.[37] Another system using nitromethane–acetic acid–ethanol–water saturated with boric acid (8:1:1:1, v/v/v/v at 37°C) will separate several monosaccharides and their alditols.[38] Table 4-3 lists R_{Glc} values for these two solvent systems. For the separation of mixtures of oligosaccharides containing up to 12 residues, 1-propanol-water (7:3, v/v) can be used at 37°C.

Multiple-ascent paper chromatography can be used to separate oligosaccharides of two to eight residues. Two solvent systems, with three ascents, are commonly used: 1-butanol–pyridine–water (6:4:3, v/v/v) or 1-propanol–water (65:35, v/v). Separations are achieved by using three ascents at 65°C. The elevated temperature gives a reduced running time (2–3 hours per ascent) and better resolution. Both reducing and nonreducing sugars may be visualized on the paper by reaction with alkaline silver nitrate, although the sensitivity is less for the nonreducing sugars.

VISUALIZING SUGARS ON PAPER

Reagent A: 1 mL of saturated silver nitrate is added to 200 mL of acetone; water is added dropwise with mixing to dissolve any precipitated silver nitrate.

Reagent B: 2 mL of 40% sodium hydroxide is added to 200 mL of methanol.

Reagent C: 240 g of sodium thiosulfate, 10 g of sodium sulfite, and 25 g of sodium bisulfite are added to 1 L of water.

The dried chromatogram is evenly dipped into reagent A and air-dried. It is then dipped into reagent B and air-dried. Brown to black spots appear where there are sugars. The chromatogram is rapidly dipped into reagent C and immediately washed for 15 minutes in running water and air-dried.

4.10.4 Thin-Layer Chromatography of Carbohydrates

Carbohydrates may be separated on commercial silica gel plates using a variety of solvents to achieve specific separations. The results of the separation depend on the particular plate used. Whatman K5 silica gel and Merck silica gel 60 plates give good results. The following solvents may be used for the separations illustrated in Figure 4-25.

SOLVENTS FOR TLC SEPARATIONS OF CARBOHYDRATES

Solvent A: Acetonitrile–water (85:15, v/v) with four ascents (45 minutes each for a 20-cm plate) will separate mono-, di-, and trisaccharides (Fig. 4-25A).

Solvent B: Nitromethane–acetonitrile–1-propanol–water (3:4:5:7, v/v) with three ascents will separate a series of isomaltodextrins [(α-1 → 6)-linked glucose dextrins] down to 14–15 glucose residues (Fig. 4-25B).

Solvent C: Nitroethane–nitromethane–ethanol–water–1-propanol (1:2:3:4:5, v/v) with four ascents will separate a series of isomaltodextrins and(α-1→3)-branched isomaltodextrins having 10–11 glucose residues (Fig. 4-25C).

Solvent D: Ethylacetate–methanol–water (37:40:23, v/v/v) with one ascent will separate maltodextrins [(α-1→4)-linked glucose dextrins] having 8-9 glucose residues (Fig. 4-25D).

Solvent E: Nitroethane–ethanol–water (1:3:1, v/v/v) with three ascents will separate isomaltodextrins from maltodextrins down to 8 glucose residues (Fig. 4-25E).

Solvent F: Nitromethane–water–1-propanol (2:3:5, v/v/v) with three ascents will separate a series of maltodextrins or isomaltodextrins down to 25 and 12 glucose residues, respectively, with R_{Glc} values about equivalent for each dextrin size (Fig. 4-25F).

The solvent series *B–F* (Fig. 4-25) illustrates the effect of solvent composition on the separation of similar carbohydrate structures.

The visualization of carbohydrates on thin-layer silica gel plates is obtained by spraying with sulfuric acid–methanol (1:3, v/v) followed by heating for 10 minutes at 110–120°C. Most carbohydrates give black to brown spots on a white background.

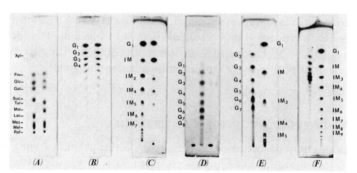

FIGURE 4-25. Separation of carbohydrates by silica gel TLC (Whatman K5 silica gel plates). (*A*) Separation of mono-, di-, and tri-saccharides with four ascents of acetonitrile–water (85:15, v/v). (*B*) Separation of glucose and isomaltodextrins with three ascents of nitromethane–acetonitrile–1-propanol–water (3:4:5:7, v/v). (*C*) Separation of glucose, isomaltodextrins (left side), and isomaltodextrins with one α-1→3 branch linkage (right side) with four ascents of nitroethane–nitromethane–ethanol–water–1-propanol (1:2:3:4:5, v/v). (*D*) Separation of maltodextrins (maltose–maltooctaose) with one ascent of ethylacetate–methanol–water (37:40:23, v/v). (*E*) Separation of glucose, maltodextrins (left side), and isomaltodextrins (right side) with three ascents of nitroethane–ethanol–water (1:3:1, v/v). (*F*) Separation of glucose, maltodextrins (left side), and isomaltodextrins (right side) with three ascents of nitromethane–water–1-propanol (2:3:5, v/v).

4.10.5 Gel Permeation Chromatography of Carbohydrates

The Sephadex and Bio-Gel P series are capable of fractionating polysaccharides up to molecular weights of 200,000. Agarose gels (Sepharose and Bio-Gel A series) may be used to fractionate polysaccharides of molecular weights between 10^4 and 10^7. See Table 4-2 for the exclusion limits of the various gels. Porous glass[39,40] and porous silica[41-43] have also found uses in fractionating polysaccharides.

The highly cross-linked polyacrylamide gel Bio-Gel P-2 ($>$ 400 mesh) in 1–2 m columns is capable of resolving maltodextrins, isomaltodextrins, and their reduced analogues down to 15 glucose residues.[44] For these neutral dextrins, water is used as eluant; 0.1–1.0 M sodium chloride solutions are used for dextrins with acidic or basic groups.

4.10.6 Affinity Chromatography of Carbohydrates

Lectins (hemagglutinating glycoproteins) with different specificities for binding different types of carbohydrate residues may be coupled as ligands to agarose or polyacrylamide gels to form affinity materials. Polysaccharides or glycoproteins containing the specific residues that the particular lectin will bind may be separated from other molecules containing carbohydrate residues that are not bound. Concanavalin A, a lectin from jack beans, has a strong affinity for α-D-mannose and α-D-glucose residues and will bind polysaccharides containing a sufficient number of nonreducing ends with these residues. Table 4-4 lists several lectins along with their carbohydrate specificities.

TABLE 4-4. Carbohydrate binding lectins

Lectin	Carbohydrate specificity
Jack bean concanavalin A	α-D-Man$p$ > α-D-Glc$p$ > α-D-Glc$p$-NAc
Wheat germ	β-D-Glc$p$-NAc (chitin oligosaccharides)
(Edible snail) (*Helix pomatia*)	α-D-Gal$p$-NAc $\gg$ α-D-Glc$p$-NAc
Soybean	α-D-Gal$p$-NAc > β-D-Gal$p$-NAc > α-D-Gal$p$
Giant clam	β-D-Gal$p$-NAc > β-D-Gal$p$
Castor bean	β-D-Gal$p$ > α-D-Gal$p$
Gorse (*Ulex europaeus*)	α-L-Fuc$p$

4.10.7 Gas–Liquid Chromatography of Carbohydrates

Gas–liquid chromatography of carbohydrates has wide application. The carbohydrates are esterified to give acetate or trimethylsilyl derivatives that are volatile at elevated temperatures. Various column supports (Chromosorb WAW DMCS or Gas-Chrom Q, 100–200 mesh) and liquid phases (OV-225: 25% cyanopropyl, 25% phenyl, 50% methyl silicone, or OV-17: 50% phenyl, 50% methyl silicone) are used at column temperatures between 140 and 250°C. The carrier gas is usually helium or nitrogen and the detector is the flame ionization type. Gas chromatography has been effectively applied to the analysis of the different partially methylated carbohydrates that result from the methylation

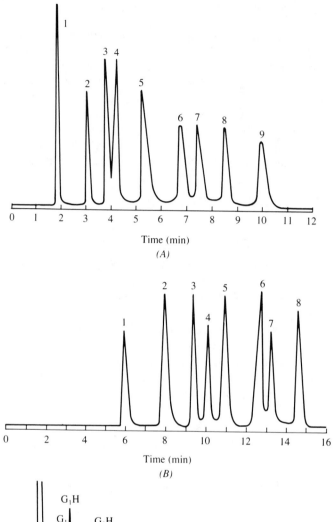

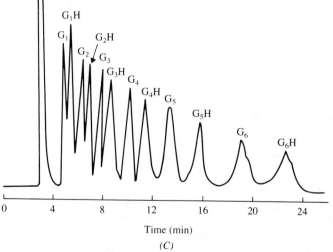

analysis of polysaccharides.[45] For the details of methylation analysis, see Section 10.1.3.1.

The separation of 17 trimethylsilyl disaccharides has been described,[46] including the α- and β-anomers of the reducing disaccharides, using a fused silica capillary column coated with SE-54 and operated isothermally at 240°C. The trimethylsilyl ethers of carbohydrates up to tetrasaccharides were found to be sufficiently volatile for gas chromatography.[47]

Trimethylsilyl derivatives of carbohydrates are prepared by dissolving the carbohydrate (10–70 mg) into 1 mL of pyridine. Hexamethyldisilazane (0.9 mL) and trifluoroacetic acid (0.1 mL) are added and the solution is stirred at room temperature. After 15 minutes of stirring, the solution should be clear with no precipitate and the carbohydrate should be completely derivatized, ready for gas–liquid chromatography.

4.10.8 High Performance Liquid Chromatography of Carbohydrates

HPLC columns and methods have been developed for the separation of carbohydrates. The columns, packings, and conditions for three combinations of carbohydrate mixtures and column systems are given in Figure 4-26. The carbohydrates can be detected by a refractive index detector. The relative quantitative amounts of the carbohydrates may be determined by measuring the areas under the curves with an integrator; and by using various concentrations of maltose as a standard, the absolute amounts may be obtained.[48]

4.11 AMINO ACIDS

4.11.1 Ion-Exchange Chromatography of Amino Acids

Ion-exchange chromatography on sulfonated polystyrene resins is used in the separation and quantitation of amino acids obtained from the acid hydrolysis of proteins and peptides. The methods of separation are based on the pioneering work of Moore and Stein,[49,50] who in 1972 received the Nobel Prize.

FIGURE 4-26. Separation of carbohydrates by HPLC. (*A*) Separation on Alltech 600CH column (300 mm × 4.1 mm) with a flow rate of 3 mL/min of acetonitrile–water (4:1, v/v) at 25°C. Peaks: 1 = solvent, 2 = xylose, 3 = fructose, 4 = glucose, 5 = sucrose, 6 = maltose, 7 = lactose, 8 = melezitose, 9 = raffinose. (*B*) Reversed phase separation on Alltech 700CH column (300 mm × 6.5 mm) with a flow rate of 0.5 mL/min of water at 90°C. Peaks: 1 = maltotriose, 2 = sucrose, 3 = glucose, 4 = galactose, 5 = fructose, 6 = mannitol, 7 = xylitol, 8 = sorbitol. (*C*) Separation of glucose (G_1), sorbitol (G_1H), cellodextrins (cellobiose–cellohexaose, G_2–G_6), and reduced cellodextrins (G_2H–G_6H) on Whatman PXS-1025 PAC column using acetonitrile–water (71:29, v/v) with a flow rate of 1.5 mL/min.[(*C*) Reproduced by permission of authors and publisher, *Anal. Biochem. 82:* 372 (1977).]

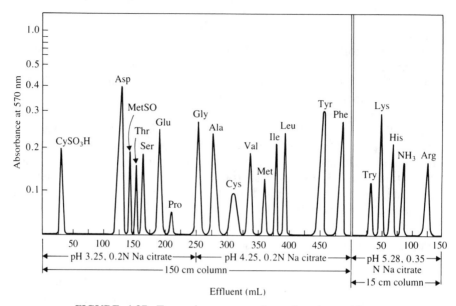

FIGURE 4-27. Two-column separation of amino acids on sulfonated polystyrene resin in the sodium form. The amino acids were detected by reaction with ninhydrin and measurement of the absorbance at 570 nm.

The original procedure used a jacketed column (0.9 cm × 100 cm) of Dowex 50 in the sodium form. The amino acids were eluted with various buffers (pH 3.4–11.0) and at temperatures ranging from 25 to 75°C. The fractions emerging from the column were reacted with ninhydrin and the absorbance was measured at 570 nm.[49] An improved method has been developed using two columns of Amberlite IR-120: a 0.9 cm × 150 cm column run at 50°C with two buffers of pH 3.25 and 4.25 for the acidic and neutral amino acids, respectively, and a 0.9 cm × 15 cm column run at 50°C and pH 5.28 for the basic amino acids.[50] The method has been automated, and several commercial amino acid analyzers are available. A typical analysis is shown in Figure 4-27.

4.11.2 Thin-Layer Chromatography of Amino Acids

Amino acids may be separated by two-dimensional TLC using either silica gel or cellulose as the separating medium. Two different solvents are used for each type of TLC plate and a different type of separation is achieved for each type (see Fig. 4-28 for the two-dimensional amino acid "maps" obtained for these systems[51,52]). The amino acids are visualized with two types of ninhydrin spray for the silica gel and the cellulose gel media.

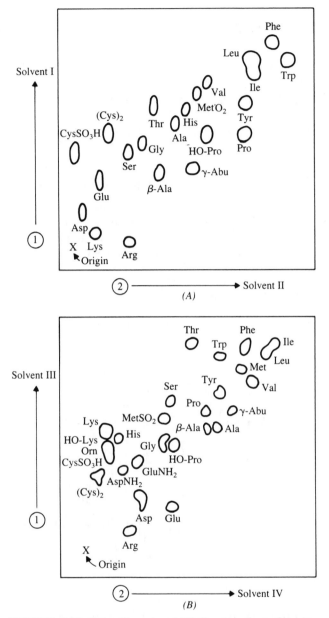

FIGURE 4-28. Two-dimensional TLC separation of amino acids. (*A*) On silica gel G with solvent I, chloroform–17% methanol (v/v)–ammonia (2:2:1, v/v/v) and solvent II, phenol–water (75:25, v/v). From reference (51). (*B*) On cellulose MN 300 with solvent III, 1–butanol–acetone–diethylamine–water (10:10:2:5, v/v/v/v, pH 12.0) and solvent IV, 2–propanol–formic acid (99%)–water (40:2:10, v/v/v, pH 2.5). From Reference (52).

NINHYDRIN SPRAYS FOR AMINO ACID DETECTION

For silica gel TLC: The plate is sprayed with a solution of 300 mg of ninhydrin + 3 mL of glacial acetic acid + 100 mL of butyl alcohol and heated for 10 minutes at 110°C.

For cellulose TLC: The plate is sprayed with a solution of 500 mg of ninhydrin + 350 mL of absolute ethanol + 100 mL of glacial acetic acid + 15 mL of 2,4,6-trimethylpyridine and heated for 10 minutes at 110°C.

4.11.3 High Performance Liquid Chromatography of Amino Acids

With the development of the HPLC technique, the quantitative analysis of a mixture of amino acids has become much faster and easier than was possible with the automated ion-exchange procedures or gas chromatography. The amino acids are precolumn derivatized by phenylthioisocyanate,[53] dimethylamino-

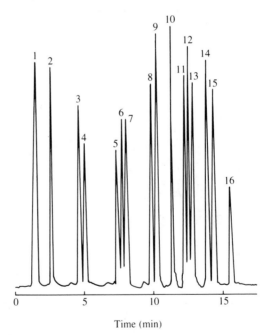

Time (min)

FIGURE 4-29. HPLC analysis of *o*-phthalaldehyde amino acids on a reversed-phase Microsorb Short-one C18 column (Rainin, Woburn, MA) using a 1.7 mL/min gradient elution of solvent A (0.1 M sodium acetate adjusted to pH 6.2 with acetic acid) and solvent B (methanol) with fluorescence detection. Peaks: 1 = Asp, 2 = Glu, 3 = Ser, 4 = His, 5 = Gly, 6 = Thr, 7 = Arg, 8 = Ala, 9 = Tyr, 10 = Abu, 11 = Met, 12 = Val, 13 = Phe, 14 = Ile, 15 = Leu, 16 = Lys.

azobenzene sulfonyl chloride (DABS-Cl),[54] or o-phthalaldehyde[55] for a greater sensitivity of detection. A separation profile of a mixture of the o-phthalaldehyde derivatives of amino acids is given in Figure 4-29. The separation of 16 amino acids is obtained by using a gradient elution with 0.1 M sodium acetate adjusted to pH 6.2 with acetic acid (solvent A) and methanol (solvent B). The amino acids are detected by fluorescence with a sensitivity of 5–10 pmol each.

High performance liquid chromatography is rapidly becoming the method of choice for the analysis of amino acids, displacing older methods that are much more time-consuming, require larger samples, and cost more. The HPLC instrument is less expensive compared with automated ion-exchange systems, and substances of many different types may be analyzed by a simple change of a column, eliminating the necessity of dedicating the instrument for one type of analysis. The derivatives required for HPLC are easily made and are applicable to most of the amino acids; moreover, volatility is not an essential property, as it is for gas chromatography.

4.12 PROTEINS

4.12.1 Adsorption Chromatography of Proteins

Hydroxylapatite, a crystalline form of calcium phosphate, is one of the oldest and most commonly used adsorbents in the purification and separation of proteins. However, preparations vary considerably in their properties. This variability combined with the development of the cellulose ion exchangers led many investigators to ignore hydroxylapatite as an adsorbent for a time. But with the development of standardized commercial preparations, hydroxylapatite is again being considered as a valuable adsorbent for protein purification chromatography. Bio-Rad Laboratories markets four types of hydroxylapatite, and LKB markets a hydroxylapatite-coated gel bead under the trade name HA-Ultrogel.

The mechanism of hydroxylapatite adsorption is complex. The proteins are usually adsorbed at low potassium phosphate concentrations and are eluted from the hydroxylapatite by increasing the potassium phosphate concentration in the eluting buffer. Other salts such as NaCl, KCl, or $CaCl_2$, up to concentrations of 3 M, have very little effect in eluting acidic proteins, but basic proteins are readily eluted by dilute chloride salts, especially $CaCl_2$. Acidic proteins, which are negatively charged at the usual operating pH values, are adsorbed by the Ca^{2+} sites on the crystal surfaces and the chloride anion is not effective in eluting them because chloride has little affinity for Ca^{2+}. The acidic proteins can be eluted by phosphate ions, which have a high affinity for calcium ions. Basic proteins, which are positively charged at the usual operating pH values, are bound to the phosphate sites of the hydroxylapatite and are readily eluted by calcium ions, which have a high affinity for phosphate.

4.12.2 Ion-Exchange Chromatography of Proteins

Ion-exchange chromatography is probably one of the most widely used procedures in purifying proteins. Several hundred proteins have at least one step in their purification that is an ion-exchange procedure. The most commonly used exchangers are DEAE–cellulose, DEAE–Sephadex, DEAE–Bio-Gel, CM–cellulose, CM–Sephadex, and CM–Bio-Gel. The DEAE–anion exchangers are more commonly employed than are the CM–cation exchangers because there are many more acidic proteins than there are basic proteins.

The purification of *E. coli* glutamate decarboxylase on DEAE–cellulose is illustrated in Figure 4-30. The specific activity of this enzyme is increased several-fold by a gradient elution, which elutes the enzyme as a single, relatively sharp peak, which separates it from contaminating protein. Figure 4.31 illustrates the separation of four isozymes of rabbit muscle aldolase on a DEAE–Sephadex column. This chromatographic material provides separation simultaneously by ion-exchange and gel permeation chromatography.

4.12.3 Gel Permeation Chromatography of Proteins

As discussed in Section 4.6, gel permeation chromatography separates molecules according to size: the large molecules, which cannot enter the pores of the gel, are excluded and emerge in the void volume, and the smaller molecules, which can

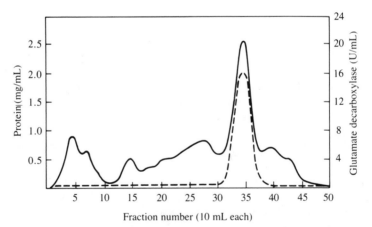

Fraction number (10 mL each)

FIGURE 4-30. Chromatographic purification of *E. coli* glutamate decarboxylase on a DEAE–cellulose column (2.2 cm × 25 cm) that was equilibrated with 0.05 M phosphate buffer (pH 6.0). Crude enzyme was added to the column in the same 0.05 M phosphate buffer, and elution was carried out with a linear gradient using 250 mL of 0.05 M phosphate buffer (pH 6.0) as solvent A and 250 mL of 0.35 M phosphate buffer (pH 6.0) as solvent B; 10-mL fractions were collected and analyzed for protein (solid curve) and glutamate decarboxylase (dashed curve) activity.

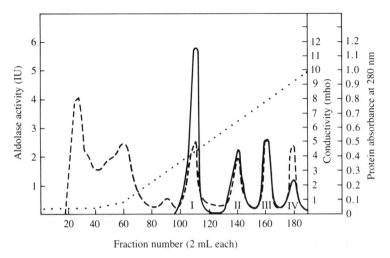

FIGURE 4-31. Chromatographic separation of rabbit muscle aldolase isozymes (I–IV) on a DEAE–Sephadex column (1 cm × 25 cm). The 2-mL fractions were eluted with a linear gradient from 0 to 0.35 M NaCl in 0.01 M Tris, 0.001 M EDTA buffer (pH 7.5). The salt gradient was measured by conductivity (dotted curve) and protein (dashed curve) was measured by absorbance at 280 nm; aldolase activity (solid curve) is given in Iu/mL.

enter the gel, are retarded. Accordingly, gel permeation chromatography can be used to purify proteins. In practice the differences between the molecular weights of two proteins must be at least twofold to obtain sufficient separation by gel chromatography. Figure 4-32 illustrates the separation of two proteins with molecular weights of 50,000 and 25,000 on Bio-Gel P-150. This type of

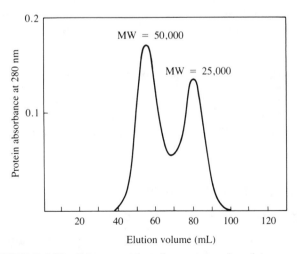

FIGURE 4-32. Gel permeation chromatography of two proteins on Bio-Gel P-150.

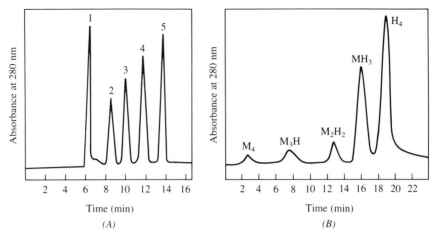

FIGURE 4-33. High performance liquid chromatography of proteins. (*A*) Separation by gel permeation chromatography using Bio-Sil TSK-250 (300 mm × 7.5 mm column), sodium sulfate (0.1 M) and sodium phosphate (0.02 M) solvent at pH 6.8, at a flow rate of 1 mL/min. Peaks: 1 = thyroglobulin, 2 = IgG (immunoglobulin), 3 = ovalbumin, 4 = myoglobin, and 5 = cyanocobalamin. (*B*) Separation by ion-exchange chromatography of heart tissue lactic acid dehydrogenase isozymes on SynChropak AX300 (250 mm × 4.1 mm i.d.), gradient elution (solvent A: 0.02 M Tris–acetate, pH 7.9; solvent B: 0.02 M Tris–acetate, 0.4 M sodium acetate, pH 7.9) at a flow rate of 1 mL/min.

chromatography is most effective when a protein of high molecular weight is to be separated from substances of much lower molecular weight. For example, it might be used to separate an enzyme from its coenzyme or to separate an enzyme from its substrate or its product. Gel permeation chromatography is very mild and rarely, if ever, produces denaturation and loss of biological activity.

4.12.4 High Performance Liquid Chromatography of Proteins

Most of the usual types of chromatography of proteins (adsorption, ion exchange, and gel permeation) may also be obtained with HPLC columns. Figure 4-33 illustrates the HPLC separation of proteins by means of a gel permeation system and an ion-exchange system.

4.13 NUCLEOTIDES AND NUCLEIC ACIDS

4.13.1 Ion-Exchange Chromatography of Nucleic Acids

The separation of the components of nucleic acids was among the earliest applications of ion-exchange chromatography. Cohn in 1953 reported the separation of mono- and oligonucleotides using anion-exchange chromatography.[56]

Strongly basic anion exchangers, such as quaternary amines, covalently bound to divinylbenzene-cross-linked polystyrene, were used to separate products of nucleic acid hydrolysis. The negatively charged nucleotides, at pH 7 or above, bind to the charged amine groups and can be sequentially eluted by decreasing the pH and increasing the salt concentration of the elution buffer.

Currently, DEAE–cellulose ion exchangers are used to give excellent separation of ribonucleotides. The four major ribonucleotides and several minor ribonucleotides have been separated on microgranular DEAE–cellulose, using 70 mM sodium acetate, pH 4.1.[57] Separation of oligonucleotides can be carried out on DEAE–cellulose or DEAE–Sephadex, by using eluants containing 7 M urea. Resolution of either deoxyribo- or ribo-oligonucleotides containing up to 20 residues has been achieved.[58]

4.13.2 Gel Permeation of Nucleic Acids

Intact DNA molecules can be separated from deoxyribonucleotides by gel filtration chromatography. DNA that has been end-labeled with ^{32}P deoxyribonucleotides can be separated from nucleoside triphosphates by passing the mixture through Sephadex G-50, equilibrated with Tris–EDTA buffer (pH 8.0). The intact DNA is excluded from the gel and is eluted in the void volume. Labeled fractions can be detected by Cerenkov counting (Section 6.9.4).

DNA and RNA can be separated from mononucleotides by preparing a column of Sephadex gel and centrifuging the sample solution through the gel to accelerate the separation.[59] First the bottom of a 1-mL disposable plastic syringe is plugged with a small amount of sterile glass wool. The syringe is filled with a Sephadex G-50 slurry equilibrated in Tris–EDTA buffer (pH 8) containing 0.1 M NaCl. The syringe is placed in a centrifuge tube and centrifuged at full speed in a table-top centrifuge ($2000g$) for 4 minutes. The Sephadex packs in the syringe. More Sephadex is added during centrifugation until the column volume is 0.9 mL. The column is washed twice with buffer containing NaCl. The DNA or RNA sample is applied in 0.1-mL increments to the column. A small plastic vial (0.5 mL) is placed under the syringe to collect the effluent and the column is recentrifuged as before. The first 0.1 mL of effluent, which should contain the DNA or RNA, is collected; the mono- and oligonucleotides remain in the syringe (Fig. 4-34).

4.13.3 Affinity Chromatography of Nucleic Acids

After the demonstration of a polyadenylic acid (polyA) segment at the 3' end of most messenger RNA, oligo(dT)–cellulose or polyU–Sephadex have been used as affinity ligands to purify mRNA fractions from total cellular RNA extracts. The RNA molecules with polyA tails are selectively bound to the oligo(dT) moiety and then eluted with buffer of low salt concentration.[60] RNA is dissolved in sterile water and heated to 65°C for 5 minutes to inactivate any ribonucleases present. An equal volume of loading buffer (20 mM Tris–HCl, pH 7.6; 0.5 M

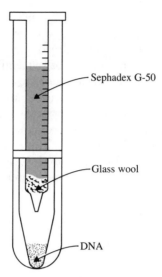

Sephadex G-50

Glass wool

DNA

FIGURE 4-34. Sephadex-filled syringe column for rapid separation of nucleic acids.

NaCl; 1 mM EDTA; 0.1% SDS) is added to the sample, and the solution is added to a 1-mL column of oligo(dT)–cellulose. The affinity support is prepared by washing sequentially with sterile water, 0.1 M NaOH containing 5 mM EDTA, and sterile water until the pH of the effluent is 8 or less. After the sample has been added, the column is washed with loading buffer to elute molecules not bound to the oligo(dT). PolyA containing RNA is eluted with 10 mM Tris–HCl (pH 7.6), 1 mM EDTA, and 0.05% SDS. The eluted mRNA can be rechromatographed on the same column by adjusting the NaCl concentration of the eluate to 0.5 M and reloading the sample onto the column.

4.13.4 High Performance Liquid Chromatography of Nucleic Acids

Nucleic acids have been separated primarily on ion-exchange and size-exclusion packings. Four transfer RNAs were separated by HPLC on a weak anion-exchange packing in less than 50 minutes. RNAs were eluted with a linear salt gradient of 0.25–0.66 M NaCl in 5 M urea, 20 mM potassium phosphate (pH 6.5), and 0.2 mM EDTA.[61]

The preparative separation of high molecular weight RNA by gel filtration, under denaturing and nondenaturing conditions, has been accomplished on an LKB-TSKG 4000 SW column. Samples were applied in nondenaturing buffer (25 mM KCl, 5 mM MgCl$_2$, 50 mM Tris–HCl, pH 7.5) or denaturing buffer (6 M urea, 0.1% SDS, 1 mM EDTA, 75 mM Tris–HCl, pH 7.5). Up to 50 μg of RNA could be separated without loss of resolution on a 17 mm × 600 mm column.[62]

4.14 LIPIDS

4.14.1 Adsorption Chromatography of Lipids

Alumina and silica are widely used in the separation of lipids. Neutral lipids can be obtained on a preparative scale by elution from silicic acid with chloroform, and polar lipids can be recovered by elution with increasingly polar solvent mixtures.[63] Triglycerides (triacylglycerols) from naturally occurring sources are normally eluted together even though they differ in their fatty acid composition. Florisil, a coprecipitate of silica and magnesia, is also used for separation of lipids.

Fractionation of polar lipids into subclasses can be achieved by gradual increases of the polarity of the eluting solvent. Choline phosphoglycerides can be eluted first, followed by phosphatidylethanolamines and phosphatidylserines, and finally by sphingomyelin.[64]

4.14.2 Ion-Exchange Chromatography of Lipids

The use of ion-exchange chromatography for lipid separations has been limited mainly because of the low solubility of lipids in polar solvents. However, Rouser *et al.* have developed a method for separation of acidic lipids from neutral lipids.[65] DEAE–cellulose is packed in glacial acetic acid and washed with methanol. Then the lipid sample is applied in methanol and the nonacidic components are eluted from the column with methanol. Elution of the acidic lipids is accomplished with methanol–chloroform in differing proportions, and finally with glacial acetic acid.

4.14.3 Paper Chromatography of Lipids

Early attempts at lipid separations on cellulose papers were of limited success. Modification of the stationary phase (e.g., Whatman No. 1 paper) by impregnating with silicic acid allows the separation of phospholipids. Excellent separations on Whatman SG-81 silica gel-loaded papers have been achieved in both one- and two-dimensional chromatography.[66] Detection of lipids on paper can be made with high sensitivity and the separated materials can be recovered by simply cutting out the desired spots. Development time of the paper chromatograms is somewhat longer than with TLC plates, and certain corrosive reagents, such as 50% sulfuric acid, cannot be used on paper.

4.14.4 Thin-Layer Chromatography of Lipids

TLC separation of lipids is fast and has excellent resolving power. Silica gel is the most common stationary phase. Chloroform–methanol–water (65:25:4, v/v/v) separates phospholipids, neutral lipids, and cholesterol esters.

A wide variety of reagents have been developed for detection of lipids on TLC. Some of the commonly used reagents include the following.

DETECTION OF LIPIDS ON THIN LAYER CHROMATOGRAMS

1. *Sulfuric acid (40 or 50% aqueous sulfuric acid)*. The chromatogram is sprayed lightly, then heated at 100°C. All organic compounds form dark brown to black spots on a white background.
2. *Iodine vapor*. Most lipids can be detected with I_2 vapor. Crystals of iodine are placed in a jar or tank, and the chromatogram is placed in the I_2 vapor for several minutes. The lipids appear as yellow spots on a white background. The reaction with I_2 is reversible, so the spots will fade when the chromatogram is removed from the I_2 vapor.
3. *2',7'-Dichlorofluorescein (0.2% w/v in 95% ethanol)*. Nonpolar lipids are visible when irradiated with short-wavelength UV light after the chromatogram has been sprayed with dichlorofluorescein.[67]
4. *Rhodamine B (0.05% w/v in 95% ethanol)*. Most neutral lipids and phospholipids can be detected after spraying the chromatogram with ethanolic rhodamine B and irradiation with UV light. Spots show red-purple on a pink background.[68]

DETECTION METHODS FOR SPECIFIC LIPIDS

1. *Cholesterol and cholesterol esters*
   A. *Antimony trichloride in chloroform-saturated solution*. After spraying the TLC, the chromatogram is heated at 100° for 10 min. Cholesterol and cholesterol esters and other steroids yield blue-violet spots.
   B. *Sulfuric acid–acetic acid (1:1, v/v)*. The chromatogram is sprayed and heated at 90°C for 15 minutes. Cholesterol and cholesterol esters yield red to purple spots. Other lipids show pink-brown spots.
2. *Lipids containing free amino groups*. The chromatogram is sprayed with 0.2% ninhydrin in ethanol and heated for 5 minutes at 100°C. Lipids containing primary amines yield purple spots.
3. *The Dragendorf test for choline-containing phospholipids*.
   Reagent I: 1.7 g of basic bismuth nitrate [$Bi(NO_3 \cdot 10H_2O)$] dissolved in 100 mL of 20% acetic acid
   Reagent II: 40 g of potassium iodide dissolved in 100 mL of water
   Reagent III: 20 mL of reagent I is mixed with 5 mL of reagent II and 7.0 mL of water; the solution is filtered to remove any precipitate that forms
   The chromatogram is sprayed with reagent III and heated at 60°C for 2 minutes. Choline-containing lipids appear as orange to orange-red spots on a yellow background.

4.14.5 Gas–Liquid Chromatography of Lipids

Gas–liquid chromatography is well suited to the analysis of fatty acid mixtures or fatty acid methyl esters. Cholesterol esters and some nonpolar lipids can also be separated using GLC. Mixtures of saturated and unsaturated C_{18} fatty acid methyl esters can be quickly and quantitatively separated using high-boiling silicone oils coated on fire brick with helium as the carrier gas (Fig. 4-35).

Acylglycerols can be resolved on the basis of molecular weight and the degree of unsaturation.[69] The column packings used for separation of neutral acylglycerols consist of a porous support and a thin film of liquid phase. SE-30, JXR, or OV-1, all nonpolar siloxane polymers, provide separation based on molecular weight. These liquid phases are stable up to temperatures of 350–375°C. QF-1, a fluoroalkyl-methylsiloxane, exhibits selectivity for steroids. With flame ionization detection, 0.5–2.0 µg of a single component can be detected. The sensitivity limit is about 0.01 µg.

4.14.6 High Performance Liquid Chromatography of Lipids

The separation of total lipid mixtures, achieved with TLC, can more efficiently be carried out with HPLC using silica columns.[70] Free fatty acids have been separated on ion-exchange resin or by reversed-phase chromatography on a C_{18}-bonded material at 50°C using methanol–water mixtures as eluant and using the refractive index (RI) method of detection.[71] Methyl esters of C_{16} and C_{18} fatty acids can be separated by Bondapak C_{18}/Corasil with acetonitrile–water and RI detection.[72] Phospholipids can be separated on MicroPak SI-10 using a

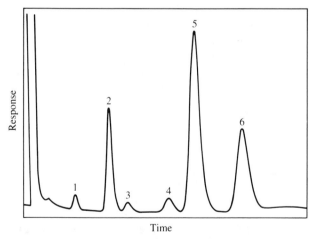

FIGURE 4-35. Separation of fatty acid methyl esters from egg yolk phosphatidylcholine on Silar 10C, using flame ionization detection. Peaks: 1 = methyl myristate, 2 = methyl palmitate, 3 = methyl palmitoleate, 4 = methyl stearate, 5 = methyl oleate, and 6 = methyl linoleate.

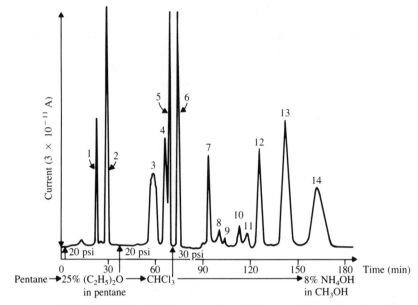

FIGURE 4-36. HPLC separation of fatty acid esters and phospholipids on Corasil II, with a three-step gradient. Peaks: 1 = methyl oleate, 2 = triolein, 3 = cholesterol, 4 = 1,3-diolein, 5 = 1,2-diolein, 6 = monoolein, 7–11 = cerebrosides, 12 = phosphatidylethanolamine, 13 = phosphatidylcholine, and 14 = sphingomyelin. [From A. Stolyhwo and O. Pruett, *J. Chromatog. Sci. 11*: 20 (1973).]

gradient elution of chloroform added to hexane. Lecithin (phosphatidylcholine) can be separated on microparticulate silica using acetonitrile–methanol–water 75:21:4.[73] A three-step gradient from pentane to pentane–diethyl ether to chloroform has been used to separate components ranging from fatty acid esters to phospholipids on pellicular silica columns (Fig. 4-36).

4.15 LITERATURE CITED

1. A.J.P. Martin and R.L.M. Synge, "A new form of chromatogram employing two liquid phases: 1. Theory of chromatography. 2. Application to the microdetermination of the higher monoamino acids in proteins," *Biochem. J. 35*: 91, 1358 (1941).
2. R. Consden, A. H. Gordon, and A.J.P. Martin, "Qualitative analysis of proteins: A partition chromatographic method using paper," *Biochem. J. 38*: 224 (1944).
3. E. A. Peterson and H. A. Sober, in *A Laboratory Manual of Analytical Methods of Protein Chemistry*, P. Alexander and R. J. Block, Eds. Pergamon Press, New York, 1960, pp. 88–102.
4. E. A. Peterson, "Cellulosic ion exchangers," in *Laboratory Techniques in Biochemistry and Molecular Biology*, Vol. 2, Part II, T. S. Work and E. Work, Eds. Elsevier, Amsterdam, 1970, pp. 228–400.

5. D. French, A. O. Pulley, M. Abdullah, and J. C. Linden, "Two-dimensional paper chromatography interspersed with reaction on the paper," *J. Chromatog. 24*: 271 (1966).

6. D. French and G. M. Wild, "Correlation of carbohydrate structure with papergram mobility," *J. Am. Chem. Soc. 75*: 2612 (1953).

7. J. Porath and P. Flodin, "Gel filtration: A method for desalting and group separation," *Nature, 183*: 1657 (1959).

8. J. R. Whitaker, "Determination of molecular weights of proteins by gel filtration on Sephadex," *Anal. Chem. 35*: 1950 (1963).

9. P. Andrews, "The gel filtration behaviour of proteins related to their molecular weight over a wide range," *Biochem. J. 96*: 595 (1965).

10. G. A. Locasio, H. A. Tigier, and A. M. del C. Battle, "Estimation of molecular weights of proteins by agarose gel filtration," *J. Chromatogr. 40*: 453 (1969).

11. R. Axen, J. Porath, and S. Ernback, "Chemical coupling of peptides, and proteins to polysaccharides by means of cyanogen halides," *Nature, 214*: 1302 (1967).

12. J. Porath, R. Axen, and S. Ernback, "Chemical coupling of proteins to agarose," *Nature, 215*: 1491 (1967).

13. J. Porath, "Molecular sieving and adsorption," *Nature, 218*: 834 (1968).

14. J. K. Inman and H. M. Dintzis, "The derivatization of cross-linked polyacrylamide beads. Controlled introduction of functional groups for the preparation of special-purpose, biochemical adsorbents," *Biochemistry, 8*: 4074 (1969).

15. C. F. Lane, "Sodium cyanoborohydride—A highly selective reducing agent for organic functional groups," *Synthesis, 135* (1975).

16. P. Cuatrecasas, "Protein purification by affinity chromatography: Derivatives of agarose and polyacrylamide beads," *J. Biol. Chem. 245*: 3059 (1970).

17. "Coupling reactions of ligands and general methodology of affinity chromatography," in *Methods in Enzymology*, Vol. 34, B. Jakoby and M. Wilchek, Eds. Academic Press, New York, 1974, pp. 13–305.

18. P. Cuatrecasas and C. Anfinsen, "Affinity chromatography," in *Methods in Enzymology*, Vol. 22, W. B. Jakoby, Ed. Academic Press, New York, 1971, pp. 355–378.

19. S. Shaltiel, "Hydrophobic chromatography," in *Methods in Enzymology*, Vol. 34, B. Jakoby and M. Wilchek, Eds. Academic Press, New York, 1974, pp. 126–139.

20. L. G. Butler, "Enzyme immobilization by adsorption on hydrophobic derivatives of cellulose and other hydrophobic materials," *Arch. Biochem. Biophys. 171*: 645 (1975).

21. C. Horvath, B. Preiss, and S. Lipsky, "Fast liquid chromatography: An investigation of operating parameters and the separation of nucleotides on pellicular ion exchangers," *Anal. Chem. 39*: 1422 (1967).

22. J. Kirkland, "High-speed liquid chromatography with controlled surface porosity supports," *J. Chromatogr. Sci. 7*: 7 (1969).

23. R. L. Whistler and D. F. Durso, "Chromatographic separation of sugars on charcoal," *J. Am. Chem. Soc. 72*: 677 (1950).

24. D. French, J. F. Robyt, M. Weintraub, and P. Knock, "Separation of maltodextrins by charcoal chromatography," *J. Chromatogr. 24*: 68 (1966).

25. W. J. Whelan, "Isomaltodextrins," *Methods Carbohydr. Chem. 1*: 321 (1962).

26. G. L. Miller, J. Dean, and R. Blum, "A study of methods for preparing oligosaccharides from cellulose," *Arch. Biochem. Biophys. 91*: 21 (1960).

27. J. Havlicek and O. Samuelson, "Chromatography of oligosaccharides from xylan by various techniques," *Carbohydr. Res. 22*: 307 (1972).

28. S. Dziedzic and M. W. Kearsley, "Preparation of macro quantities of glucose oligomers," *J. Chromatogr. 154*: 295 (1978).

29. J. X. Khym and L. P. Zill, "The separation of monosaccharides by ion exchange," *J. Am. Chem. Soc. 73*: 2399 (1951).

30. J. X. Khym and L. P. Zill, "The separation of sugars by ion exchange," *J. Am. Chem. Soc. 74*: 2090 (1952).

31. E. F. Walborg, Jr. and L. E. Kondo, "Automated system for ion-exchange chromatography of saccharides," *Anal. Biochem. 37*: 320 (1970).

32. G. O. Aspinall, J. A. Malloy, and C. C. Whitehead, "*Araucaria bidwillii* gum. Part III. Partial acid hydrolysis of the gum," *Carbohydr. Res. 12*: 143 (1970).

33. M. Duckworth, K. C. Hong, and W. Yapha, "The agar polysaccharides of *Gracilaria* species," *Carbohydr. Res. 18*: 1 (1971).

34. H. Neukom and W. Kuendig, "Fractionation on diethylaminoethylcellulose columns. Fractionation of neutral and acidic polysaccharides by ion-exchange column chromatography on diethylaminoethyl(DEAE)–cellulose," *Methods Carbohydr. Chem. 5*: 14 (1965).

35. G. O. Aspinall and G. R. Sanderson, "Plant gums of the genus *Sterculia*. Part IV. Acidic oligosaccharides from *Sterculia urens* gum," *J. Chem. Soc.* 2256 (1970).

36. D. F. Smith, D. A. Zopf, and V. Ginsburg, "Fractionation of sialyl oligosaccharides of human milk by ion-exchange chromatography," *Anal. Biochem. 85*: 602 (1978).

37. L. Hough and J. K. N. Jones, "Chromatography on paper," *Methods Carbohydr. Chem. 1*: 21 (1962).

38. J. F. Robyt, "Paper chromatographic solvent for the separation of sugars and alditols," *Carbohydr. Res. 40*: 373 (1975).

39. A.C.M. Wu, W. A. Bough, E. C. Conrad, and K. E. Alden, Jr., "Determination of molecular-weight distribution of chitosan by high-performance liquid chromatography," *J. Chromatogr. 128*: 87 (1976).

40. A. M. Basedow, K. H. Ebert, H. Ederer, and H. Hunger, "Die Bestimmung der Molekulargewichtsverteilung von Polymeren durch Permeationschromatographie an porösem Glas. 1. Dextrane bis zur einem Molekulargewicht von 600,000," *Makromol. Chem. 177*: 1501 (1976).

41. S. A. Barker, B. W. Hatt, J. B. Masters, and P. J. Somers, "Fractionation of dextran and hyaluronic acid on porous silica beads," *Carbohydr. Res. 9*: 373 (1969).

42. S. A. Barker, B. W. Hatt, and P. J. Somers, "Automated, continuous analysis for concentration of total carbohydrate and reducing end-groups in the fractionation of dextran on porous silica beads," *Carbohydr. Res. 11*: 355 (1969).

43. F. A. Buytenhuys and F.P.B. van der Maeden, "Gel-permeation chromatography on unmodified silica using aqueous solvents," *J. Chromatogr. 149*: 489 (1978).

44. E. Schmidt and B. S. Enevoldsen, "Comparative studies of malto- and isomalto-oligosaccharides and their corresponding alditols," *Carbohydr. Res. 61*: 197 (1978).

45. G.G.S. Dutton, "Applications of gas–liquid chromatography to carbohydrates. Part II," *Adv. Carbohydr. Chem. Biochem. 30*: 9–110 (1974).

46. Z. L. Nikolov and P. J. Reilly, "Isothermal capillary column gas chromatography of trimethylsilyl disaccharides," *J. Chromatogr. 254*: 157 (1983).

47. C. C. Sweeley, R. Bentley, M. Makita, and W. W. Wells, "Gas–liquid chromatography of trimethylsilyl derivatives of sugars and related substances," *J. Am. Chem. Soc. 85*: 2497 (1963).

48. K. Kainuma, T. Nakakuki, and T. Ogawa, "High-performance liquid chromatography of maltosaccharides," *J. Chromatogr. 212*: 126 (1981).

49. S. Moore and W. H. Stein, "Chromatography of amino acids on sulfonated polystyrene resins," *J. Biol. Chem. 192*: 663 (1951).

50. S. Moore, D. H. Spackman, and W. H. Stein, "Chromatography of amino acids on sulfonated polystyrene resins: An improved system," *Anal. Chem. 30*: 1185 (1958).
51. A. R. Fahmy, A. Niederwieser, G. Pataki, and M. Brenner, "Dünnschicht-Chromatographie von Aminosäuren auf Kieselgel G. 2. Eine Schnellmethode zur Trennung und zum qualitätiven Nachweis von 22 Aminosäuren," *Helv. Chim. Acta 44*: 2022 (1961).
52. E. von Arx and R. Neher, "Eine multidimensionale Technik zur chromatographischen Identifizierung von Aminosäuren," *J. Chromatogr. 12*: 329 (1963).
53. R. Somack, "Complete phenylthiohydantoin amino acid analysis by high-performance liquid chromatography on ULTRASPHERE–octadecyltri-methyloxysilane," *Anal. Biochem. 104*: 464 (1980).
54. J.-y. Chang, R. Knecht, and D. G. Braun, "Amino acid analysis at the picomole level: Application to the C-terminal sequence analysis of polypeptides," *Biochem. J. (Molecular Aspects) 199*: 547 (1981).
55. K. Blau and G. S. King, *Handbook of Derivatives for Chromatography*. Wiley, New York, 1979, pp. 378–379.
56. W. E. Cohn, "The separation of biochemically important substances by ion-exchange chromatography," *Ann. N.Y. Acad. Sci. 57*: 204 (1953).
57. N. Holmes and G. Atfield, "Chromatographic separation of mononucleotides derived from transfer ribonucleic acids," *Biochem. J. 121*: 371 (1971).
58. G. M. Tener, "Ion-exchange chromatography in the presence of urea," *Methods Enzymol. 12A*, L. Grossman and K. Moldave, Eds. Academic Press, New York, 1967, p. 398.
59. T. Maniatis, E. F. Fritsch, and J. Sambrook, *Molecular Cloning*. Cold Spring Harbor Laboratory, Cold Spring Harbor, NY, 1983, p. 466.
60. H. Avi and P. Leder, "Purification of biologically active globin messenger RNA by chromatography on oligothymidylic acid–cellulose," *Proc. Natl. Acad. Sci. U.S.A. 69*: 1408 (1972).
61. M. Colpan, "Large-scale purification of viroid RNA using Cs_2SO_4 gradient centrifugation and high-performance liquid chromatography," *Anal. Biochem. 131*: 257 (1983).
62. L. Graeve, W. Golmann, P. Foldi, and J. Kruppa, "Fractionation of biologically active messenger RNAs by HPLC gel filtration," *Biochem. Biophys. Res. Commun. 107*: 1559 (1982).
63. C. C. Sweeley, "Chromatography on columns of silicic acid," *Methods Enzymol. 14*, J. M. Lowenstein, Ed. Academic Press, New York, 1969, p. 254.
64. K. Carroll, "Column chromatography of neutral glycerides and fatty acids," in *Lipid Chromatographic Analysis*, Vol. I, 2nd ed., G. Marinetti, Ed. Dekker, New York, 1976.
65. G. Rouser, G. Kritchevsky, D. Heller, and E. Lieber, "Lipid composition of beef brain, beef liver, and the sea anemone: Two approaches to the quantitative fractionation of complex lipid mixtures," *J. Am. Oil Chem. Soc. 40*: 425 (1969).
66. R. E. Wuthier, "Paper for the microanalysis of polar lipids," *J. Lipid Res. 7*: 544 (1966).
67. H. K. Mangold and D. C. Malins, "Fractionation of fats, oils, and waxes on thin layers of silicic acid," *J. Am. Oil Chem. Soc. 37*: 383 (1960).
68. H. Wagner, L. Horhammer, and P. Wolff, "Dünnschichtchromatographie von Phosphatiden und Glykolipiden," *Biochem. Z. 334*: 175 (1961).
69. A. Kukses, "Gas chromatography of neutral acylglycerols" in *Lipid Chromatographic Analysis*, Vol. I, 2nd ed. G. Marinetti, Ed. Dekker, New York, 1976, Chapter 6.

70. A. Pryde and M. T. Gilbert, *Applications of High Performance Liquid Chromatography*. Chapman & Hall, London, 1979.
71. P. Pei, R. S. Henley, and S. Ramachandran, "New applications of high pressure reversed-phase liquid chromatography in lipids," *Lipids, 10*: 152 (1975).
72. G. Arvidson, "Separation of naturally occurring lecithins according to fatty-acid chain-length and degree of unsaturation on a lipophilic derivative of Sephadex," *J. Chromatogr. 103*: 201 (1975).
73. F. Jangalivala, J. Evans, and R. McClure, "High-performance liquid chromatography of phosphatidylcholine and sphingomyelin with detection in the region of 200 nm," *Biochem. J. 155*: 55 (1976).

4.16 REFERENCES FOR FURTHER STUDY

Books
1. I. Chaiken, M. Wilchek, and I. Parikh, *Affinity Chromatography and Biological Recognition*. Academic Press, New York, 1984.
2. P. Cuatracasas, "Affinity chromatography of macromolecules," in *Advances in Enzymology*, Vol. 36, A. Meister, Ed. Wiley, New York, 1972.
3. P.D.G. Dean, W. S. Johnson, and F. A. Middle, Eds., *Affinity Chromatography*. IRL Press, Washington, DC, 1985.
4. E. Heftman, Ed., *Chromatography—A Laboratory Handbook of Chromatographic and Electrophoretic Methods*, 3rd ed. Van Nostrand Reinhold, New York, 1975.
5. W. B. Jakoby, and M. Wilchek, Eds., *Affinity Chromatography Techniques*, Vol. 34 in *Methods in Enzymology* series. Academic Press, New York, 1974.

Manufacturers' Literature
1. *A Laboratory Manual on Gel Chromatography*. Bio-Rad Laboratories, Richmond, CA, 1975.
2. *Gel Filtration: Theory and Practice*. Pharmacia Fine Chemicals, Uppsala, Sweden, 1975.

Volumes in the Journal of Chromatography Library
1. L. S. Ettre and A. Zlatkis, Eds., *75 Years of Chromatography—A Historical Dialogue*, Vol. 17. Elsevier, Amsterdam, 1979.
2. E. Heftman, Ed., *Chromatography*, Part A: *Fundamentals and Techniques*, Vol. 22A. Elsevier, Amsterdam, 1983.
3. E. Heftman, Ed., *Chromatography. Fundamentals and Applications of Chromatographic and Electrophoretic Methods*, Part B: *Applications*, Vol. 22B. Elsevier, Amsterdam, 1983.
4. J. Turkova, Ed., *Affinity Chromatography*, Vol. 12. Elsevier, Amsterdam, 1978.

CHAPTER 5

ELECTROPHORETIC TECHNIQUES

Electrophoresis is the movement of charged molecules in an electric field. It is an important method for the separation of biological molecules because it usually does not affect the native structure of biopolymers and because it is highly sensitive to small differences in both charge and mass. Early references to electrophoretic separations appear in the 1900s. Arne Tiselius did pioneering work on moving boundary electrophoresis (1927) and later developed a zone method for the purification of biomolecules.

5.1 THEORY

The movement of a molecule in an electric field is a function of the net charge q and the frictional coefficient f. The velocity v of a charged molecule in an electric field is a function of the field strength E, thus giving the relationship:

$$v = \frac{Eq}{f} \qquad (5\text{-}1)$$

The ratio of velocity to field strength is defined as *mobility* μ, which equals the charge (q) divided by the frictional coefficient (f):

$$\mu = \frac{v}{E} = \frac{q}{f} \qquad (5\text{-}2)$$

The frictional coefficient is a measure of the hydrodynamic size of a molecule. Increasing molecular size increases the frictional coefficient. Small molecules, such as amino acids, tend to have spherical shapes in solution, and their shape is not as important as size in defining the frictional coefficient. Larger molecules, such as nucleic acids and proteins, have distinctive shapes, and their length-to-width ratios can vary widely. Thus for these molecules both size and shape are important in defining frictional coefficients. Therefore, the mobility of a molecule is a function of size, shape, and charge. Although the movement of complex biological molecules in aqueous solution or in gel media is much more complex than indicated in Equation 5-2, it is still useful in understanding electrophoretic separations.

The most important type of electrophoresis today is **zone electrophoresis**. A spot or band of sample is applied to a solid support, an electric field is applied, and sample molecules move through the support toward the positive or negative electrode, depending on the net charge of the molecule. Support media include paper and cellulose acetate, starch, agarose, polyacrylamide, and silica gels. The primary function of the support is to prevent convection or other disturbances of the sample molecules, but in special cases such as polyacrylamide or agarose gels it may also function as a molecular sieve or as a stabilizing medium for a pH gradient.

The voltage used depends on the nature of the sample and the type of support medium. Higher voltages are used for paper supports than for gels because of the much higher resistance of the paper. Higher voltages are also used to overcome diffusion, which can lead to the spreading of the molecules. With higher voltages, higher currents are created and heat is generated, requiring the cooling of the system.

After separation, compounds may be located on the support by staining with specific dyes, by autoradiography of radioactively labeled compounds, or by observation of fluorescence after irradiation. For preparative separations, compounds can be eluted from the support medium and recovered in purified form.

5.2 TYPES OF ELECTROPHORESIS

5.2.1 Paper and Cellulose Acetate Electrophoresis

Many samples can be separated on a paper support, but the best separations are obtained with compounds of medium to low molecular weight ($\leq 10,000$). The paper is moistened with a minimum of buffer, spotted with the sample, and placed between two compartments. The spot is kept as small as possible by using a small volume of a concentrated sample solution. Since each of the compartments contains buffer and an electrode, the paper forms a bridge through which charged molecules will move when a potential is applied. Buffers should be of relatively low ionic strength and composed of volatile components, such as ammonium acetate or formate, which are easily removed if the separated substances are to be eluted from the solid support and concentrated after electrophoresis. Standards are usually electrophoresed along with the samples to facilitate identification of sample components. A convenient method to monitor the progress of electrophoresis is to include a charged dye, which also moves in the electric field.

Because of the high resistance of paper as a support, relatively high voltages are required (200–300 V), generating heat in the sample. It is, therefore, necessary to cool the system. Figure 5-1 shows a system for electrophoresis with cooling.

FIGURE 5-1. High voltage electrophoresis chamber for paper strips. Cooling water circulates through upper and lower metal plates, which act as efficient heat sinks. Courtesy of Shandon Southern Instruments. Reprinted by permission.

There are several special separation techniques using two-dimensional electrophoresis or combining electrophoresis with chromatography. A sample can be separated first on paper by chromatography, and then in the second dimension with electrophoresis (Fig. 5-2). Alternatively, a sample can be separated by electrophoresis at one pH in the first dimension and at a different pH in the second dimension. Hartley introduced a special technique for proteins and peptides in which two-dimensional electrophoresis is carried out with an oxidation treatment in between. This technique is known as **diagonal electrophoresis** because peptides not changed by the oxidation step remain on a diagonal, while peptides with sulfhydryl groups oxidized to cysteic acid, $R-SO_3^-$, acquire a new negative charge and move away from the diagonal.[1]

Papers for electrophoresis should be of high wet strength and free from contaminants. Since their fibers are oriented in the direction of machining, sample separation is better when the flow of current is parallel to the fiber axis, rather than perpendicular. It is advisable to prewash these papers to remove metal ions and free amino acids, especially if samples are to be eluted and concentrated after electrophoretic separation. Whatman 3MM paper has high wet strength and intermediate thickness (0.33 mm). Whatman No. 1 is a smoother, thinner paper (0.17 mm), but it tears easily when wet.

Paper sheets allow several samples and controls to be simultaneously separated under identical conditions, or a single band of sample can be applied across the sheet for preparative electrophoresis. After separation, components can be visualized using any of a wide variety of staining techniques compatible with the paper support. Primary amines can be located by spraying or dipping in ethanolic ninhydrin (Section 7.2.1). Carbohydrate-containing compounds can be detected by reaction with alkaline silver nitrate (Section 4.10.3). Nucleotide-containing substances can be located by fluorescence upon irradiation with

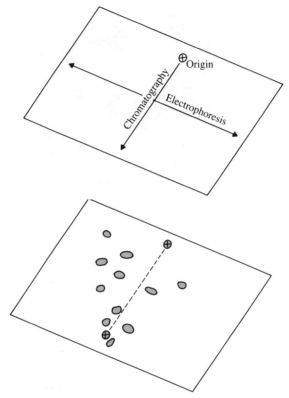

FIGURE 5-2. Two-dimensional separation on paper. The sample is subjected to chromatography in the first dimension and then electrophoresis at 90°.

short-wavelength UV light. Lipids can be detected by exposure to iodine vapor or specific lipid reagents (Section 7.4).

Separated compounds can be recovered from paper after electrophoresis by elution with an appropriate solvent. Sections of paper can be cut out and placed in a test tube of solvent, or compounds can be chromatographed off the paper section by applying solvent at one end of the paper and collecting the solvent as it elutes from the other end (see also Section 4.4).

A modification of paper electrophoresis, introduced by Kohn,[2] is obtained with cellulose acetate produced from the treatment of cellulose with acetic anhydride. This modification decreases the nonspecific interactions between the support and the molecules to be separated, such as proteins and glycoproteins. The cellulose acetate supports are stable to dilute acid and base and so can be used over a wide pH range. Cellulose acetate is commonly used in strips rather than sheets, and is ideal for staining and direct scanning of the stained compounds.

5.2.2 Thin-Layer Electrophoresis

Use of thin layers of support material in electrophoresis leads to high-resolution separation in relatively short time periods, and requires microamounts of sample. Plates are coated with thin layers of cellulose, polyacrylamide gel, or silica gel.

Electrophoresis through thin layers of cellulose or gels is generally carried out on plates 5–10 cm long. The thin-layer plates are connected to the anode and cathode buffer compartments by means of paper wicks (Whatman 3MM), which serve to make contact between buffer on the plates and the buffer troughs. Silica gel on glass or plastic backing allows the use of more aggressive spray reagents such as sulfuric acid in methanol. Ultraviolet-absorbing substances can be detected directly on the thin-layer plate without destruction, and the sample can then be eluted by scraping the thin layer material off the backing into a small test tube or Pasteur pipet. The compound can then be extracted from the thin-layer material with a suitable solvent.

5.2.3 Gel Electrophoresis

The ideal supporting medium for electrophoresis is chemically inert and easy to handle and has controllable porosity. Although paper is the support of choice for some applications such as amino acids, small peptides, or preparative electrophoresis, it has limitations because the porosity cannot be controlled. High molecular weight compounds, proteins and nucleic acids, do not separate well on paper because the cellulose fibers offer too much resistance to movement. Gels with controllable or variable porosity are ideal for the separation of high molecular weight compounds because they permit easy movement of macromolecules while preventing convection and minimizing diffusion. Gels also add the possibility of combining electrophoretic separation with molecular sieving effects—gel permeation—as described in Section 4.6. Gels that effect sieving can be prepared with pore sizes approximately the same as the size of the molecules to be separated by controlling the gel concentration and the degree of cross-linking.

Three types of gels are commonly used—starch, polyacrylamide, and agarose—and the properties, uses, and preparation of each of these are discussed next.

5.2.3.1 STARCH

Electrophoresis in starch gels, the first method in which the sieving effect played a significant role in the separation, was introduced in 1955 by Smithies.[3] Although starch gels generally have been replaced by other, more reproducible materials, they retain some special and useful applications. To prepare the gel, reagent-grade soluble starch is heated in buffer until the starch granules rupture, and a viscous, homogeneous solution (13% starch w/v) is obtained. The solution is degassed and poured into a gel form, usually a thin rectangular slab. The starch is allowed to solidify and cool before the gel is prepared for sample application.

Slots are cut near one end of the gel with a razor blade or knife and samples are added to the slots. After electrophoresis, the gel can be stained intact or it can be cut into layers for multiple staining, as is necessary for determinations of both protein and enzymatic activity (Fig. 5-3). Because soluble starch varies in composition from preparation to preparation, it is difficult to prepare gels of reproducible pore size. Starch gels have been used for separation of hemoglobins and plasma proteins, and for isoenzyme studies.

5.2.3.2 AGAROSE

Agarose gels are formed by suspending dry agarose in aqueous buffer, heating until a clear solution is obtained, and cooling the solution to room temperature. Agarose, the neutral polysaccharide fraction of agar, is composed of a linear polymer of D-galactopyranose linked β-1$\rightarrow$4 to 3,6-anhydro-L-galactopyranose, which is linked α-1$\rightarrow$3 (Fig. 4-15). The concentration of agarose varies from 0.5 to 8% w/v.[4] The agarose gel network that forms on cooling has large pore sizes, which are stabilized by hydrogen bonds. The pore size is controlled by the initial

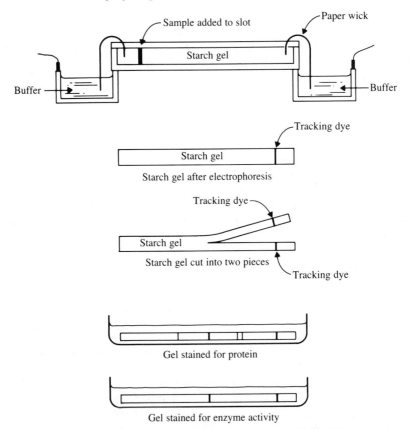

FIGURE 5-3. Starch gel electrophoresis and parallel staining. The gel slab is cut horizontally into two sections for different staining procedures. Courtesy of Bio-Rad Laboratories. Reprinted by permission.

concentration of agarose: large pore sizes are formed from low concentrations and smaller pore sizes are formed from the higher concentrations. The large pore size of the low-concentration gels makes possible the separation of the very large molecules such as DNA,[5] lipoproteins,[6] immunoglobulins,[7] and enzyme complexes. Agarose is the support of choice for separation of DNA strands in preparation for sequencing (Section 10.3.4). DNA fragments can be extracted directly from low-melting agarose ($62-65°C$) after liquefying the agarose gel by heating to $65°C$. Agarose gels are frequently used for immunoelectrophoresis, in which large antigen–antibody complexes are separated, and also for horizontal slab gel isoelectric focusing.

Agarose slurries require no polymerizing or cross-linking agents, which could affect the substances to be separated. Agarose is easily poured, and the gels are physically firmer than low-percentage polyacrylamide gels.

5.2.3.3 POLYACRYLAMIDE

Acrylamide and bisacrylamide can be polymerized in the presence of free radicals to generate gels of varying but controlled pore size. This permits the electrophoretic medium to have a direct effect on the electrophoretic separation. If the pore sizes are approximately the same as the size of the molecules, smaller molecules will move more freely in an electric field and larger molecules will be restricted in their migration.

Raymond and Weintraub[8] introduced polyacrylamide gel layers, and Ornstein and Davis[9] first used electrophoresis in gel rods. Polyacrylamide gel is formed by the polymerization of acrylamide, $CH_2{=}CH{-}C({=}O){-}NH_2$, with the cross-linking agent N,N'-methylene bisacrylamide, $CH_2{=}CH{-}C({=}O){-}NH{-}CH_2{-}NH{-}C({=}O){-}CH{=}CH_2$. The polymerization is catalyzed by agents that generate free radicals, such as ammonium persulfate or riboflavin. The formation of free radicals from ammonium persulfate is catalyzed by N,N,N',N'-tetramethylethylenediamine (TEMED), while visible light causes the photodecomposition of riboflavin to generate free radicals. Oxygen will inhibit the free-radical polymerization, so degassing of solutions used in the polymerization is advised. The pore size in polyacrylamide gel is influenced by the total acrylamide concentration, whereas the stiffness and swelling properties and to a lesser degree the pore size are more affected by the amount of cross-linking agent. A 2.5% gel sieves molecules of 10^5 to 10^6 molecular weight; a 7% gel sieves molecules of 10^4 to 10^5 molecular weight; and a 30% gel sieves molecules of 2×10^2 to 2×10^3 molecular weight. The structure of cross-linked polyacrylamide is shown in Figure 4-14.

Polyacrylamide gels can be formed in cylinders or in flat slabs.[10] Cylindrical gels are formed in glass tubes, 7 mm in diameter, and of varying length (10–20 cm or more). The tubes are easy to load with sample and easy to run. However, only one sample can be run per gel tube, and since conditions of separation may vary from tube to tube, comparison with standards is not always accurate. Gel slabs allow one to run several samples and standards in a single gel under uniform conditions. Heat is more easily dissipated from a thin flat slab than from a cylinder. However, handling of the slab, usually between 0.7 and 1.5 mm thick, can be more difficult than handling gel cylinders.

a. Continuous and Discontinuous Conditions. Gel separations can be accomplished with continuous variables (one gel density, one buffer, and one pH) or with discontinuous conditions (different gel densities, buffers, and/or pH's within the same gel). Discontinuities produce sharp, highly concentrated zones of material and make possible the separation of mixtures of large numbers of different molecular species, or the separation of mixtures of highly similar substances.

Discontinuous polyacrylamide gel electrophoresis is carried out with a stacking gel and a separating gel. The stacking gel, which makes up 10% of the total volume of the gel system, is of a lower polyacrylamide concentration and at a pH different from the separating gel. In the stacking gel, charged molecules move freely under the influence of an electric field, and molecules with similar charge and size accumulate in stacks of closely spaced bands or zones. Let us consider in detail one of the causes of the stacking effect. When a charged protein molecule M lags behind the zone containing like protein molecules, it enters a slower moving zone of different molecules. The molecules in the slower zone have less net charge, and the zone has a lower ionic strength, but a higher field strength (V/cm). Because of the higher field strength, the velocity of the lagging protein molecule M will increase in this zone, and it will catch up to its original zone. If the charged molecule M outruns its own zone and moves into a faster moving zone of molecules with greater net charge, the velocity of M will decrease as a result of the lower potential, and M will be overrun by its own zone. The net result is the stacking of molecules with similar charge and size in bands or zones only a few micrometers thick.

Stacking is increased at each discontinuity in pH, buffer composition, or gel density. In the separating gel, bands separate into discrete areas based on molecular size and shape or on the charge-to-mass ratio.

REAGENTS FOR DISCONTINUOUS POLYACRYLAMIDE
GEL ELECTROPHORESIS

The following systems can be used for the separation and localization of proteins in polyacrylamide gels, using a 7% separating gel (pH 8.8–9.0) and a 2.5% stacking gel (pH 6.6–6.8).

Stock Solutions

A. Acrylamide*	28 g	
Bisacrylamide	0.735 g	
Water	to make 100 mL	
B. Acrylamide*	10 g	
Bisacrylamide	2.5 g	
Water	to make 100 mL	

* *Caution.* Acrylamide has been reported to be a neurotoxin and should be handled with care.

C. *N,N,N',N'*-tetramethylethylene diamine (TEMED) in 3 M Tris–HCl buffer (pH 8.8)

TEMED	0.23 mL
1 N HCl	48 mL
Tris	36.3 g
Water	to make 100 mL, pH 8.8–9.0

D. TEMED in 0.5 M Tris–HCl buffer (pH 6.6)

TEMED	0.46 mL
1 N HCl	48 mL
Tris	6 g
Water	to make 100 mL, pH 6.6–6.8

E. Riboflavin catalyst

Riboflavin	4 mg
Water	to make 100 mL

F. Ammonium persulfate catalyst (prepare fresh for each use)

Ammonium persulfate	0.14 g
Water	to make 100 mL

G. Buffer (10 ×) Tris–glycine

Tris	6 g
Glycine	29 g
Water	to make 1 L

H. Tracking dye

Bromophenol blue	0.005% in water

I. 20% Sucrose

Sucrose	20 g
Water	80 mL

J. Staining solution

Coomassie brilliant blue

R250	0.25 g
Methanol	45 mL
Acetic acid	9 mL
Water	45 mL

K. Destaining solution

Acetic acid	7 mL
Methanol	7 mL
Water	to make 100 mL

Preparation of Acrylamide Solutions

1. Solution for separating gel, 7% acrylamide (pH 8.8–9.0): 1 part A, 1 part C, and 2 parts F are mixed thoroughly. This solution will polymerize in approximately 30 minutes.

2. Solution for stacking gel, 2.5% acrylamide (pH 6.6–6.8): 2 parts B, 1 part D, 1 part E, and 4 parts water are mixed thoroughly.

GEL PREPARATION AND ELECTROPHORESIS

Casting the Gels

Cylindrical Gels   Gels are cast in scrupulously cleaned and rinsed glass tubes (7–8 mm i.d.), covered with a layer of Parafilm on one end. A constant amount of the 7% solution is delivered into each tube with a Pasteur pipet, filling the tubes approximately two-thirds full, and taking care not to trap air bubbles. Water is applied carefully to the top of the gel solution, without mixing, to a depth of 3–4 mm, to give a level gel surface. Complete polymerization, which may require up to 30 minutes, is indicated by the appearance of a thin refractile line between the gel and the water.

The water is poured from the top of the separating gel, and 0.2 mL of the stacking solution is pipetted onto the separating gel. Water is carefully layered onto the top of the stacking solution to a height of 3–4 mm. The riboflavin-containing solution is polymerized by placing the tubes close to a fluorescent light for approximately 20 minutes. The water is removed from the top of the gel just before the sample is added.

Slab gels   Glass plates are washed with detergent, rinsed thoroughly with water, then with ethanol, and wiped dry with tissue. Plastic spacers are placed on three edges of the plate, leaving a small gap (2 mm) along the outside edge of the plate to accommodate the agarose seal. Two glass plates are clamped together with two bulldog clamps on each side.

A 1% agarose gel is warmed until the gel liquifies and then a bead of agarose is applied on the sides and across the bottom of the glass plates to make a seal. To form the separating gel, degassed acrylamide solution is added between the glass plates with a Pasteur pipet or a syringe to about 20 mm from the top of the plates. Water-saturated 1-butanol is layered carefully across the top of the acrylamide solution while polymerization is occurring. The aqueous butanol solution is poured or aspirated off the gel after polymerization, and the gel is rinsed once with water. To form the stacking gel, 1 mL of acrylamide solution is added between the plates and spread across the top of the gel with a rocking motion. Then the plates are filled with stacking solution and the sample-well-comb is inserted into position at the top of the plates. After polymerization, the comb is very carefully removed by lifting from each end with a rocking motion. Just before electrophoresis, the spacer across the bottom of the gel is removed and the gap is filled using a Pasteur pipet with 1% agarose solution to exclude air bubbles. Gels can be stored in the refrigerator for 2–3 days before being used.

Applying the Sample

The sample is mixed with an equal volume of 20% sucrose or 50% glycerol. For a mixture of proteins, the best resolution occurs with a maximum of 200 μg per gel; for highly purified protein samples, 10 μg of the protein component is necessary to obtain a clearly visible band with protein-staining dye. The sample is applied to the top of the stacking gel with a Pasteur pipet. The sample well in a slab gel holds up to 50 μL; cylinder gels can accommodate sample volumes as large as 200 μL.

Electrophoresis

Cylindrical gels   The gel tubes are placed in the electrophoresis apparatus and filled with buffer G, diluted 1:10, by carefully layering the buffer onto the dense sample solution in the gel tube (see Fig. 5-4). The upper and lower reservoirs are filled with buffer G, diluted 1:10. Air bubbles at the top of the sample and at the bottom of the gel must be removed before electrophoresis begins. A few drops of tracking dye solution H are added to the individual tubes, or 4 mL of dye solution H is added to the upper reservoir. The anode is connected to the bottom reservoir and the cathode to the upper reservoir. The high voltage power supply is turned on and the current adjusted to 2–3 mA per gel. Electrophoresis is allowed to proceed until the tracking dye has almost reached the bottom of the gel. Then the power is turned off and the gel tubes are removed. The gel tubes are placed in ice water for 10 minutes to aid in the removal of the gel from the tube. The gels are loosened from the tubes with a hypodermic needle inserted carefully between the gel and the side of the tube. The gel can be displaced into a small tube by squeezing water from a dropper bulb into the gel tube. Because gels are fragile and broken pieces are useless, care should be taken to obtain the separating gel as an intact cylinder.

Slab gels   Gels are run in the vertical dimension. Glass plates (0.3 mm thick, 17 cm × 19 cm) rest on ledges in the lower buffer compartment. The upper

FIGURE 5-4. A typical apparatus for cylinder gels. The upper buffer chambers can accommodate 12 or 18 cylinders. The central core provides cooling of the buffer during electrophoresis. Courtesy of Hoefer Scientific Instruments, San Francisco, CA, designer and manufacturer. Reprinted by permission.

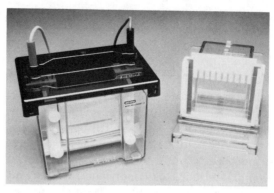

FIGURE 5-5. A typical apparatus for vertical slab gels. The number of sample wells is determined by the number of teeth in the plastic comb used to form the wells. A central cooling core acts as a heat exchanger.

electrode is the cathode and the lower is the anode. Contact with the upper electrode compartment is made by the buffer covering the top of the gel itself (Fig. 5-5). Air bubbles in the sample wells and at the bottom of the gel must be removed before application of samples and electrophoresis. A drop of tracking dye is added to each sample, and the sample is transferred to the well with a Pasteur pipet or drawn-out capillary tube. The voltage is adjusted to give 25–30 mA per gel slab. Electrophoresis is allowed to proceed until the tracking dye has almost reached the bottom of the gel. The power is turned off, the gel plates are removed from the tank, and the top glass plate is carefully separated from the gel with a thin spatula. The gel slab is carefully peeled off the remaining plate and floated into a shallow tray containing buffer or staining solution.

Staining the Proteins in the Gel

The gels are suspended in staining reagent J for 30–180 minutes. Background stain is removed by suspending the stained gels in reagent K. The destaining reagent is replaced every 30 minutes until a clear background is obtained. The period in the destaining reagent need not be timed precisely and may be distributed over several days if necessary. Proteins will appear as dark blue bands on a light blue background.

b. Dissociating Conditions. High molecular weight proteins, protein complexes, or nucleic acids are frequently electrophoresed under dissociating conditions—that is, in the presence of agents that disrupt the native structure. Sodium dodecyl sulfate (SDS), is a detergent that has both polar and nonpolar properties. SDS binds to most proteins, with the nonpolar hydrophobic portion buried in the nonpolar regions of the protein and the negatively charged sulfate portion exposed to the solvent. Binding of SDS leads to two effects. The first is interference with native hydrophobic and ionic interaction, causing the dissociation of most oligomeric proteins into their monomer subunits and the disruption of their secondary structure. If the polypeptide chains are joined by covalent disulfide linkages, additional treatment such as heating in the presence of disulfide reducing agents (i.e., β-mercaptoethanol or 1,4-dithiothreitol) is necessary to dissociate the protein into its smallest subunits. The second effect occurs when proteins become saturated with negatively charged SDS molecules (1.4 g of SDS per gram of protein), and the electrophoretic separation then depends exclusively on the molecular weight: low molecular weight monomers move faster because the frictional coefficient is smaller.

Electrophoresis under these dissociating conditions can be used to estimate the molecular weight of proteins. Weber and Osborn[11] showed that molecular weights of most proteins could be determined by measuring the mobility in polyacrylamide gels containing SDS. Protein standards of known molecular weight are electrophoresed and their mobilities measured and plotted as a function of the log of the molecular weight (Fig. 5-6). This graph yields a straight-line relationship. The mobility of an unknown protein can be determined under identical conditions, and the molecular weight can be obtained from the plot of

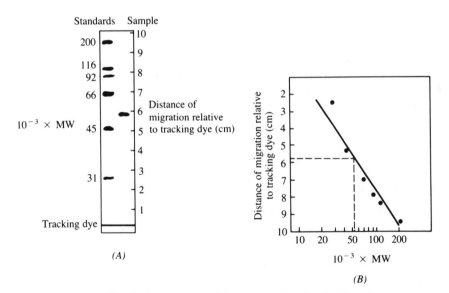

FIGURE 5-6. Molecular weight estimation by SDS–polyacrylamide gel electrophoresis (SDS–PAGE). (A) Separation of molecular weight standards and sample in 7% SDS–PAGE. (B) Calibration curve for molecular weight estimation. From the migration of the sample band, a molecular weight of about 54,000 is obtained.

the mobilities of the standards versus their respective log molecular weights. Molecular weights determined by electrophoresis under dissociating conditions can be determined with an accuracy ranging between 5 and 10%.

SDS–PAGE CONTINUOUS ELECTROPHORESIS: WEBER–OSBORN PROCEDURE[11]

Stock Solutions

A. Acrylamide-bisacrylamide*

Acrylamide	28 g
Bisacrylamide	0.735 g
Water	to make 100 mL

B. Sodium phosphate buffer, 0.2 M, pH 7.2

$NaH_2PO_4 \cdot H_2O$	7.8 g
$Na_2HPO_4 \cdot 7H_2O$	38.6 g
Water	to make 1 L

C. Sodium dodecyl sulfate, 10%

SDS, electrophoresis quality	10 g
Water	to make 100 mL

D. Catalyst (prepared fresh for each use)

Ammonium persulfate	0.14 g
Water	to make 100 mL

E. TEMED (N,N,N',N'-tetramethylethylenediamine)

To prepare 7.5% separating gel solutions (100 mL total volume), mix the following together thoroughly:

Solution A	26.7	mL
Solution B	10	mL
Solution C	1	mL
Solution D	3	mL
Solution E	0.05	mL

Water to make 100 mL

The gels are cast either as cylinders or slabs (Section 5.2.3.3).
The reservoir buffer is a 1:10 dilution of B, with 0.1 volume of C.
To prepare the sample, the following are mixed thoroughly:

Protein (1–2 mg/mL in 10 mM phosphate buffer, pH 7.2)	10–50 μL
10% SDS	5 μL
2-Mercaptoethanol	5 μL

The mixture is heated in a boiling water bath for 3–5 minutes and then cooled, and an equal volume of 20% sucrose or 20% glycerol is added, as well as 10 μL

* *Caution.* Acrylamide has been reported to be a neurotoxin and should be handled with care.

of 0.005% bromophenol tracking dye. Insoluble material is removed by centrifugation. Commercially available protein molecular weight standards should be prepared for electrophoresis under conditions identical to the sample proteins.

SDS–PAGE DISCONTINUOUS ELECTROPHORESIS: LAEMMLI PROCEDURE[10]

Stock Solutions

A. Acrylamide-bisacrylamide*
Acrylamide	30 g
Bisacrylamide	0.8 g
Water	to make 100 mL

B. Stacking gel buffer, 0.5 M Tris–HCl, pH 6.8
Tris	6.0 g
HCl, 1 M	48 mL
Water	to make 100 mL

C. Resolving gel buffer, 3.0 M Tris–HCl, pH 8.8
Tris	36.3 g
HCl, 1 M	48 mL
Water	to make 100 mL

D. Sodium dodecyl sulfate, 10%
SDS	10 g
Water	to make 100 mL

E. Ammonium persulfate, 1.5%
Ammonium persulfate	1.5 g
Water	to make 100 mL

F. Riboflavin, 0.004%
Riboflavin	4 mg
Water	to make 100 mL

G. TEMED (N,N,N',N'-tetramethylethylenediamine)

H. 0.25 M Tris, 1.92 M glycine, 1% SDS, pH 8.3
Tris	30.3 g
Glycine	144 g
SDS	10 g
Water	to make 1000 mL

To prepare separating gel (30 mL, 7.5%), mix the following thoroughly:

Solution A	7.5	mL
Solution C	3.75	mL
Solution D	0.3	mL
Solution E	1.5	mL

* *Caution.* Acrylamide has been reported to be a neurotoxin and should be handled with care.

Water   16.95 mL
Solution G   0.015 mL

To prepare stacking gel (20 mL, 3.75%), mix the following thoroughly:

Solution A   2.5 mL
Solution B   5.0 mL
Solution D   0.2 mL
Solution E   1.0 mL (or 2.5 mL F)
Water   11.3 mL (or 9.8 mL water)
Solution G   0.015 mL

The gels are cast as either cylinders or slabs (Section 5.2.3.3).
The reservoir buffer is a 1:10 dilution of solution H.
To prepare the sample, the following are mixed:

Protein (1–2 mg/mL in 62.5 mM Tris–HCl, pH 6.8)   10–50 μL
10% SDS   10 μL
2-Mercaptoethanol   3 μL

The mixture is heated in a boiling water bath for 3 min, cooled, and an equal volume of 20% sucrose or 20% glycerol added. 10 μL of bromophenol (0.005%) is added, and any insoluble material removed by centrifugation.

Electrophoresis

The upper and lower reservoirs are filled with a 1:10 dilution of solution H. The samples and molecular weight standards are applied to the gel. The anode is connected to the bottom reservoir, the power supply is turned on, and the current is adjusted to 6 mA per cylindrical gel or 25 mA per slab gel. Electrophoresis is continued until the tracking dye has reached the bottom of the gel.

STAINING METHODS FOR GELS

(1) Coomassie Blue Stain for Proteins. Coomassie blue is a general stain for proteins in gels. Gels are stained for 20–30 minutes at room temperature, then removed from the stain and destained for 48 hours. Protein bands appear as dark blue on a light blue background.

Staining Solution

Coomassie brilliant blue R 250   1.25 g
Methanol   227 mL
Glacial acetic acid   46 mL
Water   to make 500 mL

Destaining Solution

Methanol	7 mL
Glacial acetic acid	7 mL
Water	to make 100 mL

(2) Amido Black for Proteins. Amido black is a general protein stain but is less sensitive than Coomassie blue. Gels are stained for 0.5–2 hours at room temperature, then destained for 48 hours. Protein bands appear dark blue.

Staining Solution

Amido black 10B (naphthalene black 10B)	5 g
Ethanol	450 mL
Glacial acetic acid	100 mL
Water	450 mL

The staining solution must be filtered before use.

Destaining Solution

Ethanol	20 mL
Glacial acetic acid	20 mL
Water	160 mL

(3) Silver Stain for Proteins and Nucleic Acids in Polyacrylamide Gels.[12] Silver stain is suitable for both proteins and nucleic acids. The method is up to 100 times more sensitive than Coomassie blue stain.

1. Gels are soaked for 30 minutes in 10% unbuffered glutaraldehyde.
2. Gels are rinsed twice in 1 L of distilled, deionized water for 10 minutes each. The gel is added to 1 L of distilled deionized water and left to soak for at least 2 hours.
3. The water is drained off. Freshly made ammoniacal silver solution is added. To prepare 100 mL of the solution, 1.4 mL of 28%, reagent grade NH_4OH is added to 21 mL of 0.36% NaOH. With stirring, 4 mL of 19.4% $AgNO_3$ is added slowly (20 g of $AgNO_3$ to 100 mL of deionized water). A transient brown precipitate may form but will dissolve when water is added to the final volume. Enough solution is used to allow the gel to float freely and stain for 15 minutes.
4. The gel is removed from the silver solution and washed for 2 minutes in deionized water. Excess silver can be precipitated from the staining solution as AgCl by acidifying with HCl. The ammoniacal silver solution should not be stored because it may become explosive.
5. The gel is transferred to a freshly prepared solution containing 0.005% citric acid, 0.019% formaldehyde (made by dilution of a 38% formaldehyde solution containing 10–15% methanol). The stained bands become visible at this

step. After staining the gel should be washed for at least 1 hour in deionized water with agitation. The minimum protein concentrations detectable are in the range of 0.05–0.25 ng/mm^2. Both single- and double-stranded DNA and RNA can be stained by this method. However, the method is not recommended for agarose gels.

(4) Silver Stain in Agarose Gels. After electrophoresis, gels are fixed for 30 minutes in a mixture of 500 mL of methanol, 120 mL of glacial acetic acid, and 50 g of glycerol made up to 1 L with water. Gels are dried for 90 minutes in a 37°C oven. Oven-dried gels are soaked in distilled water three times for 20 minutes, then placed in a solution of 20 g of $K_4Fe(CN)_6 \cdot 3H_2O$ (ferrocyanide) per liter of water for 5 minutes. Gels are washed three times in deionized water.

To prepare the silver stain, equal volumes of solutions A and B (below) are mixed immediately before use. Solution B must be added to solution A with vigorous stirring.

Solution A	5% sodium carbonate, aqueous	
Solution B	NH_4NO_3	2 g
	$AgNO_3$	2 g
	Tungstosilicic acid $(SiO_3 \cdot 12WO_3 \cdot 26H_2O)$	10 g
	37% formaldehyde	6.7 mL
	Water	to make 1 L

Gels are stained for 10 minutes or until dark bands appear. Staining is stopped by transferring the gel to a 1% acetic acid solution for 5 minutes. Gels are rinsed in distilled water and allowed to air-dry.

(5) Ethidium Bromide* for DNA and RNA. Gels are stained for 45 minutes at room temperature in solutions of ethidium bromide (0.5 μg/mL) in buffer or deionized water. Destain gels in 1 mM $MgSO_4$ for 1 hour at room temperature. DNA and RNA are visualized by irradiating the gel with 254-nm light.

* *Caution.* Ethidium bromide is a powerful mutagen. Always wear gloves and work in a hood when using solid ethidium bromide.

5.2.4 Immunoelectrophoresis

Immunoelectrophoresis[13] exploits the separation of proteins on the bases of their charge-to-mass ratio and their antigenicity. In practice, a mixture of proteins is separated in one dimension in an agarose gel and the proteins are allowed to interact by diffusion with a specific antibody preparation. The antibody diffuses through the gel and forms a visible precipitate with the electrophoresed proteins for which the antibody has affinity (Fig. 5-7). Microscope slides are coated with agarose, and a Pasteur pipet is used to cut two circular sample wells near the outside edges in the center of the slide. Samples of antigen are placed in the wells and separated by electrophoresis. After electrophoresis, a trough in the center of

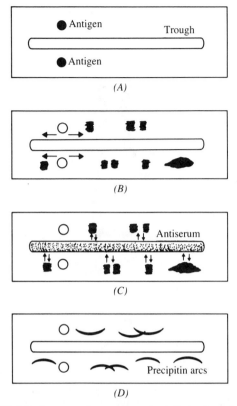

FIGURE 5-7. Immunoelectrophoresis. (A) Antigen is placed in wells on either side of trough. (B) Antigen is electrophoresed. (C) Antiserum (antibody) is placed in trough. Antigen and antiserum diffuse. (D) Insoluble antigen–antibody complexes form precipitin arcs.

the gel, running parallel to the long side of the slide, is cut with a razor blade, and an antibody preparation is added to the trough and allowed to diffuse for 48 hours. If an antigen–antibody reaction occurs, precipitation lines appear in the area of antigen–antibody equivalence. This technique allows the identification of the number of antigens in a sample and indicates the relative amounts of immunoreactive material (Fig. 5-8).

A modification of immunoelectrophoresis is the determination of antigen concentration by means of **rocket electrophoresis**, so called because the patterns that form on the electrophoresis gel look like rockets. Varying amounts of antigen are placed in the sample wells and electrophoresed into the agarose gel, which contains antibody. The antigens are electrophoresed at a pH above their isoelectric pH values (pH$_I$: see Section 5.2.5), where they exist as anions and all migrate in the same direction toward the anode, while the antibodies tend to exist as cations and migrate toward the cathode. As the antigen moves into the gel, it

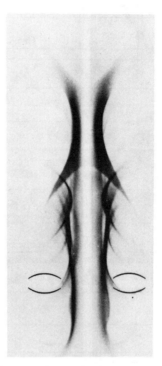

FIGURE 5-8. Immunoelectrophoresis of serum. Equine serum was placed in two wells and after electrophoresis the precipitin pattern was developed with whole equine antiserum. The size of the precipitin arcs indicates the relative amounts of protein present. Photo courtesy of Joann Kinyon, Iowa State University.

encounters antibody, but it does not precipitate until the antigen–antibody concentrations are equivalent. After the precipitate has formed, more antigen will be electrophoresed into the position of the precipitin band, causing the precipitate to dissolve because of excess antigen. The complex will then migrate farther toward the anode. Eventually the antigen concentration becomes constant and the rocket precipitate becomes stationary. The distance traveled by the tip of the rocket is proportional to the amount of antigen in the sample[14] (Fig. 5-9). A standard curve is prepared by plotting known concentrations of the antigen against the distance traveled by the tip of the rocket.

5.2.5 Isoelectric Focusing

If a protein is placed in a pH gradient, it will move under the influence of an electric field until it reaches a place in the pH gradient at which the protein has no net charge. The pH at which the molecule has no net charge is called the **isoelectric pH** or pH_I. With a mixture of proteins, each species will migrate until it

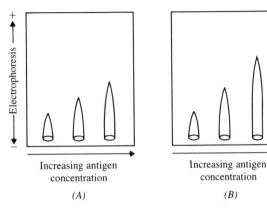

FIGURE 5-9. Rocket immunoelectrophoresis. (*A*) Antigen is placed in the wells of antibody-containing agarose gel. After 60 minutes of electrophoresis, free antigen remains at the leading edge. (*B*) After 300 minutes the antigen concentration is reduced to equivalence with the antibody and the rocket precipitate becomes stationary.

The distance traveled by the tip of the rocket is proportional to the concentration of the antigen in the sample. It should be noted that the lowest antigen concentration travels the same distance in 300 minutes as it did in 60 minutes. This indicates that for this amount of antigen equivalence with antibody was reached after 60 minutes of electrophoresis, whereas the tips of the other two antigen concentrations migrated farther in 300 minutes than they did in 60 minutes.

reaches its own pH_I, at which point it has no net charge and does not migrate farther, giving distinct separation bands for each protein according to their respective pH_I values. This method of separation is termed **isoelectric focusing**.

A stable pH gradient is necessary for isoelectric focusing. This is accomplished with special **ampholytes**—compounds containing both positively and negatively charged groups. Each ampholyte must have buffering capacity and adequate conductivity. Carrier ampholytes are incorporated into the gel solution (2% w/v) before polymerization. The pH gradient is established by adding a buffer of lower pH to the anode chamber and a buffer of higher pH to the cathode chamber. When the current passes through the gel, the ampholyte carriers distribute themselves according to their pH_I values, giving a continuous variation in pH. Ampholyte mixtures usually cover a pH range of 2–3 units, but can also cover the broad range of pH 3–10. The carrier with the most carboxyl groups and the lowest pH_I will be negatively charged and will migrate toward the anode. Conversely, carriers with high pH_I values will be positively charged and will migrate toward the cathode. Gels and density gradients are used to stabilize the pH gradients and give resistance to mixing. The gels should be of high porosity to ensure that the separation is based on net charge only, not on molecular size. Agarose, polyacrylamide and Sephadex are commonly used.

5.3 SPECIAL APPLICATIONS

5.3.1 Proteins and Peptides

Ingram[15] first used combined paper electrophoresis and chromatography to investigate the structures of hemoglobin A and hemoglobin S. He treated hemoglobins A and S with trypsin and used this two-dimensional separation method to look for differences in their primary structure. Ingram called this separation method **peptide fingerprinting**. A sample of peptides is placed on one corner of a paper sheet, which is moistened with buffer with a pH at which all the molecules have a net positive charge, hence will all migrate in the same direction. The high voltages used (up to 2500 V) effect faster separations, decreasing diffusion and sample spreading. The paper is dried after electrophoresis, turned 90°, and subjected to descending chromatography. The procedure can be modified by carrying out chromatography first and then performing electrophoresis. Anfinsen developed this method using high voltage electrophoresis and called it **peptide mapping**.[16] He used potentials of up to 2500 V, with current flow between 75 and 150 mA for a 56-cm sheet of paper (Fig. 5-10). Because this amount of current causes much heating of the system, an efficient method of cooling the paper is essential.

O'Farrell[17] developed a two-dimensional method for separating proteins by combining isoelectric focusing in a polyacrylamide cylinder gel in the first dimension and polyacrylamide slab gel electrophoresis in the presence of SDS in the second dimension. Proteins from *E. coli*, labeled with ^{14}C, were electrophoresed in a long (130 mm) tube of isoelectric focusing polyacrylamide gel. The gels were electrophoresed at 400 V for 12 hours and then 800 V for 1 hour. The gel was extruded from the glass tube and soaked in SDS-containing buffer. The second dimension was a discontinuous SDS–gel system, using an exponential gradient

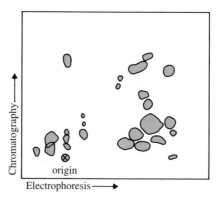

FIGURE 5-10. Two-dimensional separation of tryptic phosphopeptides from transforming protein P-130. Peptides were separated in the first dimension by electrophoresis and in the second by chromatography. [G. Weinmaster and T. Pawson, *J. Biol. Chem. 281*: 328 (1986).]

FIGURE 5-11. Two-dimensional separation of the proteins of *Escherichia Coli* labeled with ^{14}C-amino acids. Proteins were subjected to isoelectric focusing in the first dimension and to electrophoresis in a gradient acrylamide gel containing sodium dodecyl sulfate in the second dimensions. The proteins were detected by autoradiography. Photo courtesy of P. H. O'Farrell.

from 5 to 22% acrylamide, or a 10% running gel and a 4.75% stacking gel. The isoelectric cylinder gel was loaded onto the SDS-polyacrylamide slab gel in 1% agarose to keep the cylinder in place and to prevent the proteins from mixing during their migration out of the cylinder into the slab. The slab gel (164 cm × 146 cm) was electrophoresed at 20 mA for 5 hours or until the tracking dye reached the bottom. The labeled proteins were detected by autoradiography. More than 1000 radioactive spots were detected in the original autoradiogram, indicating the extraordinary resolving power of this two-dimensional system (Fig. 5-11).

Gel electrophoresis of normal human serum is an important clinical method for separating and estimating the amounts of the various protein components. Originally five major fractions were observed in free-boundary electrophoresis, but starch gel and immunoelectrophoresis are capable of resolving serum into 16 different protein components (Fig. 5-12).

5.3.2 Nucleic Acids and Oligonucleotides

5.3.2.1 DNA SEPARATING GELS

The standard method for separating and purifying large DNA fragments is electrophoresis in agarose gels. This simple method is capable of effecting the rapid separation of DNA molecules in a wide range of sizes. DNA bands can be detected in the gel with the fluorescent dye ethidium bromide. As little as 1 ng of DNA can be detected by examination of the gel under UV light. Large DNA

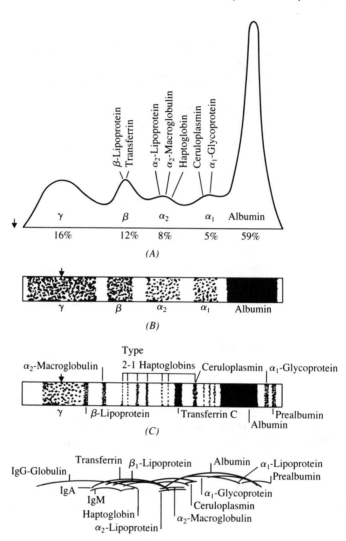

FIGURE 5-12. Schematic representation of the electrophoretic patterns of normal human serum in pH 6.8 buffer as obtained by four methods. (*A*) Tiselius or in free-boundary electrophoresis. (*B*) Paper electrophoresis. (*C*) Starch gel electrophoresis. (*D*) Immunoelectrophoresis. In free-boundary and paper electrophoresis, the serum is resolved into 5 fractions. In starch gel and immunoelectrophoresis, up to 16 components are revealed. From F. W. Putnam, Ed. *The Plasma Proteins*, 2nd Ed., Vol. 1, p. 18. Copyright © 1975 by Academic Press. Reprinted by permission.

molecules, 1–20 kilobases [1 kilobase (kb) = 10^3 nucleotides], are separated in 0.6% gel, and small DNA molecules, 0.1–3 kb, are separated in 2% gel. The maximum resolution is obtained with a field strength of 5 V/cm.

**PREPARATION OF AGAROSE GELS WITH ETHIDIUM BROMIDE
FOR HORIZONTAL ELECTROPHORESIS**

1. The correct amount of powdered agarose, type II, is added to a measured volume of electrophoresis buffer, usually Tris–acetate or Tris–borate, 50 mM, pH 8.2.
2. The slurry is heated in a boiling water bath until the agarose dissolves.
3. The solution is cooled to 50°C and ethidium bromide is added to a final concentration of 0.5 μg/mL.
4. A clean, dry glass plate is sealed around the edges with tape to form a mold (Fig. 5-13, page 154).
5. With a Pasteur pipet, a bead of agarose is spread around the edge of the mold to form a seal.
6. A plastic comb is clamped across the plate to form sample wells, and when the seal is set, the warm agarose solution is carefully poured into the mold.
7. When the gel is completely set, the comb is removed, and the tape is carefully peeled away. The gel is placed horizontally in the electrophoresis chamber, and the electrophoresis buffer is added to cover the top of the gel to a depth of about 1 mm.
8. Samples are loaded through the buffer into the sample wells. The sample contains added sucrose or glycerol, and tracking dye. Since it is more dense than the buffer, the sample will not float out of the sample well before it moves into the gel bed. Several samples and controls can be separated simultaneously. The length of time for electrophoresis will depend on the percentage of gel and the size of the DNA to be separated.
9. Detection of separated bands is accomplished by irradiating the gel with 254-nm light, if ethidium bromide has been incorporated into the gel and the buffer. Alternatively, the electrophoresed gel can be immersed in buffer containing ethidium bromide (0.5 μg/mL) for 30 minutes. The gel is removed from the staining solution and irradiated with UV light.

An alternative method for separation of DNA in polyacrylamide gels and recovery of the DNA from the gel is given in Chapter 10 (Section 10.3.4).

5.3.2.2 DNA SEQUENCING GELS

Polyacrylamide gels are used to analyze and prepare fragments of DNA less than one kilobase long. The gels range in concentration from 3.5% polyacrylamide for separation of large DNA fragments (500–1000 nucleotides) to

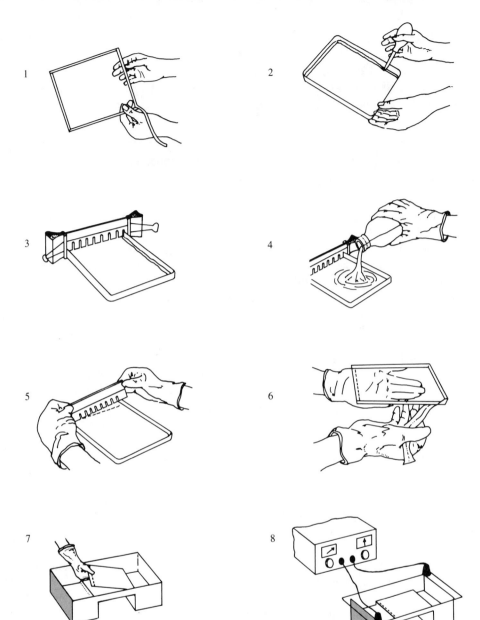

FIGURE 5-13. Steps in the preparation of horizontal agarose gels:

1. A glass plate is sealed around edges with waterproof tape.
2. A bead of agarose is spread around the edge of the plate.
3. A plastic comb is clamped across the plate to form sample wells.
4. Warm (50°C) agarose is poured into the mold.
5. The comb is removed when the gel is set.
6. The tape is peeled away.
7. The gel is placed horizontally in the electrophoresis chamber and covered with buffer. Samples are loaded through the buffer.
8. Gels are electrophoresed at constant current.

20% polyacrylamide for separation of small DNA fragments (2–50 nucleotides). Slab gels are formed between two glass plates to exclude oxygen during the polymerization. The gels can vary in length from 10 to 100 cm, depending on the number of oligonucleotides to be separated, and are almost always run in the vertical position. Scrupulously clean glass plates are separated by thin spacers (≤ 1 mm) along the long edge and taped along the entire length of two sides and the bottom with waterproof tape.

Acrylamide gel solution is poured carefully, or injected from a syringe, between the glass plates. To avoid trapping any air bubbles, the plates are filled until a small excess of liquid is present at the top. A plastic comb, used to form the sample wells, is carefully inserted into the solution at the top of the plates and clamped in position during the polymerization of the acrylamide. The plates are placed horizontally during the 30-minute polymerization period. The tape and comb are removed after polymerization and the gels are pre-electrophoresed to remove excess catalyst and potential oxidizing agents.

Immediately before sample loading, the wells are rinsed with buffer. Drawn-out glass capillaries are used to load the sample into the wells of the gel. Samples should contain 1–10 μg of DNA in 2–3 μL. If a number of samples are to be separated, a few samples can be loaded and subjected to electrophoresis for 5 minutes; then additional samples can be applied and the electrophoresis completed. Electrophoresis is carried out at 1–8 V per centimeter of gel length. At higher voltages, heating of the gel occurs and sample bands may be distorted or the glass plates may break. Marker dyes (bromophenol blue and xylene cyanol) travel in the gel at different rates and are used to measure the progress of separation of the DNA fragments.

The DNA cleavage products from the Maxam–Gilbert DNA sequencing protocol (Section 10.3.4) are separated on ultrathin polyacrylamide gels containing 8.3 M urea. The electrophoresis is carried out at high temperature to ensure denaturing conditions and for periods of 1–8 hours, depending on the size and number of fragments. Fragments are visualized using highly sensitive techniques because each fragment is present in minute amounts. Detection by autoradiography of ^{32}P-labeled nucleotides is the most common method of visualizing the separated DNA fragments (Fig. 5-14). (See also Section 10.3.4.1.)

5.3.2.3 RNA SEPARATION IN HORIZONTAL AGAROSE GELS

RNA species of varying molecular size, from 5s to 30s can be separated and detected in low percentage agarose gels. Agarose (1.5 %) is dissolved by heating (80–90°C) in pH 7.4 buffer containing 40 mM Tris, 36 mM NaH_2PO_4 and 1 mM EDTA. The solution is cooled to 50°C and poured onto a 5 × 8 cm plate as shown in figure 5-13. RNA samples, containing 10–20 μg, are mixed with one-third volume of 20 % sucrose containing 0.1 % pyronin Y dye and added to the sample wells of a gel submerged just under the surface of the buffer in the electrophoresis cell. Electrophoresis is carried out at 10–20 V/cm for 30–45 minutes. RNA binds the pyronin Y and the bands are visible during electrophoresis. The RNA bands can be intensified by placing the gel into a solution of 1% pyronin Y in 15% acetic acid for 60 minutes. The gel is destained in 7% acetic acid until the background has been sufficiently decreased.

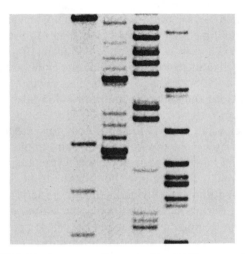

FIGURE 5-14. An autoradiograph of a DNA sequencing gel showing DNA fragments labeled at the 5' end with ^{32}P.

5.3.3 Lipoprotein Separation

Since lipids are not soluble in aqueous solvents, electrophoresis is not commonly used as a separation method. However, lipid–protein complexes are soluble in polar solvents and are characterized by electrophoretic separation in gels.

Serum lipoprotein complexes can be separated into four major classes by polyacrylamide gel electrophoresis. The fastest moving fraction is high density lipoprotein (HDL), followed by low density lipoprotein (LDL) and very low density lipoprotein and chylomicrons. The HDL fraction contains the smallest diameter particles and the chylomicrons the largest diameter particles, hence the observed electrophoretic mobilities. The lipoprotein fractions can be visualized by adding Sudan black B dye, which binds to the lipids and moves through the gels with the various lipoprotein complexes.[18]

5.4 LITERATURE CITED

1. B. S. Hartley, "Strategy and tactics in protein chemistry," *Biochem. J. 119*: 805 (1970).
2. J. Kohn, "A new supporting medium for zone electrophoresis," *Biochem. J. 65*: 9P (1957).
3. O. Smithies, "Zone electrophoresis in starch gels: Group variations in the serum proteins of normal human adults," *Biochem. J. 61*: 629 (1955).
4. H. Hock and C. G. Lewallen, "High concentration agarose gel: A new medium for high resolution electrophoresis," *Anal. Biochem. 78*: 312 (1977).
5. T. Maniatis, E. F. Fritsch, and J. Sambrook, *Molecular Cloning*. Cold Spring Harbor Laboratory, Cold Spring Harbor, NY, 1983, pp. 150–171.
6. A. Van Melsen et al., "A modified method of phenotyping of hyperlipoproteinaemia on agarose electrophoresis," *Clin. Chim. Acta, 55*: 225 (1974).

7. (a) C.-B. Laurell, "Antigen–antibody crossed electrophoresis," *Anal. Biochem. 10*: 358 (1965). (b) R. Verbruggen, "Quantitative immunoelectrophoretic methods," *Clin. Chem. 21*: 5 (1975).

8. S. Raymond and L. Weintraub, "Acrylamide gel as a supporting medium for zone electrophoresis," *Science, 130*: 711 (1959).

9. (a) L. Ornstein, "Disk electrophoresis. I. Background and theory," *Ann. N.Y. Acad. Sci. 121*: 321 (1964). (b) B. Davis, "Disk electrophoresis. II. Method and application to human serum proteins," *Ann. N.Y. Acad. Sci. 121*: 404 (1964).

10. U. K. Laemmli, "Cleavage of structural proteins during the assembly of the head of bacteriophage T4," *Nature, 227*: 680 (1970).

11. K. Weber and M. Osborn, "The reliability of molecular weight determinations by dodecyl sulfate–polyacrylamide gel electrophoresis," *J. Biol. Chem. 244*: 4406 (1969).

12. B. Oakley, D. Kirsch, and N. Morris, "A simplified ultrasensitive silver stain for detecting proteins in polyacrylamide gels," *Anal. Biochem. 105*: 361 (1980).

13. P. Grabar and C. A. Williams, "Méthode permettant l'étude conjuguée des propriétés électrophorétiques et immunochimiques d'un mélange de protéines. Application au sérum sanguin," *Biochim. Biophys. Acta, 10*: 193 (1953).

14. C.-B. Laurell, "Quantitative estimation of proteins by electrophoresis in agarose gel containing antibodies," *Anal. Biochem. 15*: 45 (1966).

15. V. Ingram, "Abnormal human haemoglobins. I. The comparison of normal human and sickle-cell haemoglobins by 'fingerprinting,'" *Biochim. Biophys. Acta, 28*: 539 (1958).

16. A. Katz, W. Dreyer, and C. Anfinsen, "Peptide separation by two-dimensional chromatography and electrophoresis," *J. Biol. Chem. 234*: 2897 (1959).

17. P. O'Farrell, "High-resolution two-dimensional electrophoresis of proteins," *J. Biol. Chem. 250*: 4007 (1975).

18. H. K. Naito, M. Wada, L. Ehrhart, and L. Lewis, "Polyacrylamide-gel electrophoresis as a screening procedure for serum lipoprotein abnormalities," *Clin. Chem. 19*: 228 (1973).

5.5 REFERENCES FOR FURTHER STUDY

1. Z. Deyl, "Electrophoresis—A survey of techniques and applications." *Journal of Chromatography Library*, Vol. 18, Part A. *Techniques*. Elsevier, Amsterdam, 1969.

2. B. D. Holmes and D. Rickwood, Eds., *Gel Electrophoresis of Proteins*. IRL Press, Oxford, 1981.

3. H. R. Maurer, *Disc Electrophoresis*, 2nd ed. de Gruyter, New York, 1971.

4. D. Freifelder, *Physical Biochemistry*, 2nd ed. Freeman, San Francisco, 1982.

5. O. Gaál, G. Medgyesi, and L. Vereczkey, *Electrophoresis in the Separation of Biological Macromolecules*, Wiley-Interscience, New York, 1980.

6. I. Smith, *Chromatographic and Electrophoretic Techniques*, 4th ed, Vol. 2, Heinemann Medical Books, London, 1976.

CHAPTER 6

THEORY, MEASUREMENT, AND
USE OF RADIOISOTOPES

Radioisotopes are atoms with unstable nuclei that undergo transformation into other atoms, which have more stable nuclei. The transformation takes place by the release of energetic particles or radiant energy, which can be quantitatively measured. Radioisotopes participate in chemical and biological reactions just like nonradioactive atoms of the same element. Thus they can be used as quantitative tracers for following a reaction in a biochemical process without disrupting the chemistry of the process.

6.1 NATURE OF ATOMS, ISOTOPES, AND RADIOISOTOPES

Atoms are made up of a dense nucleus of positive charge surrounded by a cloud of negatively charged electrons moving in orbits around the nucleus. The nucleus is an assembly of particles of approximately equal mass called neutrons and protons. The proton carries a positive charge and the neutron is neutral. Electrons are very small particles, each having a mass 1/1836 that of the proton or neutron. The amount of negative charge on the electron equals the amount of positive charge on the proton, and the number of orbital electrons in an atom equals the number of protons in the nucleus; thus the atom as a whole is electrically neutral. The number and arrangement of the orbital electrons determines the chemical properties of the atom. This number is equal to the number of protons, is characteristic of the particular type of atom, and is called the **atomic number**. The weight of the atom is essentially the sum of the weights of the protons and neutrons, since the weight of the electrons contributes only a very small fraction of the weight. The sum of the number of protons and neutrons, thus, makes up the **mass** or **mass number**, or **atomic weight**, of the atom. The predominant form of carbon has 6 protons, 6 neutrons, and 6 electrons and has an atomic number of 6 and an atomic mass of 12; it is symbolized as $^{12}_{6}C$.

For many years, it was believed that all atoms of a particular element were identical in size and mass. Early in the twentieth century, however, atoms were found that had identical chemical properties but different atomic weights. These atoms of different masses but identical chemistry were called **isotopes**. Since their chemistry is the same, all isotopes of a single element have identical numbers of protons and orbital electrons; their different masses are due to a different number of neutrons. Most isotopes of the lighter elements will differ in mass number by one, two, or three mass units or neutrons.

It was further observed that some of the isotopes of an element were unstable and underwent disintegration. This property was correlated with elements having an unequal number of neutrons and protons in the nucleus. For example, the isotopes of carbon—^{11}C, ^{12}C, ^{13}C, and ^{14}C—have 6 protons and 5 neutrons, 6 protons and 6 neutrons, 6 protons and 7 neutrons, and 6 protons and 8 neutrons, respectively. Three of these isotopes have a number of neutrons different from the number of protons. Two, ^{11}C and ^{14}C, are so unbalanced in the number of neutrons to protons that their nuclei are radioactive and eject charged particles in an attempt to achieve more balanced nuclei. Carbon-14 emits an electron and an antineutrino (a very small uncharged particle of negligible mass) converting a neutron into a proton, yielding a nucleus with 7 protons and 7 neutrons, which is nitrogen, $^{14}_{7}N$. Carbon-11, on the other hand, emits a positron, a particle having the same mass as an electron but with a positive charge instead of a negative charge, and a neutrino (see Section 6.4.1). This process converts a proton into a neutron, reduces the atomic number by 1 and produces boron, $^{11}_{5}B$. These processes may be written as radiochemical disintegration reactions:

$$^{14}_{6}C \rightarrow \beta^{-} + ^{14}_{7}N + \bar{\nu}_{e}$$
$$^{11}_{6}C \rightarrow \beta^{+} + ^{11}_{5}B + \nu_{e}$$

where β^{-} and β^{+} are electrons and positrons, respectively, and ν_{e} and $\bar{\nu}_{e}$ are neutrinos and antineutrinos. For an explanation of the role of the antineutrino and neutrino, see Section 6.4.1.

For each element, there is a specific ratio of neutrons to protons that makes for the greatest stability. In the lighter elements, this ratio is approximately 1:1, but as we progress through the periodic table to the heavier elements, the ratio for maximum stability approaches 1.5 neutrons to 1 proton. As we have seen for carbon, when there are more neutrons than protons, the radioactive decay process decreases the number of neutrons and increases the number of protons; and when there are more protons than neutrons, the radioactive decay decreases the number of protons and increases the number of neutrons. Thus, the type of radioactive decay observed is related to the type of nuclear instability.

6.2 TYPES OF RADIOACTIVE DECAY

6.2.1 Decay by Negative-Beta (Electron) Emission

As seen with $^{14}_{6}C$, nuclei with excess neutrons may reach stability by the ejection of an electron and an antineutrino from a neutron, giving a proton. The ejected electron originates in the nucleus and should not be confused with the orbital electrons. Otherwise, however, the negative β-particle is identical with the orbital electron in mass and charge. Many of the radioactive elements of biological interest are β^--emitters. The following equations give the radiochemical disintegrations of these elements.

$$^{3}_{1}\text{H} \rightarrow \,^{3}_{2}\text{He} + \beta^- + \bar{\nu}_e$$
$$^{14}_{6}\text{C} \rightarrow \,^{14}_{7}\text{N} + \beta^- + \bar{\nu}_e$$
$$^{24}_{11}\text{Na} \rightarrow \,^{24}_{12}\text{Mg} + \beta^- + \bar{\nu}_e$$
$$^{32}_{15}\text{P} \rightarrow \,^{32}_{16}\text{S} + \beta^- + \bar{\nu}_e$$
$$^{33}_{15}\text{P} \rightarrow \,^{33}_{16}\text{S} + \beta^- + \bar{\nu}_e$$
$$^{35}_{16}\text{S} \rightarrow \,^{35}_{17}\text{Cl} + \beta^- + \bar{\nu}_e$$
$$^{36}_{17}\text{Cl} \rightarrow \,^{36}_{18}\text{Ar} + \beta^- + \bar{\nu}_e$$
$$^{45}_{20}\text{Ca} \rightarrow \,^{45}_{21}\text{Se} + \beta^- + \bar{\nu}_e$$
$$^{59}_{26}\text{Fe} \rightarrow \,^{59}_{27}\text{Co} + \beta^- + \bar{\nu}_e$$
$$^{131}_{53}\text{I} \rightarrow \,^{131}_{54}\text{Xe} + \beta^- + \bar{\nu}_e$$

Negative β-emitting isotopes have found the greatest use in biochemical research because the emitted electron is relatively easy to detect and measure.

6.2.2 Decay by Positive-Beta (Positron) Emission

As seen in the case of $^{11}_{6}\text{C}$, nuclei with excess protons may reach stability by the ejection of a positron and a neutrino from a proton, giving a neutron. Positrons have the mass of an electron but equal and opposite charge. Any excess energy in the nucleus after positron emission is emitted as gamma (γ) radiation. The positron is an example of an antiparticle that is part of antimatter. When a positron is emitted from a nucleus, it encounters the orbiting electrons of the atoms in its surroundings. When a positron collides with an electron, both are rapidly annihilated and are converted into energy (γ-radiation) of about 1.02 million electron volts (MeV):

$$\beta^+ + e^- \rightarrow \gamma \,(1.02 \text{ MeV})$$

Some examples of positron emitting nuclei are:

$$^{11}_{6}\text{C} \rightarrow \,^{11}_{5}\text{B} + \beta^+ + \nu_e$$
$$^{13}_{7}\text{N} \rightarrow \,^{13}_{6}\text{C} + \beta^+ + \nu_e$$
$$^{65}_{30}\text{Zn} \rightarrow \,^{65}_{29}\text{Cu} + \beta^+ + \nu_e$$

6.2.3 Decay by Electron Capture

In another mechanism for the conversion of a nuclear proton into a neutron, the nucleus captures one of the two electrons orbiting in the innermost K-shell. This process is sometimes called K-capture (KC) and gives the same result as the emission of a positron — the formation of a neutron and the release of an electron neutrino. The only particle directly ejected in electron capture is the electron neutrino. However, because of the electron vacancy in the K-shell, an outer electron moves into the vacant position and X-rays are emitted at wavelengths characteristic of the newly formed atom. Sometimes an unstable nucleus will decay alternately by two routes, electron capture or positron emission:

$$\begin{array}{c} \nearrow \quad {}^{22}_{10}\text{Ne} + \beta^+ + \nu_e \\ {}^{22}_{11}\text{Na} \\ \searrow_{\text{KC}} \quad {}^{22}_{10}\text{Ne} + \text{X-ray} + \bar{\nu}_e \end{array}$$

Positron emission will dominate as the decay pathway over electron capture when the atomic number is low or the decay energy is high. The following isotopes undergo K capture:

$$^{55}_{26}\text{Fe} \xrightarrow{\text{KC}} {}^{55}_{25}\text{Mn} + \text{X-ray} + \bar{\nu}_e$$

$$^{57}_{27}\text{Co} \xrightarrow{\text{KC}} {}^{57}_{26}\text{Fe} + \text{X-ray} + \bar{\nu}_e$$

$$^{125}_{53}\text{I} \xrightarrow{\text{KC}} {}^{125}_{52}\text{Te} + \text{X-ray} + \bar{\nu}_e$$

6.2.4 Decay by γ-Radiation

Certain excited nuclei get rid of their excess energy and return to the ground state by the emission of γ-rays, which all have the same discrete energy. In γ-ray emission, the number of neutrons and protons in the nucleus does not change, only the energy of an excited nucleus changes to a lower value. γ-Ray emission can occur after β^- or β^+ emission. Sometimes, however, the γ-ray emission does not occur immediately; in such cases, there are two forms of the isotope, an excited form and a ground-state form. The more excited form is indicated by the letter "m" following the mass number.

$$^{131}_{53}\text{I} \rightarrow \beta^- + {}^{131m}_{54}\text{Xe}^* \rightarrow {}^{131}_{54}\text{Xe} + \gamma$$

$$^{137}_{55}\text{Cs} \rightarrow \beta^- + {}^{137m}_{56}\text{Ba}^* \rightarrow {}^{137}_{56}\text{Ba} + \gamma$$

$$^{60}_{27}\text{Co} \rightarrow \beta^- + {}^{60}_{28}\text{Ni} + \gamma$$

6.2.5 Decay by α-Particle Emission

Heavy atoms with more than 82 protons attain a stable nucleus by ejecting an α-particle, a helium nucleus of atomic mass number of 4 (2 protons and 2 neutrons) with a charge of $+2$. α-Particles are ejected by $^{226}_{88}\text{Ra}$, $^{232}_{90}\text{Th}$, $^{235}_{92}\text{U}$, and $^{238}_{92}\text{U}$. In general, α-particle emitters are not used in biochemical studies but are of interest for their radiological toxicity.

6.3 OCCURRENCE AND ARTIFICIAL PRODUCTION OF RADIOACTIVE ISOTOPES

In 1896 Antoine Henri Becquerel found that uranium emits penetrating rays spontaneously and continuously. Pierre and Marie Curie later introduced the term **radioactivity** to designate this process and showed that it is a property of the uranium atoms themselves and is not related to a particular chemical combination of the uranium. Several other naturally occurring radioactive isotopes were recognized in the early part of the twentieth century. Today about 1500 radioactive isotopes are known, most of them artificially produced. In 1934 Irène Curie and Frédéric Joliot-Curie produced the first artificial radioactive isotope, phosphorus-30, by bombarding aluminum-27 with α-particles. Thereafter many other artificial radioactive isotopes were produced.

Enrico Fermi found that slow-moving neutrons are very effective bombarding agents because they are uncharged and readily enter a nucleus, becoming part of it, and producing a new element that is radioactive. One of the common methods of isotope production, thus, uses neutron bombardment of elements in a nuclear reactor. The production of some biochemically important radioactive isotopes by neutron bombardment is shown in the following reactions:

$$^{6}_{3}\text{Li} + ^{1}_{0}\text{n} \rightarrow ^{3}_{1}\text{H} + ^{4}_{2}\text{He}$$

$$^{14}_{7}\text{N} + ^{1}_{0}\text{n} \rightarrow ^{14}_{6}\text{C} + ^{1}_{1}\text{H}$$

$$^{32}_{16}\text{S} + ^{1}_{0}\text{n} \rightarrow ^{32}_{15}\text{P} + ^{1}_{1}\text{H}$$

$$^{35}_{17}\text{Cl} + ^{1}_{0}\text{n} \rightarrow ^{35}_{16}\text{S} + ^{1}_{1}\text{H}$$

Other bombarding agents, such as α-particles, protons (hydrogen nuclei), and deuterons (nuclei of the hydrogen isotope deuterium containing one proton and one neutron) are also used at high velocities in particle accelerators to obtain radioactive isotopes.

In 1940 Ruben and Kamen produced carbon-14 by bombarding graphite with deuterons ($^{12}_{6}\text{C} + ^{2}_{1}\text{H} \rightarrow ^{14}_{6}\text{C} + \beta^{+}$) and started the radioactive era in biochemical research. Now, however, ^{14}C is produced primarily by bombarding compounds containing ^{14}N (e.g., Be_3N_2) with neutrons: $^{14}_{7}\text{N} + ^{1}_{0}\text{n} \rightarrow ^{14}_{6}\text{C} + ^{1}_{1}\text{H}$.

Several properties of ^{14}C result in an isotope that can be detected and safely handled: a very long half-life (> 5000 years), the ease with which its emitted electron can be detected by gas ionization, and the relatively low energy of the emitted electron. Thus, the availability of ^{14}C has played, and continues to play, a major role in advancing knowledge of biochemical processes. Other negative β-emitters commonly used in studying biochemical reactions are ^{32}P, ^{35}S, and ^{3}H (tritium). The latter was actually produced before ^{14}C, but because its negative β-particle is of very low energy, it is difficult to detect and measure. It was not until 1954, when liquid scintillation counters became available, that tritium could be readily measured and quantitated.

Of the naturally occurring radioactive isotopes of low atomic number, only ^{40}K is biologically significant. Potassium-40 has an extremely long half-life (see Section 6.4.4) of 1.26×10^9 years and because of the high water solubility of potassium compounds is widely distributed in the crust and water of the earth. In addition, K^+ is a common ion in living cells and a small, but measurable, amount of ^{40}K is present in all cells and, thus, contributes to the "background" radiation to which all living things are exposed.

A few biologically important radioisotopes are constantly being produced in the earth's upper atmosphere as a result of cosmic ray bombardment of atmospheric nitrogen. When ^{14}N is hit by neutrons, ^3H and ^{14}C are produced in two different reactions:

$$^{14}_{7}N + ^{1}_{0}n \rightarrow ^{3}_{1}H + ^{12}_{6}C \quad \text{and} \quad ^{14}_{7}N + ^{1}_{0}n \rightarrow ^{14}_{6}C + ^{1}_{1}H$$

The ^3H so produced combines with oxygen to form tritiated water and the ^{14}C combines with oxygen to form $[^{14}C]$ carbon dioxide. Water and carbon dioxide are readily incorporated into living cells. In the case of tritium and ^{14}C, the rate of decay has reached an equilibrium with the rate of formation, so the abundance of these isotopes is relatively constant in nature. As a result, during the life of an organism there are defined concentrations of tritium and ^{14}C in its tissues. It is the decrease of the concentration of ^{14}C, by radioactive decay, after the death of the organism that makes possible radiocarbon (i.e., ^{14}C) dating.

6.4 PROPERTIES OF RADIOACTIVE EMISSIONS

6.4.1 Energy of β-Particle Emissions

Beta particles, positrons and electrons, are ejected from the nucleus with velocities approaching the speed of light, 3×10^{10} cm/s. They are ejected with a continuous range of kinetic energies up to a maximum value E_{max}, which is characteristic of the particular radioisotope. Figure 6-1 shows typical energy distributions for the emission of the negative β-particle from ^3H, ^{14}C, ^{32}P, and ^{35}S. The number of electrons emitted with E_{max} is very small, and a much larger number are emitted with the average energy E_{avg}. The average energy for most β^--emissions is approximately $\frac{1}{3}E_{max}$.

Since it was known that discrete energy levels existed within the nucleus, the emission of β-particles with a continuous energy spectrum was an enigma. In 1931, Pauli suggested that each β-emission actually occurred with an energy equivalent to E_{max}. He postulated that an undiscovered particle, the neutrino, shares the total energy with the β-particle in varying proportions for each β-particle emission. For example, a β-particle emitted from ^{35}S with an energy of 0.1 MeV would be accompanied by a neutrino with an energy of 0.067 MeV, giving a total emission of 0.167 MeV, the E_{max} for ^{35}S. The neutrino was postulated to have no charge and negligible mass; because theoretically it had virtually no interaction with matter, investigators realized it would be very difficult to detect. In 1956 the neutrino's existence was finally demonstrated by a complicated and elaborate experiment, which verified Pauli's hypothesis and explained the continuous energy spectrum observed for β-emissions.

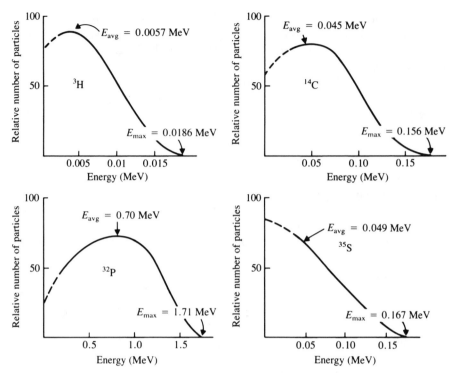

FIGURE 6-1. β-particle energy spectra for four radioisotopes commonly used in biochemical studies. The dashed portions of the low energy curves of the spectra indicate theoretical values of E, since experimental determinations in these regions are not readily made.

6.4.2 Interactions of β-Particles with Their Environment

An electron ejected from a nucleus into a medium will encounter and interact with both orbital electrons and nuclei in the medium. During these interactions, the ejected electron will knock orbital electrons out of their orbit to produce ion pairs. Another, but less prominent effect is electron excitation. Orbital electrons are excited to higher energy states and then emit ultraviolet or visible light as they return to their ground state. During these interactions, the ejected electrons lose energy and are deflected or scattered from their original direction of emission. Ejected electrons are scattered with a probability that increases with the electron density (electrons per gram) of the medium through which they are passing. In air, about 34 eV is required to produce one ion pair. A particle loses this amount of energy at each ionizing event and will eventually lose all its energy and come to rest. An ejected electron, with an energy of 1.5 MeV, will travel about 2 meters in air and about 1 centimeter in tissue. Eventually the ejected electron will be captured by an empty electron orbit of a positive ion.

An ejected positron interacts with its environment in a different manner. When a positron encounters an electron, the two particles annihilate each other and their masses are converted into electromagnetic γ-radiation, which is usually in the form of two 0.51-MeV photons moving in opposite directions. This interaction is termed **pair annihilation**, and the resulting photons are referred to as **annihilation photons**.

6.4.3 Gamma Radiation and Its Interaction with Matter

Gamma rays are high energy photons or "light particles" that are electrically neutral with zero mass and consequently can penetrate matter readily. Gamma rays are frequently emitted from a nucleus immediately after electron or positron emission and represent a readjustment of the energy content in the nucleus from an excited state to a more stable state. Another source of these rays is positron annihilation, just mentioned.

In contrast to β-particle emissions, which have a continuous energy spectrum, a γ-ray is emitted from the nucleus with discrete energy. This probably reflects discrete energy levels that exist in the nucleus. The discrete energies of γ-rays range between 10 keV and 3 MeV. Many isotopes emit γ-rays of several energies in discrete steps; the ^{131}I nucleus dissipates its excess energy after β^- emission by a series of γ-rays (Fig. 6-2). The partial multiple decay of ^{131}I is shown for the major β^--emissions giving E_{max} for the formation of $\beta^- + \bar{\nu}_e$ and an excited Xe, which undergoes further decay by γ-ray emission of discrete energy. The overall energy of each of the decay reactions ($\beta^- + \gamma$) is 0.97 MeV.

The principal interactions of γ-rays with matter result in the photoelectric effect, the Compton effect, and pair production.

In the **photoelectric effect**, a γ-ray of relatively low energy (≤ 0.1 MeV) interacts with an orbital electron, transferring all its energy to the electron and disappearing in the process. The electron is displaced from the atom with a certain amount of kinetic energy. This electron then interacts with other atoms in its path, just as a β^--particle would, leading to further ionization. The photoelectric effect can be accompanied by X-ray emission when the other orbital electrons move to fill the vacant orbital caused by the loss of the photoelectron.

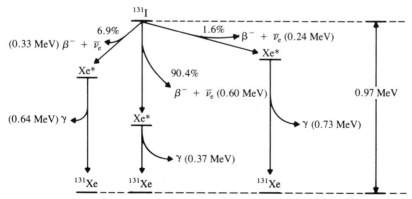

FIGURE 6-2. Products of the decay of the ^{131}I nucleus; Xe* = xenon in the excited state.

The **Compton effect** results when γ-rays of medium energy (≥ 0.1 MeV) undergo elastic collisions with loosely bound orbital electrons and a portion of the γ-ray energy is transferred to the ejected electron. The γ-ray photon is deflected in a new direction with reduced energy and may undergo several more similar collisions before losing all its energy. Because different amounts of the γ-ray energy can be imparted to the electrons, Compton electrons have a wide range of energies. These electrons will produce secondary ionizations as they travel through the medium, dissipating their energy.

Pair production results when a relatively energetic (≥ 1.02 MeV) γ-ray interacts with a nuclear force field. The γ-ray photon has all its energy converted into a positron and an electron, which are ejected from the site with varying energy. For pair production to occur, the γ-ray must have an energy equal to, or greater than, 1.02 MeV, which is the energy equivalent to the rest mass of one positron and one electron. *This is essentially the conversion of energy into mass.* The opposite conversion occurs when a positron is annihilated by colliding with an electron. If the energy of the γ-ray exceeds 1.02 MeV, the excess goes into the kinetic energies of the ejected positron and electron. Both the ejected particles (positron and electron) may cause ionization of atoms in their paths, and eventually as the positron loses its kinetic energy, it will be annihilated by a collision with an electron.

The ionization effects of the secondary electrons produced by the three modes of γ-ray interaction with matter provide the means of detecting γ-radiation.

6.4.4 Kinetics of Radioactive Decay

The process of decay of a radioactive nucleus into a stable, nonradioactive nucleus is irreversible. The rate of particle emission is equal to the rate of disintegration of the radioactive nuclei, which is proportional to the number of radioactive nuclei present. The number of nuclei in a radioactive sample decreases exponentially with time (Fig. 6-3). The rate of disintegration is indepen-

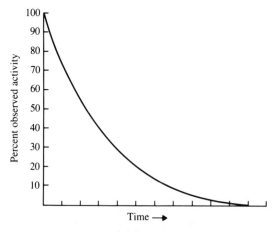

FIGURE 6-3. Exponential decay curve for radioisotopes.

dent of pressure, temperature, and mass action effects that are characteristic of most chemical reactions or physical changes. Radioactive decay is therefore an excellent example of a first-order reaction. That is, the rate of radioactive decay is directly proportional to the number of radioactive nuclei present in a given sample. So, if the number of radioactive nuclei in a sample is doubled, the number of particles emitted by the sample per unit time is doubled.

The decay of nuclei in a sample does not occur at regular intervals but occurs randomly. For any given radioactive nucleus, the probability that it will disintegrate in time t is given by the decay constant, λ. The radioactive process may be represented by the following differential equation in which N represents the number of radioactive nuclei present at time t:

$$-\frac{dN}{dt} = \lambda N \tag{6-1}$$

where N is constantly decreasing with time. Rearrangement of Equation 6-1 gives:

$$\frac{dN}{N} = -\lambda \, dt \tag{6-2}$$

If we integrate the differential equation (6-2) between t_0 and t, where $t_0 = 0$,

$$\int_{N_0}^{N} \frac{dN}{N} = -\lambda \int_{t_0}^{t} dt \tag{6-3}$$

we obtain:

$$\ln \frac{N}{N_0} = -\lambda t \quad \text{or} \quad N = N_0 e^{-\lambda t} \tag{6-4}$$

Equation 6-4 is known as the **radioactive decay law**. It gives the number of radioactive nuclei N at time t when the number N_0 is known at t_0.

Since the rate of disintegration of the radioactive nuclei is equal to the rate of particle emission and since the emitted particles can be detected and quantitatively measured as the **activity** A of the radioactive sample, the measurement of the "counting rate" of the detector (A) divided by the efficiency (E) of the detecting system is equal to the rate of disintegration of the nuclei (N).

$$\frac{A}{E} = N \tag{6-5}$$

Substituting A/E for N in Equation 6-2, we have:

$$\frac{(dA)/E}{A/E} = -\lambda\,dt \qquad (6\text{-}6)$$

Then we can write:

$$\frac{dA}{A} = -\lambda\,dt \qquad (6\text{-}7)$$

and after integration of Equation 6-7, we have:

$$\ln\frac{A}{A_0} = -\lambda t \qquad (6\text{-}8)$$

Conversion of Equation 6-8 into $\log_{10}$, gives:

$$2.303\log\frac{A}{A_0} = -\lambda t \qquad (6\text{-}9)$$

The measurement of the activity A_0 at t_0 and the use of Equation 6-9 will give the activity A at time t.

   The use of the radioactive decay law, Equation 6-8 or 6-9, can tell us how much activity a sample has at present or at some date in the future, if we know how much activity the sample had at some previous time. For example, if a sample of tritium had 495,000 disintegrations per minute (dpm) on February 18, 1987, we can determine how much activity the sample will have on February 18, 1994. The decay constant λ for tritium is 1.52×10^{-4}/day.

Assuming there are 365.25 days/year, the number of days that will have passed by 1994 are 7 years $\times$ 365.25 days/year = 2556.75 days. Now applying Equation 6-9, with $A_0 = 495{,}000$ dpm, we have:

$2.303\log\dfrac{A}{A_0} = -\lambda t$

$2.303\,(\log A - \log A_0) = -\lambda t$

$2.303\log A - 2.303\log(4.95\times 10^5) = -(1.52\times 10^{-4}/\text{day})(2556.75\ \text{days})$

$2.303\log A = 13.115 - 3.89\times 10^{-1} = 1.273\times 10^{1}$

$$\log A = \frac{1.273 \times 10^1}{2.303} = 5.5276$$

$$A = 337{,}000 \text{ dpm}$$

The decay constant λ may be obtained experimentally for a radioactive sample by plotting the log of the observed activity (in percent) versus time (Fig. 6-4). The slope of the resulting line is equal to the value $-(\lambda/2.303)$, from which the decay constant may be obtained. The limitation of this method of determining λ is that to give a significant slope over the period during which the activity is measured, the isotope must be undergoing decay at an appreciable rate. Table 6-1 gives decay constants for several radioactive isotopes of interest.

The **half-life** $t_{1/2}$ is another expression of the decay rate of a radioactive isotope. It is simply the time required for the radioactivity or number of radioactive nuclei to decrease by half. After one half-life, 50% of the initial activity remains; after two half-lives, 25% remains; after three half-lives, 12.5% remains, and so forth. The value of the half-life can be derived from Equation 6-9 as follows.

When half of the initial activity has disappeared,

$$\frac{A}{A_0} = \frac{1}{2} \qquad \text{and} \qquad t = t_{1/2}$$

Substituting into Equation 6-9, we can write:

$$2.303 \log \frac{1}{2} = -\lambda t_{1/2}$$

or

$$2.303 \log 2 = \lambda t_{1/2}$$

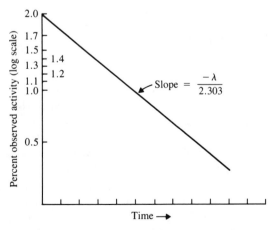

FIGURE 6-4. Semilogarithmic linear decay curve for radioisotopes.

TABLE 6-1. **Half-life, decay constant, type of radiation, and maximum energy of radioisotopes important in biochemistry**

Isotope	Half-life	Decay constant (λ)	Type of radiation	Maximum energy (MeV)
^{3}H	12.26 yrs	1.55×10^{-4}/day	β^-	0.018
^{14}C	5730 yrs	1.21×10^{-4}/year	β^-	0.156
^{22}Na	2.62 yrs	7.24×10^{-4}/day	$\beta^+ + \gamma$	0.55 (1.28)[a]
^{32}P	14.3 days	4.85×10^{-2}/day	β^-	1.71
^{33}P	25 days	2.77×10^{-2}/day	β^-	0.25
^{35}S	87 days	7.97×10^{-3}/day	β^-	0.167
^{36}Cl	3×10^5 yrs	2.31×10^{-6}/year	β^-	0.71
^{40}K	1.3×10^9 yrs	5.33×10^{-10}/year	$\beta^- + \gamma$	1.4 (1.5)
^{45}Ca	165 days	4.2×10^{-3}/day	$\beta^- + \gamma$	0.26 (0.013)
^{59}Fe	45 days	1.54×10^{-2}/day	$\beta^- + \gamma$	0.46 (1.1)
^{60}Co	5.3 yrs	3.58×10^{-4}/day	$\beta^- + \gamma$	0.318 (1.33)
^{65}Zn	245 days	2.83×10^{-3}/day	$\beta^+ + \gamma$	0.33 (1.14)
^{90}Sr	29 yrs	6.54×10^{-5}/day	β^-	0.54
^{125}I	60 days	1.16×10^{-2}/day	γ	0.036
^{131}I	8.06 days	8.60×10^{-2}/day	$\beta^- + \gamma$	0.61 (0.36)
^{137}Cs	30.2 yrs	6.28×10^{-5}/day	$\beta^- + \gamma$	0.51 (0.66)
^{226}Ra	1620 yrs	4.28×10^{-4}/year	$\alpha + \gamma$	4.78 (0.19)

[a] Where two types of radiation occur, the number in parentheses is the maximum energy for the second type of radiation.

Therefore,

$$t_{1/2} = \frac{0.693}{\lambda} \tag{6-10}$$

Equation 6-10 is called the **half-life equation**. The half-life for an isotope can be obtained by preparing a semilog plot similar to that of Figure 6-4 for the determination of λ and using Equation 6-10. The half-life has been determined for many isotopes and for all the radioactive isotopes commonly used in biochemical research (see Table 6-1). The half-life is characteristic of the particular isotope and ranges from a microsecond (10^{-6} s) to billions of years (10^{10} yr).

The half-life must be considered when using a particular radioisotope in experiments. An isotope with a short half-life is usually difficult to use because it must be prepared and shipped immediately and rapidly to the experimenter, who then must perform and analyze the experiment rapidly before too much of the isotope has disintegrated. Furthermore, for an isotope with a short half-life, the amount of disintegration that has occurred during the course of the experiment must be determined, to permit the quantitative evaluation of the amount of isotope involved in the experiment. Fortunately many isotopes of biochemical interest have reasonably long half-lives. For example, ^{14}C has a half-life of approximately 5730 years and its decay during an experiment is not a concern.

The much shorter half-life of tritium, 12.26 years, is still sufficiently long that ^{3}H decay during an experiment usually is not significant. ^{32}P, however, is an isotope that has a relatively short half-life of 14.28 days, which must be considered during the course of an experiment.

6.5 UNITS USED IN RADIOACTIVE MEASUREMENTS

6.5.1 The Electron Volt

As we have seen in Section 6.4.1, particles are emitted with energies given in electron volts (eV), which are frequently expressed as millions of electron volts (MeV). One electron volt (1 eV) is the kinetic energy that an electron develops when it is accelerated through a potential difference of 1 volt in a vacuum. The electron volt is a very small energy unit, which is equivalent to 3.85×10^{-20} calorie or 1.602×10^{-19} joule. The amount of energy that accompanies chemical reactions, which involve the outermost electron orbits, is about 10 eV per atom, whereas the amount of energy of an electron emitted from a nucleus is much greater. For example, the electron emitted from ^{14}C has an average energy of 4500 eV per nucleus (atom).

6.5.2 The Curie: The Unit of Radioactive Disintegration

The **curie** (Ci) was originally defined as the number of disintegrations occurring per second in 1 gram of pure radium 226. One mole or 226 grams of pure radium-226 contains Avogadro's number of atoms (i.e., 6.02×10^{23} atoms). One gram of radium would have $(1/226)(6.02 \times 10^{23}) = 2.66 \times 10^{21}$ atoms. The half-life of radium has been determined to be 1600 years, and there are 3.16×10^7 seconds in a year. Employing Equation 6-10 to determine the decay constant λ gives:

$$\lambda = \frac{0.693}{(1600 \text{ yr})(3.16 \times 10^7 \text{ s/yr})} = 1.37 \times 10^{-11}/\text{s}$$

Employing Equation 6-1, the decay rate of 1 g of radium (dN/dt) would equal the product of the decay constant times the number of atoms in a gram, or:

$$\frac{dN}{N} = \lambda N = (1.37 \times 10^{-11}/\text{s})(2.66 \times 10^{21})$$

$$= 3.70 \times 10^{10} \text{ disintegrations per second (dps)}$$

This calculation of the value of the curie depends on the determination of the half-life of ^{226}Ra, which because of its magnitude is subject to experimental error and subsequent refinement. Indeed, the half-life was later found to be 1620 years instead of 1600 years. To avoid problems caused by defining the curie on the basis of the decay constant of radium, the International Union of Pure and Applied Chemistry (IUPAC) and the Union of Pure and Applied Physics (UPAP) defined the curie in 1950 to be 3.70×10^{10} dps. The curie is also given

TABLE 6-2. Units of radioactivity

Unit name	Fraction of a curie	Activity dps	Activity dpm
curie (Ci)	—	3.70×10^{10}	2.22×10^{12}
millicurie (mCi)	10^{-3}	3.70×10^{7}	2.22×10^{9}
microcurie (μCi)	10^{-6}	3.70×10^{4}	2.22×10^{6}
nanocurie (nCi)	10^{-9}	3.70×10^{1}	2.22×10^{3}
picocurie (pCi)	10^{-12}	3.70×10^{-2}	2.22×10^{0}

as 2.22×10^{12} disintegrations per minute (dpm). Because the curie is a large unit, several smaller units are frequently used; Table 6-2 defines millicurie through picocurie.

The MKS system (Table 1-3) defines the basic unit of radioactivity as one disintegration per second and calls this a **becquerel**. One curie equals 3.70×10^{10} becquerels.

6.6 SPECIAL TECHNIQUES AND SAFETY IN HANDLING RADIOISOTOPES

The weak β-emitters (^{3}H, ^{14}C, ^{35}S) pose little, if any, health hazards for the handler; stronger β-emitters and γ-ray emitters (^{32}P, ^{60}Co, ^{125}I, ^{131}I) should be respected for their potential health hazard. Regardless of the type of isotope being used, however, great care should be taken to prevent ingestion, inhalation, or absorption of radioactivity. For example, eating, drinking, smoking, or placing items in the mouth should not be done while handling radioisotopes or while in the radioisotope laboratory. Contamination of the laboratory and personnel must be prevented.

A laboratory coat should be worn as protective clothing, and when contamination of the hands is a possibility, disposable plastic gloves should be used. Proper procedures for using radioactive isotopes in the laboratory are the same as those of usual, good laboratory technique: the working area should be neat and clean; equipment and supplies not required for the experiment should be removed; and the experiment should be confined to an area as small as possible. In addition, the laboratory bench should be covered with plastic-backed absorbent paper and the experiment should be performed in a tray lined with the same paper, to contain and facilitate disposal of any spills that might occur. All radioactive samples should be stored in an appropriate container that is marked "*radioactive*," with the *type* and *amount* of the isotope and the *date*. Containers used briefly in the experiment need not be labeled if they are cleaned and decontaminated immediately after use. One radioactive waste container for solids (paper, disposables, etc.) and another for liquids should be available near the experimental area. Radioactive samples should *never* be pipetted by mouth. A rubber bulb, a syringe pipette, or an automatic pipette (Section 2.2) with a disposable tip should be used to transfer radioactive liquids. Experiments in which there is the possibility for the release of radioactivity into the air should be conducted in a filtered, ventilated fume hood.

The laboratory should be routinely monitored for possible contamination of bench, sinks, hood, refrigerators, and so on. This may be rapidly and easily done for strong β-emitters with a thin-window Geiger–Müller survey monitor equipped with an audible counter. In addition, damp filter paper swipes of the laboratory area can be made and the paper counted by heterogeneous liquid scintillation to especially detect the weak β-emitters.

If strong, penetrating radiation is being used, such as X-rays, γ-rays, or high energy β-particles, the investigator should wear a personal dosimeter, which is a badge consisting of two or more pieces of X-ray film in a light-tight plastic case. After the badge has been worn for 1–4 weeks at the level of the belt or shirt pocket, the film is developed and its optical density compared with films that have received known exposures. The radiation dose may then be estimated for the badge and, thus, for the individual.

If a radioactive substance is spilled, the area should be secured immediately and the type of radiation and its extent determined before decontamination is attempted. Then, wearing protective clothing (coat, mask, gloves) appropriate for the situation, the individual decontaminating the area should wash it with a detergent solution, always working toward the center, to contain the contamination in the smallest possible space. The area must be monitored frequently to determine the effectiveness of the decontamination procedure. All cleaning materials and solutions are disposed of in the radioactive waste containers.

If skin becomes contaminated, the most common area being the hands, it should be thoroughly washed with soap and warm water for at least 2 minutes and the lather thoroughly rinsed. The contaminated area should be monitored and the washing procedure repeated until the contamination has been removed. Care must be exercised that contamination does not enter the body through broken skin. The thyroid area of anyone handling ^{125}I or ^{131}I should be monitored with a survey monitor.

6.7 STATISTICS OF COUNTING RADIOISOTOPES

The radioactive decay process is random and the number of nuclei in a sample undergoing decay in any relatively short period is not fixed. In the measurement of radioactivity, the randomness of the decay often is the greatest source of indeterminate error. This randomness or variability of the count rate is readily demonstrated by making a number of short-time activity measurements with a radioactive isotope of relatively long half-life. A comparison of the measured count rates shows that the counts vary. If this experiment is repeated a large number of times and a plot of the number of times a given count rate occurs versus the count rate, a normal distribution curve (Fig. 6-5) is obtained.

It will be noted that the curve in Figure 6-5 is identical to the curve presented in Figure 1-1. It bears repeating in the present context that in any measurement with a normal distribution there is a standard deviation, s, which is given as

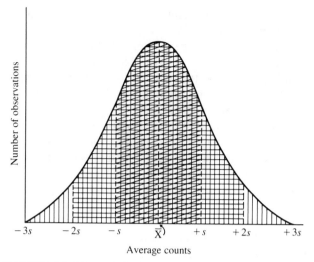

FIGURE 6-5. Normal distribution of the average counts obtained for a single sample showing the standard deviations from the mean $\bar{x}$; $\pm s$ gives 68.3% confidence, $\pm 2s$ gives 95.5% confidence, and $\pm 3s$ gives 99.7% confidence.

follows:

$$s = \sqrt{\frac{\sum (x - \bar{x})^2}{N - 1}} \qquad (6\text{-}11)$$

and can be simplified to:

$$s = \sqrt{\bar{x}} \qquad (6\text{-}12)$$

where x is the individual measurement and $\bar{x}$ is the average of N measurements. In a normal distribution, as pointed out in Chapter 1, it can be shown that 68.3% of all measurements fall within the limits of one standard deviation, s, or $\bar{x} \pm \sqrt{\bar{x}}$, and that 95.5% fall within $\pm 2s$, and 99.7% fall within $\pm 3s$.

A more common situation is a single measurement on a radioactive sample. It can be shown that s approximately equals the square root of the single determination x:

$$s = \sqrt{x} \qquad (6\text{-}13)$$

TABLE 6-3.   Determining the accuracy of a sample at 1000–40,000 cpm

x (counts)	t (min)	s	Confidence level 68.3%	Confidence level 95.5%	% RSD 1s	% RSD 2s
1,000	1	$\sqrt{1,000} = 32$	$1,000 \pm 32$	$1,000 \pm 64$	3.2	6.4
2,500	2.5	$\sqrt{2,500} = 50$	$2,500 \pm 50$	$2,500 \pm 100$	2.0	4.0
5,000	5	$\sqrt{5,000} = 71$	$5,000 \pm 71$	$5,000 \pm 142$	1.4	2.8
10,000	10	$\sqrt{10,000} = 100$	$10,000 \pm 100$	$10,000 \pm 200$	1.0	2.0
40,000	40	$\sqrt{40,000} = 200$	$40,000 \pm 200$	$40,000 \pm 400$	0.5	1.0

Thus, a statement concerning the true value from a single measurement x can be made. For example, if a sample has 3000 counts, $s = \sqrt{3000} = 54.8$. We can say with 68.3% confidence that the true count rate for the sample is 3000 ± 55 counts per minute (cpm), with 95.5% confidence it is 3000 ± 110 cpm, and with 99.7% confidence the true count rate is 3000 ± 165 cpm.

A number of parameters can be used to determine the normal distribution of the measurement in a sequence of measurements. There are two common parameters in radioactivity measurements: the number of times the sample is counted and the length of time the sample is counted. It can be shown that the width of the normal distribution curve depends on the number of determinations made. The larger the number of determinations, the narrower the distribution width. In practice, however, one cannot make an infinite number of measurements on a single sample. Usually the maximum number of measurements on a single sample is three and the average or mean of these three counts is computed.

The length of time the sample is counted also influences the accuracy of the determination. For example, if various counts are obtained for a sample by counting for various lengths of time, the value of s changes, hence the accuracy of the count changes. The longer we count the sample, the higher the accuracy. Let us say we count a sample for various lengths of time and obtain the counts given in Table 6-3.

If we counted the sample for 1 minute and obtained 1000 counts, the value of the count would be 1000 ± 32; but if we counted the sample for 10 minutes the value of the count would be $10,000 \pm 100$ or 1000 ± 10 cpm. Thus, by counting the sample for a longer period, we increase the accuracy of our measurement.

Another commonly used method for expressing the "true value" of a measurement is the **percent relative standard deviation** (% RSD), which is expressed as follows:

$$\% \ RSD = \frac{\sqrt{x}}{x} \times 100 \qquad (6\text{-}14)$$

Scintillation counters automatically compute the % RSD for the count obtained. The % RSD also can be expressed as a function of the number of standard deviations. Thus, we can see in Table 6-3 that the five counts have % RSD values of 3.2, 2.0, 1.4, 1.0, and 0.5 for one standard deviation giving 68.3% confidence, or the five counts have % RSD values of 6.4, 4.0, 2.8, 2.0, and 1.0 for two standard deviations giving 95.5% confidence. The equation expressing % RSD as a function of the number of standard deviations is:

$$\% \, \text{RSD}_{ns} = n \frac{\sqrt{x}}{x} \times 100 \qquad (6\text{-}15)$$

Regardless of how active a sample is, a 1.0% RSD (1s) is obtained by counting it to 10,000 counts and a 1.0% RSD (2s) by counting to 40,000 counts. If a sample has an activity of 2350 cpm, it would have to be counted for 4.26 minutes to obtain 10,000 counts and a 1.0% RSD (1s) and 17 minutes to obtain 40,000 counts for a 1.0% RSD (2s).

It would be ideal to count all samples to 1.0% RSD, especially for comparisons of individual counts. This, however, is not always practical. For example, let us say a sample has 230 cpm. To obtain 10,000 counts or 1% RSD (1s), this sample would have to be counted for 43.5 minutes. In itself, that is not too bad, but let us say that 50 samples of this level must be counted an average of 40 minutes each to obtain 1% RSD (1s). For a single count for each sample, we would have to count a total of 2000 minutes or 33.33 hours—nearly a day and a half. Practical considerations of tying up the counter and having to wait for the counts suggest that, in this case, we would have to accept a higher % RSD.

Scintillation counters have the capability of allowing the investigator to choose the maximum number of counts to be collected before the next sample is counted, or the maximum length of time each sample is to be counted. Practical considerations suggest that the maximum time be 10 minutes and the maximum count be 10,000 whichever comes first, with the latter being preferred.

6.8 MEASUREMENT OF RADIOACTIVITY BY GAS IONIZATION

Several radioactivity detectors use the ionizing effects of radiation on gases. When an energetically charged particle, such as β^--particle, passes through a gas, its electrostatic field displaces orbital electrons from the gas molecules close to its path. This displacement also produces a positive ion from the other part of the molecule, resulting in an **ion pair**. If an electric field is applied in the chamber where the ion pairs are formed, the displaced electrons are accelerated toward the anode and the positive ions move slowly (because of their larger mass) toward the cathode. When the applied electric field is low, single ion pairs are formed from single β-particle emissions. If higher electric fields are applied, the electron displaced by the first β-particle collision is accelerated and collides with other gas molecules to produce additional ion pairs. The result is an amplification of the number of electrons from the first electron, giving a "shower of electrons."

This type of radioactive detection is used in the Geiger–Müller (G–M) tube, which is a gas-filled tube with a cylindrical cathode and a fine wire anode (Fig. 6-6). Gases of different composition are used in the G–M tube. A common gas mixture is the so-called Q-gas, which is 1.3% isobutane in helium. One end of the tube has a "window" made up of a thin sheet of mica or plastic, through which the β-particles from the radioactive sample must enter the gas chamber. A voltage potential is applied between the electrodes of the tube, and the electron initially produced by collision with the β-particle is accelerated toward the anode. In its acceleration, the electron collides with gas molecules in the G–M tube, producing more electrons, which in turn are accelerated, to yield a "shower" of electrons from the primary β-particle coming from the radioactive sample. The electrons are collected at the anode within a microsecond or less from the time the single β-particle entered the chamber. A strong electrical pulse is formed and fed into an external circuit, where it is amplified and measured.

As the potential gradient between the electrodes of the tube is increased, the number of electrons reaching the anode increases sharply for a given single ionization event. The number of electrons collected on the anode is proportional

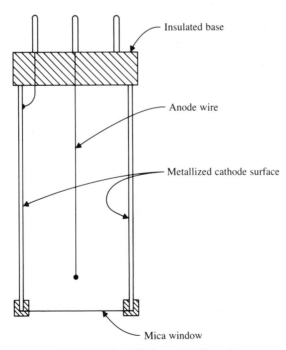

FIGURE 6-6. A Geiger–Müeller tube.

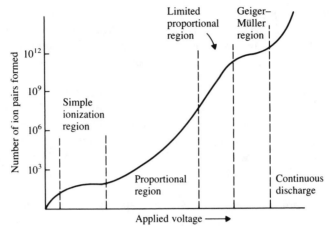

FIGURE 6-7. Relationship of the number of ion pairs produced to the voltage applied to a G–M tube.

to the number of initial ionizations, hence is proportional to the number of β-particles entering the chamber. This voltage region of the gas ionization tube is called the **proportional region** (Fig. 6-7). The avalanche of electrons produced in this region is collected on only a portion of the anode wire illustrated in Figure 6-6, and only a small fraction of the gas molecules of the chamber are involved in the formation of ions.

As the voltage is increased further, the **limited proportional region** is entered, and ionization of the gas extends throughout the entire chamber involving all the gas molecules. Electrons are collected along the entire length of the anode wire. At even higher voltages the **Geiger–Müller region**, is entered, and even weak β-particles and γ-rays are able to form sufficient numbers of electrons to completely

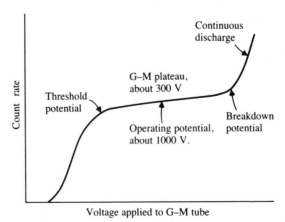

FIGURE 6-8. Operating voltage characteristics of a G–M tube.

fill the ion space in the chamber. At higher voltages, the G–M tube undergoes continuous discharge (Fig. 6-7).

The operating voltage characteristics of a G–M tube are given in Figure 6-8. The curve first shows a sharp rise in the counting rate where only the most energetic β-particles cause a response. At the **threshold potential**, this rise levels off to a plateau region, where most G–M tubes are operated, at approximately 1000 V.

The radioactive sample, usually dissolved in water for biological samples, is added to a planchet (Fig. 6-9) to produce a thin, uniform, dried film. The planchet is then placed close to the window of the G–M tube. The β-particle emissions come from point sources of the dried sample and those that pass through the window are counted. A number of factors affect the number of β-particles that come from the sample and pass through the window. In addition to the average energy of the particular β-particles, we mention the following conditions.

1. Because β-particle emissions originate from point sources, and proceed in all directions, at best only 50% of the particles have a chance to enter the tube from the planchet.
2. The distance of the planchet from the window has a major effect, since the probability that a β-particle will enter the tube falls off as a function of the inverse square of the distance $(1/d^2)$ between the planchet and the tube.
3. The character and thickness of the sample on the planchet are important because β-particles frequently are absorbed by the sample before they can penetrate the sample surface and enter the tube—so-called **self-absorption**.
4. The activity of the sample should not be too high: if a large number of emissions is allowed to saturate the system, the detection of particles ceases during the "dead time"—the time required for the system to recover.

The efficiency of the G–M counting is determined by counting a sample of known activity (i.e., a standard). The percent efficiency is the number of measured counts per unit time divided by the number of actual disintegrations per unit time times 100, or:

$$\% \text{ efficiency} = \frac{\text{counts per minute}}{\text{disintegrations per minute}} \times 100 = \frac{\text{cpm}}{\text{dpm}} \times 100 \quad (6\text{-}16)$$

FIGURE 6-9. Planchets for G–M counting. To minimize flow to the edge, planchets are manufactured with rough surfaces (e.g., anodized) or concentric rings.

The efficiency of G–M counting is relatively low for the reasons discussed above, the efficiency for ^{14}C is only 1–5%, and ^{3}H cannot be measured at all. Modifications of the G–M tube have been developed to improve the efficiency. For example, in the **gas-flow G–M** system the end window is eliminated and the sample is placed directly in the gas chamber. Because the introduction and removal of the sample deplete the gas, however, it is necessary to maintain a constant flow of counting gas through the chamber.

6.9 MEASUREMENT OF RADIOACTIVITY BY SCINTILLATION COUNTING

Scintillation counting dates back to 1903, when W. Crookes placed a source of α-radiation in front of a zinc sulfide screen in a dark room and counted small flashes of light or scintillations on the screen. The development of liquid scintillation counting occurred in the 1950s with the discovery of organic compounds called **scintillators** or **fluors** and the development of photomultiplier tubes. Liquid scintillation counting, because of its high sensitivity, is especially useful for the measurement of low energy β-emitters and has essentially replaced G–M counting of weak β-emitters (e.g., ^{14}C and ^{3}H). The first liquid scintillation counter appeared in 1953, and the automatic, multisample counter was introduced in 1957. Solid scintillation counters are used for γ-ray emitters.

6.9.1 Mechanism of Liquid Scintillation Counting

Scintillators have the property of absorbing radiant energy either in the solid state or in solution. The absorption of this energy by the scintillator results in the formation of excited atoms or molecules that rapidly return to the ground state, releasing their excitation energy as photons and heat. The number of photons emitted is directly related to the amount of radiant energy absorbed by the scintillator. Particles from nuclear disintegration, with high energy, will produce a greater number of photons than particles with lower energy. The number of photons detected during a very short period (less than a nanosecond) or the intensity of light is, therefore, directly related to the energy of the emitted radiation.

The term "liquid scintillation counting" is used because the scintillators are dissolved in a solvent, giving a "cocktail" to which the radioactive sample is added. The formation of photons results from the following sequence of events: (*a*) the energy of the β-particle emitted from the radioactive isotope is first absorbed by the solvent molecules, causing them to become excited, and (*b*) the excited solvent transfers the energy to the scintillator, causing the scintillator molecules to become excited, and when they return to the ground state they emit photons. Some scintillation cocktails use a secondary fluor that absorbs photons from the primary fluor and reemits photons at a wavelength more favorable for detection by the phototubes. The photons are then detected by a phototube–photomultiplier system, which converts the photons into amplified numbers of

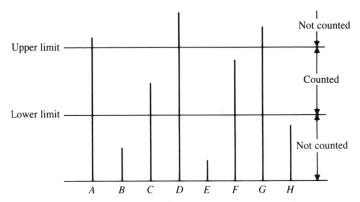

FIGURE 6-10. Principle of pulse height analysis.

electrons. The electrons produce a voltage pulse, which is measured by a scaler. Beta-particles with different energies produce voltage pulses of different sizes. Voltage pulses may be analyzed by a **pulse height analyzer**, which is an electronic sorter that can be adjusted to accept electrical pulses within a selected range of pulse heights and reject all others. The device is set by means of two discriminators (lower and upper) that determine the size of the pulses to be counted. In the example of Figure 6-10, where the bars $A-H$ represent eight pulses of varying energies generated by the photomultiplier tube, pulses B, E, and H are rejected because they are too small and pulses A, D, and G are rejected because they are too large. Only pulses C and F are passed on and counted by the scaler, which tabulates them. Figure 6-11 is a block diagram of the parts for a two-channel liquid scintillation counter. Modern instruments have two and sometimes three channels for analyzing the sample at different discriminator settings to obtain pulses of different energies.

6.9.2 Pulse Height Analysis and the Beta Energy Spectrum

There are three basic controls for each channel of a liquid scintillation counter; a gain control, which affects the voltage output of the amplifier (Fig. 6.11), and the lower and upper discriminator controls of the pulse height analyzer. There are two types of gain control: linear gain and logarithmic gain. Instruments with logarithmic gain usually are set at a fixed value, whereas instruments with linear gain usually can be varied. The optimum gain for instruments of the latter type is obtained by determining the maximum efficiency of counting for an isotope by having the lower and upper discriminator settings as far apart as possible.

At optimum gain, an energy spectrum of an isotope may be obtained by measuring the number of counts by varying the lower and upper discriminator settings at close intervals (e.g., 25 discriminator units). For example, if L is lower and U is upper, the first settings might be $L = 0$ and $U = 25$, the second $L = 25$ and $U = 50$, the third $L = 50$ and $U = 75$, and so on. By plotting the number of counts for each interval against the discriminator settings, one obtains an energy

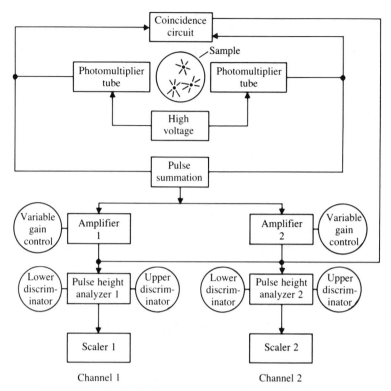

FIGURE 6-11. Block diagram of a two-channel liquid scintillation spectrometer. Two photomultiplier tubes, with high voltage power supply, detect light emissions from the sample vial. The resulting pulses from these two PM tubes are fed into a summation amplifier and then to channels 1 and 2, where they are amplified, analyzed, and recorded on a scaler.

or pulse spectrum for the isotope, from which a smooth curve can be drawn. A typical energy spectrum obtained in this way (Fig. 6-12) provides information on the choice of instrument settings to be used for optimum counting of an isotope and provides a rationale for the separation of isotopes in a dual-labeled isotope mixture.

6.9.3 Determining the Activity of Each Isotope in a Dual-Labeled Mixture

Different β-emitting isotopes have different energy spectra, as seen in Figure 6-1. The proper selection of the gain and discriminator settings allows a mixture of two isotopes, with different β-particle energies, to be counted simultaneously by a two-channel liquid scintillation counter. The hypothetical example of two isotopes with different β-energy spectra is given in Figure 6-13. For ideal detection of two isotopes, the difference in their E_{avg} should be a factor of 10. The disintegrations from two isotopes may be selected by the proper choice of gain

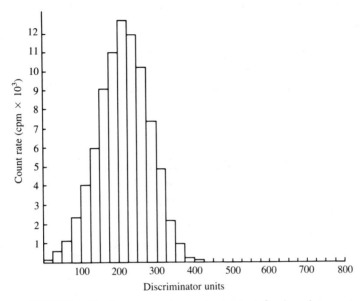

FIGURE 6-12. Energy spectrum obtained at fixed optimum gain by counting the isotope at small variable discriminator settings.

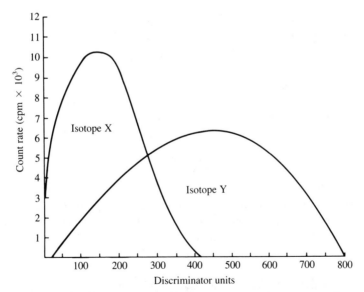

FIGURE 6-13. Liquid scintillation counter energy spectra of isotopes X and Y obtained using linear amplification.

and discriminator settings. The philosophy of counting two isotopes in a dual-labeled sample is to obtain a relatively high efficiency of counting of one of the isotopes in channel 1 and a relatively low efficiency of counting of the same isotope in channel 2, and the opposite for the other isotope (a relatively low efficiency in channel 1 and a relatively high efficiency in channel 2). From Figure 6-13, we see that the discriminator settings for channel 1 to obtain the maximum number of counts of isotope X and the minimum number of counts of isotope Y would be $L = 50$ and $U = 300$; the discriminator settings for Channel 2 to obtain the maximum number of counts of isotope Y and the minimum number of counts of isotope X would be $L = 300$ and $U = 800$. Counts would be obtained in both channels for isotope Y, but there would be a greater number of counts in channel 2 than in channel 1. To determine the actual number of dpm's for a mixture of the two isotopes, one would count standards of the two isotopes, X and Y, in both channels and determine the counting efficiencies for each of the isotopes in the two channels. One would then take the mixture of the isotopes and measure the total net counts in each of the two channels and substitute into the two simultaneous equations (6-17 and 6-18) to obtain the dpm's of isotopes X and Y in the sample:

$$C_1 = X(E_X)_1 + Y(E_Y)_1 \qquad (6\text{-}17)$$
$$C_2 = X(E_X)_2 + Y(E_Y)_2 \qquad (6\text{-}18)$$

where C_1 = net total observed counts in channel 1
$\quad\quad C_2$ = net total observed counts in channel 2
$\quad (E_X)_1$ = efficiency of counting of isotope X in channel 1
$\quad (E_Y)_1$ = efficiency of counting of isotope Y in channel 1
$\quad (E_X)_2$ = efficiency of counting of isotope X in channel 2
$\quad (E_Y)_2$ = efficiency of counting of isotope Y in channel 2
$\quad\quad X$ = dpm of isotope X in the sample
$\quad\quad Y$ = dpm of isotope Y in the sample

Some instruments come with preset gain and discriminator settings for the common isotopes (e.g., ^{3}H, ^{14}C, ^{32}P) and the common isotope pairs (^{3}H/^{14}C, ^{3}H/^{32}P, ^{14}C/^{32}P). However, the efficiencies for each of the isotopes in each channel still must be determined. It should also be mentioned that a single-channel liquid scintillation counter can be used to determine two isotopes in a mixture by counting the sample twice, using a different setting each time.

6.9.4 Cocktail Composition and Sample Preparation

The most commonly used scintillator or primary fluor is 2,5-diphenyloxazole, abbreviated as PPO, and the most commonly used secondary fluor is 1,4-bis-2-(5-phenyloxazolyl)benzene, abbreviated as POPOP.

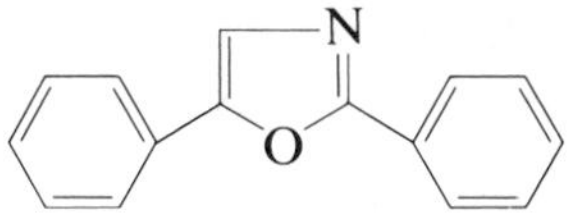

PPO

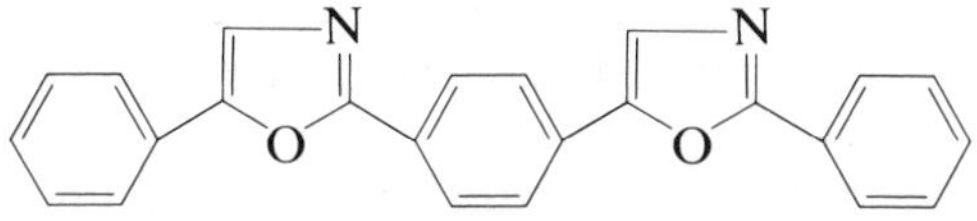

POPOP

The most common solvents are toluene and dioxane. The toluene-based cocktails are used for radioactive samples that are soluble in organic solvents and for heterogeneous liquid scintillation counting. The dioxane cocktails are used for aqueous samples. Table 6-4 gives the compositions of toluene and dioxane cocktails.

Radioactive organic compounds, soluble in organic solvents, can be added directly to the toluene cocktail. Aqueous solutions can be added directly to the dioxane cocktail up to 10% by volume. Very often it is convenient to count samples of solutions or suspensions on solid supports, which can be introduced directly into the toluene scintillation cocktail. The solid supports may be small discs or squares of filter paper, glass fiber paper, or cellulose acetate or nitrate. Aliquots of the radioactive solution are spotted and dried onto the solid support, or suspensions are filtered onto the support. The solid support is added directly to the toluene cocktail. Radioactive samples, located by autoradiography, on paper chromatograms or plastic-backed TLC sheets, may be cut out and added to the toluene cocktail. This type of counting is called **heterogeneous counting**, in

TABLE 6-4. Liquid scintillation cocktails

Solvent	Primary fluor	Secondary fluor	Additives	Comments
Toluene	PPO (5 g/L)	POPOP (0.1 g/L)	—	All samples soluble in toluene; insoluble samples adsorbed onto inert supports
Dioxane	PPO (4 g/L)	POPOP (0.2 g/L)	60 g naphthalene 100 mL methanol 20 mL ethylene glycol 880 mL dioxane	Bray's cocktail, used for aqueous samples up to 10% v/v of the cocktail volume
Dioxane	PPO (5 g/L)		100 g naphthalene	Universal cocktail, used for aqueous samples up to 10% v/v of the cocktail volume and for toluene-soluble samples

contrast to **homogeneous counting**, in which the sample is dissolved in the cocktail. The efficiency of counting samples on insoluble supports varies with the nature of the insoluble support and the particular isotope. The counting efficiency is lowered by self-absorption, which is affected by the thickness and absorptivity of the support. Very weak β-emitters, such as ^3H, are particularly affected. Although the efficiency of counting ^{14}C by heterogeneous liquid scintillation counting is decreased from the efficiency obtained by homogeneous liquid scintillation counting, it is not as drastic as it is for ^3H. The convenience and ease of heterogeneous liquid scintillation counting often compensates for the lowered efficiency, especially for ^{14}C, ^{32}P, ^{35}S, and even ^3H in some cases. To obtain reproducible results, samples on solid supports must be insoluble in the cocktail or completely extracted into the cocktail.

The polyacrylamide gels commonly used in the electrophoresis of proteins and nucleic acids (Section 5.3) are insoluble in liquid scintillation cocktails. The gels can be counted directly, however, if they are cut into thin slices (1–2 mm thick) and the separated proteins or nucleic acids are fixed by soaking the slices in 4% acetic acid. One method of solubilizing gel slices is to treat 1–2 mm slices of the gel with 0.5 mL of 30% hydrogen peroxide at 50°C until the gels dissolve. The hydrogen peroxide also decolorizes the stained gels. The digested gel can be counted in a dioxane cocktail.

The efficiency of the various cocktails is determined by using radioactive compounds of known activity. With toluene cocktails, the efficiency of a homogeneous system is most easily determined for ^3H and ^{14}C using ^3H- or ^{14}C-labeled toluene standard. For heterogeneous counting, a radioactive standard, containing the particular isotope under study, is added to the solid support and counted. The efficiency of aqueous samples in dioxane cocktails can be determined by using aqueous standard solutions of the isotope under study.

Background "counts" due to electronic noise, cosmic radiation, ^{40}K in the glass vials, and unknown sources must be measured and subtracted from the sample counts to give the net counts of the sample. In general, background counts increase with increasing the gain of the spectrometer.

In the special case of **Cerenkov counting**, no fluor is required. Cerenkov radiation is emitted when β-particles leave the nucleus at speeds approaching the speed of light in a vacuum and enter the surrounding medium, where the light moves more slowly than it does in a vacuum. The light photons of Cerenkov radiation can be detected by a photomultiplier to give a measure of the amount of radioactivity present. Only relatively high energy β-emitters, such as ^{32}P (1.72 MeV), contain enough energy to cause photons to be emitted. Therefore samples containing ^{32}P can be counted in water with relatively high efficiency (30–50%).

6.9.5 Quenching and Quench Correction

"Quenching" is any process that causes a reduction in the amount of fluorescence produced by β-emission in a liquid scintillation cocktail or a reduction in the amount of light reaching the photomultiplier tube. The effect of quenching is to produce a decrease in the counting efficiency. Two principal quenching mechanisms have been identified: chemical quenching and color quenching.

Chemical quenching is caused by the presence in the sample of materials that interfere with the transfer of energy from solvent to fluor molecules, hence reducing the amount of fluorescence from the cocktail. Chemical quenching may be caused by one or more of the following: (a) acid, resulting in the reaction of H^+ with the primary or secondary fluors, which are both bases, reducing the transfer of energy from solvent to fluor and from fluor to fluor; (b) dipole–dipole interaction of quenching agent with solvent or fluor, resulting in the loss of energy because of an increase in the vibrational energy of the excited molecule; (c) electron capture by quenching agent, resulting in loss of energy transfer from the secondary electrons to the solvent molecules.

Color quenching may result from the presence in the sample of colored compounds, which absorb the photons emitted by the fluors. This problem often is encountered in work with biological samples that contain pigments such as blood, plant extracts, and certain fermentation broths.

Quenching reduces the efficiency of counting. There are three common methods for correcting quenching by determining the counting efficiency of the sample: the internal standard method, the channels ratio method, and the external standard method.

The **internal standard method** is the classical procedure for determining the counting efficiency of a sample. First the sample is counted; then a known amount of nonquenching radioactive standard is added and the sample is recounted. The additional counts produced by the addition of standard are used to compute the counting efficiency of the sample. The actual dpm's of the sample are then computed using the determined value of the efficiency. The internal standard is added in a small volume to ensure that the counting characteristics of the original sample are not changed. The activity of the added standard should be appreciably greater than that of the sample. For toluene cocktails, labeled toluene is commonly used as an internal standard. The internal standard method is a straightforward method for correcting quenching and determining the counting efficiency of the sample. The disadvantages of the method are the need for a second counting, the tediousness of several pipetting operations when many samples are involved, and the additional cost required for the use of a labeled standard, especially for the testing of many samples.

The **channels ratio method** is based on the displacement of the pulse height spectrum to lower energies that always occurs when quenching agents are present. With a two-channel instrument one channel is set to obtain as many cpm's as possible for a nonquenching standard (A in Fig. 6-14). The second channel (B in Fig. 6-14) is set to count approximately the upper half of the β-spectrum of the nonquenching standard. The channels ratio is then defined as the ratio of the counts in the narrow window channel (channel B) divided by the counts in the wide window channel (channel A), giving a number less than 1. As quenching agents are added, the ratio of the cpm in channel B to cpm in channel A decreases because the pulse height spectrum is shifting to lower energies.

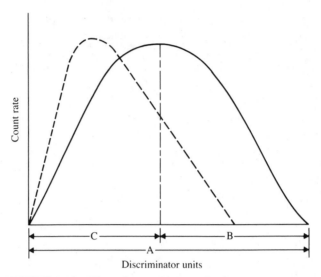

FIGURE 6-14. Effect of quenching on the β-energy spectrum. The solid curve represents unquenched isotope and the dashed curve represents quenched isotope; A, B, and C represent various window widths of the pulse height analyzers.

A correction curve relating the channels ratio to the efficiency is obtained by preparing a series of standards to which varying amounts of quenching agent are added. For each of the standards, both the efficiency and channels ratio are determined. A number of organic compounds may be used as quenching agents (e.g., nitromethane, acetic acid, acetone, carbon tetrachloride, chloroform) to prepare a chemical quench, channels ratio plot. A single quench correction curve, prepared with one chemical quenching agent, can be used for the correction of chemical quench caused by any of the chemical quenching agents. A separate chemical quench correction curve must be prepared for each isotope. A chemical quench correction curve would not be suitable for color quench correction. Nonlabeled color quenching agents, (blood, plant extract, fermentation broth, etc.) would have to be added to the standards to obtain a color quench, channels ratio plot. A typical channels ratio plot is shown in Figure 6-15. The actual dpm of a quenched sample is obtained by dividing the cpm by the efficiency obtained from the channels ratio plot.

The **external standard method** for quench correction depends on the use of a highly radioactive γ-radiation source, such as ^{226}Ra or ^{137}Cs. After the sample has been counted, the γ-source is brought near the sample vial and the sample is counted for a short additional period. The method is based on the production of Compton electrons, which are formed when the γ-rays enter the cocktail solution. It is assumed that these Compton electrons behave like the β-particles emitted by the radioactive sample in the cocktail. Thus, the presence of quenching agents in the cocktail should have the same effect on the Compton electrons as it does on the β-particles emitted by the sample.

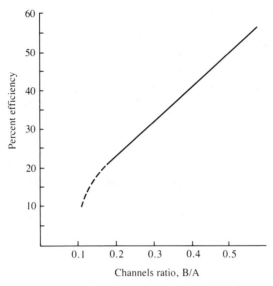

FIGURE 6-15. A typical channels ratio plot of efficiency versus channels ratio; A and B are pulse height analyzer windows (see Fig. 6-14).

In theory, the number of counts produced by the external standard (ES) is the total counts obtained when the external standard is placed next to the vial minus the counts obtained from the sample. The ratio of the net external standard counts obtained from a vial of cocktail with sample to the counts of the external standard obtained from a vial of cocktail without sample gives the efficiency of counting of the vial with sample.

$$\% \text{ efficiency by ES} = \frac{\text{for a sample containing quenching agent}}{\text{total ES counts from a vial of}} \times 100$$
$$\frac{(\text{total counts} - \text{sample counts})}{\text{cocktail without sample}}$$

(6-19)

In practice, because of the statistical variation in radioisotope counting, the external standard is counted in two channels with different energy selections (different windows) so that a ratio of net counts between two external standard channels is obtained. For example, an unquenched sample will give an ES ratio of 0.80–0.85. Quenched samples give lower ES ratios (e.g., 0.75, 0.70, 0.65). This method of quench correction is easy to perform, especially with many samples, and has very little potential for error. Consequently, most liquid scintillation counters are equipped with an external standard and the mechanical means of moving the external standard up to the sample vial in the counting well. A plot of the percent efficiency versus the ES ratio can be obtained similar to the plot of percent efficiency versus channels ratio shown in Figure 6-15.

6.10 SOLID SCINTILLATION COUNTING OF γ-RADIATION

Gamma radiation can be detected and quantitated by the use of a crystalline inorganic fluor—most commonly sodium iodide with a small amount of thallium as an activator. This material, which can be produced as large transparent crystals, has high density and high stopping power and is therefore suitable for counting γ-rays. The γ-rays interact with the crystalline fluor to give ion pairs, Compton electrons, and photoelectrons. These electrons excite the crystalline fluor, which emits photons on returning to ground state. The crystals are optically coupled to the window of a photomultiplier tube, and the whole assembly is screened from external light sources by encapsulation in a metal cylinder. The crystalline fluor must be hermetically sealed because sodium iodide is hygroscopic. The sample is inserted into a well with the solid fluor surrounding it (Fig. 6-16). The efficiency of counting is determined with the use of standards. Gamma counters with automatic multisample capability have found especially wide use in counting ^{125}I used in radioimmunoassays (Section 6.14.6). Gamma counters with two or three channels are available with preset and variable discriminator settings for the counting of γ-emitters such as ^{125}I, ^{51}Cr, and ^{60}Co, and for the selection of specific γ-energies from dual-labeled samples. Optimization of the gain and discriminator settings is obtained by the procedures outlined for liquid scintillation counting (Section 6.9.2).

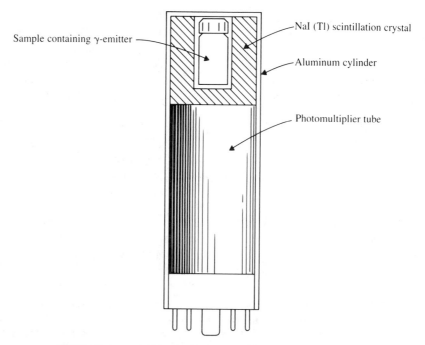

Sample containing γ-emitter

NaI (Tl) scintillation crystal

Aluminum cylinder

Photomultiplier tube

FIGURE 6-16. Solid scintillation, well-type counter for measuring samples with γ-ray emitters.

6.11 DETECTION BY AUTORADIOGRAPHY

Autoradiography is primarily a technique for determining the location of radioisotopes in tissue, tissue sections, chromatograms, and gels. It is a non-destructive method for locating radioactive compounds, which can then be isolated and counted. The radioactive sample is placed in close contact with X-ray film. Radiation from the areas of radioactivity in the sample, in the form of β-emissions, hits the silver bromide in the film emulsion reducing the silver ions to metallic silver. After a period of exposure in a light-tight container, the film is developed in the usual manner by immersing in developer, stop bath, fixer, and water wash, followed by drying. The location of the radioactivity in the sample is determined from the pattern of film darkening.

Various radioactive isotopes can be detected by autoradiography: location of ^3H in nucleic acids of chromosomes during cell division; location of ^{45}Ca in growing bone tissue; location of $[^{14}C]N$-acetyl-glucosamine in the cell wall of yeast during division; and the distribution of ^{32}P in plant tissue during growth. Figure 6-17 presents an autoradiogram of a plant leaf that has been exposed to $[^{14}C]$-CO_2.

The film employed for contact autoradiography for weak β-emitters (e.g., ^{14}C, ^{35}S, ^{45}Ca) has emulsion on one side only. The film employed for stronger β-emitters and γ-emitters (e.g., ^{32}P and ^{125}I) has emulsion on both sides. The more expensive two-sided emulsion film is unnecessary for the weak β-emitters, which do not emit β-particles with enough energy to penetrate the second emulsion. A commonly used single-sided film is Kodak SB-5 X-ray film, and Kodak also makes double-sided film, NS-5T. To have optimum darkening of the film, one needs 1×10^6 events (β-particles) reaching the film per square centimeter: 1×10^5 events/cm^2 would barely be detectable. Thus, if the sample has 1×10^3

FIGURE 6-17. Autoradiogram of a plant leaf that has been exposed to $[^{14}C]$-CO_2.

cpm/cm^2 on a paper chromatogram or tissue section, the film would have to be exposed for 1000 minutes or about 17 hours for optimum darkening.

An intensifying screen, made up of a fluorescent material, may be used for the strong β^-- and γ-emitters. The screen is placed against the back of the double-sided film. The radiation passes through the double-sided film and hits the intensifying screen, producing fluorescent light that is reflected back onto the emulsion, giving further reduction of the silver ion in the film.

The detection of the weak β-emitters can be enhanced using fluorography, which involves overcoating or impregnating the sample with a suitable fluor such as PPO. Some of the energy of the β-particles is converted to light, which exposes the film. The sample, coated with the scintillator, is placed against the film, which is exposed in a light-tight container at -50 to $-80°$C. Fluorography improves the efficiency of detection of weak β-emitters when compared with direct exposure. It is almost always used for the detection of tritium but will also greatly improve the detection of ^{14}C and ^{35}S too (Fig. 6-18). A commercial spray scintillator (DuPont-NEN EN^{3}HANCE) is convenient for use with chromatograms and electrophoresis gels.

The low temperatures used in fluorography serve to stabilize latent image formation during long exposures to light. For example, at low temperatures, the fluorographic detection of ^{3}H is increased by a factor of 4 or more over exposure at room temperature.

Since most of the film exposure in fluorography is from the light emitted by the fluor, the film must have spectral sensitivity that is optimum for the

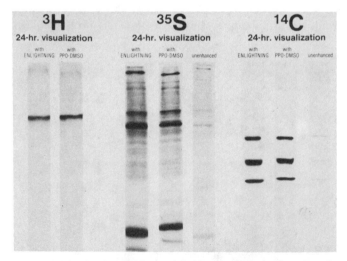

FIGURE 6-18. Comparison of fluorographic enhancement of proteins labeled with ^{3}H, ^{35}S, and ^{14}C that have been separated by polyacrylamide slab-gel electrophoresis. Enhancement was obtained with DuPont-NEN EN^{3}HANCE and PPO-DMSO, followed by exposure for 24 hours. Courtesy of E.I. duPont-NEN Products Division.

TABLE 6-5.   Estimated mean range in a photographic emulsion for β^--particle emission from different isotopes

Isotope	Average energy (MeV)	Range (μm)
^{3}H	0.0055	1
^{14}C	0.049	100
^{35}S	0.055	100
^{32}P	0.70	800

wavelength of light emitted by the fluor, which is in the ultraviolet or blue regions. A film that is sensitive for the UV and blue region is Kodak X-OMAT AR.

Resolution in autoradiographic detection is the minimum distance required between two point sources of radioactivity that can be seen as two separate images on an autoradiogram. Resolution is affected by the energy of the emitter, the thickness of the sample, and the distance between the sample source and the film emulsion.

Table 6-5 gives the estimated mean range of β^--particles in a photographic emulsion. Tritium gives the best resolution in autoradiography because the low energy β-particles travel only a short distance in the emulsion, and therefore only a small area near the source will be darkened. In contrast, β-particles from ^{32}P will travel great distances from the sample in the emulsion, causing a loss of resolution and a less distinct image. X-Ray and γ-emitters produce very diffuse images.

Resolution will also be increased by decreasing both the thickness of the sample and the distance between the sample and the film emulsion. Since radiation is emitted in all directions from a point source, the closer the source to the film, the smaller will be the image that is formed.

6.12 METHODS OF LABELING BIOCHEMICAL COMPOUNDS

There are two general methods for obtaining labeled biochemicals: biosynthesis and chemical synthesis. Each has both advantages and disadvantages.

Biosynthesis uses whole organisms or specific enzymes. The most commonly used radioisotope in biosynthetic procedures is ^{14}C, although ^{32}P and ^{35}S are also employed. The successful use of biosynthesis depends on a number of factors: an organism must be selected that synthesizes and accumulates reasonable quantities of the labeled compound(s) from available, labeled starting compounds, or a specific enzyme must be available or easily isolated, and its labeled substrate must also be available. Microorganisms, or enzymes isolated from them, are commonly used because they can be rapidly grown in large quantities, under controlled conditions, to produce a wide variety of desired, labeled compounds or desired, specific enzymes. When the labeled compound is to be obtained directly from the microorganism, the microorganism is grown in a culture containing a labeled precursor, and when the growth is complete, the desired labeled compound is isolated from the microorganism itself or from the cell-free culture supernatant fluid. When specific enzymes are used, the desired labeled product also must be isolated and purified. Whole organism biosynthesis gives a relatively large number of labeled products, whereas enzyme synthesis gives only one or a few specifically labeled products. Enzyme biosynthesis usually requires a specific, labeled substrate, whereas whole organism biosynthesis requires a labeled precursor that is much less specific (e.g., $[^{14}C]$-CO_2, $[^{32}P]$-PO_4^{-3}, ^{14}C- or ^3H-labeled acetate). Because biosynthesis is usually on a micro or semimicro scale, chromatographic methods of isolation and purification (paper, TLC, HPLC) can be used to obtain highly purified products. For a specific example, using a biological system to prepare a labeled compound, see Chapter 8 (Section 8.8.3).

Photosynthesis is commonly used to obtain ^{14}C-labeled compounds from photosynthetic organisms and $[^{14}C]$-CO_2, which is obtained from the relatively cheap starting material, ^{14}C-labeled $BaCO_3$. Starch, glucose, fructose, and sucrose labeled with ^{14}C are obtained in good yields by photosynthesis of leaves or algae in the presence of $[^{14}C]$-CO_2. Yeast grown on $[^{14}C]$D-glucose can be used to produce $[^{14}C]$L-amino acids.

More specifically labeled compounds can be obtained by enzyme biosynthesis. For example, sucrose specifically labeled with ^{14}C in either the glucose or the fructose residues can be prepared using sucrose phosphorylase, which catalyzes the following reaction:

$$\alpha\text{-D-glucose-1-phosphate} + \text{D-fructose} \rightarrow \text{sucrose} + \text{phosphate}$$

If $[^{14}C]$glucose-1-phosphate and nonlabeled fructose are used, glucose-labeled sucrose results, and if $[^{14}C]$fructose and nonlabeled glucose-1-phosphate is used, fructose-labeled sucrose results. Fructose, labeled in various specific carbons, can be obtained by the reaction of aldolase with various combinations of ^{14}C-labeled D-glyceraldehyde-3-phosphate and dihydroxyacetone phosphate. To obtain ^{32}P-labeled α-D-glucose-6-phosphate, one can react $[^{32}P]$glucose-1-phosphate with phosphoglucomutase or glucose with $[^{32}P]$ATP and hexokinase. Proteins can be enzymatically iodinated with ^{125}I under very mild conditions by the use of glucose, glucose oxidase, lactoperoxidase and $Na[^{125}I]$. The glucose oxidase generates a small, steady amount of hydrogen peroxide from the glucose and the lactoperoxidase catalyzes the peroxide oxidation of iodide-125 to iodine-125, which adds to the tyrosyl groups of the protein. These reactions illustrate a few of the numerous specific enzyme reactions used to obtain radioactive compounds.

Radioactive isotopes may also be introduced into biochemicals using the wide variety of synthetic procedures of organic chemistry. As with enzymatic methods, chemical synthesis yields products that are specifically labeled in one or a few positions. However, labeled starting materials can be scarce and expensive. An additional limitation is the relatively low yields obtained in many chemical syntheses. Furthermore, many chemical syntheses give a racemic mixture of optical isomers, only one of which has a biochemical function. For example, the organic synthesis of amino acids gives both D- and L-isomers, but usually only the L-isomer is of biochemical importance. Likewise the synthesis of carbohydrates by the Kiliani–Fischer synthesis using KCN gives two epimers. The separation of enantiomers and epimers is usually tedious and difficult, although the development of preparative high pressure liquid chromatography has made it somewhat easier.

The following are examples of some organic reactions used to introduce radioisotopes into organic compounds.

1. *Labeling of proteins.*

(a) Introduction of ^{14}C by reaction of $[^{14}C]$-formaldehyde with amino groups followed by reduction with sodium borohydride.

$$\text{protein}-NH_2 + H-^{14}CHO \rightarrow \text{protein}-N=^{14}CH_2 \xrightarrow{\text{NaBH}_4}$$

$$\text{protein}-NH-^{14}CH_3$$

(b) Introduction of ^3H by reaction of formaldehyde with amino groups followed by reduction with sodium borotritide.

$$\text{protein}-NH_2 + HCHO \rightarrow \text{protein}-N=CH_2 \xrightarrow{\text{NaB}^3\text{H}_4}$$

$$\overset{\displaystyle ^3H}{\underset{\displaystyle |}{\text{protein}-NH-CH_2}}$$

(c) Introduction of ^{125}I by reaction of $[^{125}I]$-Bolton-Hunter reagent with amino groups.

(d) Introduction of ^{125}I by reaction of $[^{125}I]$-NaI and chloramine-T with protein tyrosine groups.

2. *Labeling of nucleic acids.*

(a) Introduction of ^{14}C by reaction of $[^{14}C]$dimethyl sulfate. Nucleic acids are methylated at the N_7 position of guanine, and therefore both DNA and RNA can be labeled.

(b) Introduction of ^3H at the 3′-terminal ribonucleotide in RNA. The terminal ribose is oxidized with periodate to form the dialdehyde, which can be reduced with tritiated sodium borohydride.

(c) Introduction of ^{32}P at the 5′-terminus of DNA or RNA with T_4-polynucleotide kinase. Polynucleotide kinase catalyzes the transfer of γ-^{32}P from ATP to the 5′-OH group of DNA or RNA. Equimolar concentrations of ATP and 5′-OH nucleic acid are incubated with a few units of T_4-polynucleotide kinase in 70 mM Tris–HCl, pH 7.6, at 30°C for 30 minutes.

3. *Formation of ^{14}C carbon–carbon bonds.*
   (a) Synthesis of ^{14}C-carboxyl-labeled acids from Grignard reagent and [^{14}C]-CO$_2$.

$$R—MgCl + {}^{14}CO_2 \xrightarrow{\text{ether}} R—{}^{14}\overset{\displaystyle O}{\overset{\|}{C}}—OMgCl \xrightarrow{\text{H}_2\text{O}} R—{}^{14}\overset{\displaystyle O}{\overset{\|}{C}}—OH$$

   (b) Synthesis from aldehyde and ^{14}C labeled KCN.

$$R—CHO + K^{14}CN \rightarrow R—\overset{\displaystyle OH}{\overset{|}{C}H}—{}^{14}CN \xrightarrow{\text{H}_3\text{O}^+} R—\overset{\displaystyle OH}{\overset{|}{C}H}—{}^{14}CHO$$

4. *Synthesis of tritium-labeled compounds by reduction with NaB^3H_4.*
   (a) Synthesis of 1-[^{3}H] alcohols by reduction of aldehydes:

$$R—CHO + NaB^3H_4 \rightarrow R—\overset{\displaystyle {}^3H}{\overset{|}{C}H}—OH$$

   (b) Hydrogenation of unsaturated bonds:

$$R—CH{=}CH—R' \xrightarrow[\substack{2)\ \text{H}^+ \\ 3)\ \text{Pd/C}}]{1)\ \text{NaB}^3\text{H}_4} R—\overset{\displaystyle {}^3H}{\overset{|}{C}H}—\overset{\displaystyle {}^3H}{\overset{|}{C}H}—R'$$

5. *Synthesis of ^{35}S-labeled L-cysteine.*
   Reaction of 3-bromo-L-alanine with ^{35}S-labeled potassium thioacetate followed by deacetylation:

$$H_2N—\overset{\displaystyle |}{\underset{\displaystyle \underset{\displaystyle Br}{\overset{|}{CH_2}}}{\overset{|}{C}H}}—COOH + CH_3—\overset{\displaystyle O}{\overset{\|}{C}}—{}^{35}S^- \rightarrow H_2N—\overset{\displaystyle |}{\underset{\displaystyle \underset{\displaystyle \underset{\displaystyle O}{\overset{\|}{}}}{\overset{|}{{}^{35}S—C—CH_3}}}{\overset{|}{C}H}}—COOH \xrightarrow[\text{CH}_3\text{OH}]{\text{NaOCH}_3}$$

$$H_2N—\overset{\displaystyle |}{\underset{\displaystyle \underset{\displaystyle {}^{35}SH}{\overset{|}{CH_2}}}{\overset{|}{C}H}}—COOH$$

6.13 TYPES OF RADIOCHEMICAL LABELING

A compound can be radiolabeled *isotopically* or *nonisotopically*. In an isotopically labeled compound, one or more radioactive atoms have replaced stable atoms of the same element. For example, $^1_1$H might be replaced with $^3_1$H, $^{12}_6$C with $^{14}_6$C, or $^{31}_{15}$P with $^{32}_{15}$P. An isotopically labeled compound closely mimics the chemical and biological properties of its nonlabeled counterpart, with the possible exceptions discussed in Section 6.15.

In a nonisotopically labeled compound, a radioactive isotope has been substituted for a stable atom of another element. For example, ^{125}I might be substituted for $^1_1 H$ on the phenolic ring of tyrosine in proteins, or the $^{36}Cl_2$ might be added to the double bond of oleic acid.

Isotopic labels can be applied in three forms: specific, general (G) or random, and uniform (U). The isotope ^{14}C can be used to illustrate all three.

Specific labeling is the labeling of a selected atom. By convention, specific labels are indicated by including the number of the labeled atom in brackets with the symbol for the isotope. Thus, for example, [1-^{14}C]D-glucose and [6-^{14}C]D-glucose designate D-glucose bearing labels on C_1, the aldehyde carbon, and C_6, the primary alcohol carbon, respectively. Almost always, a specifically labeled compound is a mixture of unlabeled molecules (carrier molecules) and molecules with a radioactive atom in the single specified position.

In general labeling, the label is randomly distributed in all the carbon atoms. For example, if propionic acid (CH_3—CH_2—COOH) is randomly labeled in all three carbons, giving three types of labeled molecules (asterisks):

$$\overset{*}{C}H_3-CH_2-COOH \qquad CH_3-\overset{*}{C}H_2-COOH \qquad CH_3-CH_2-\overset{*}{C}OOH$$

we call it [^{14}C-(G)]propionic acid.

If, instead, we wanted propionic acid to be uniformly labeled, we would label all three carbons in a single molecule with the same amount of radioactivity:

$$\overset{*}{C}H_3-\overset{*}{C}H_2-\overset{*}{C}OOH$$

plus nonlabeled carrier molecules.

If the amount of label is the same in each carbon of a randomly labeled compound, it is difficult to distinguish the randomly labeled compound from the uniformly labeled compound. It is possible, however, to have a labeled compound in which the different types of carbons have different amounts of label. For example, propionic acid might be a mixture of labeled molecules in which 40% of the label is in the methyl carbon, 50% in the methylene carbon, and 10% in the carboxyl carbon. The different percentages could result from labeled carbon sources of different types in the biosynthesis of propionic acid. Such labeling can give information about the biosynthetic pathway.

Another type of specific labeling is found in molecules that are oligomers or polymers. For example, the disaccharide sucrose can be specifically labeled in three ways: with uniform labeling (i.e., all the carbons equally labeled), with specific labeling in the glucose residue only, or with specific labeling in the fructose residue. The label in the glucose or fructose residues could further be uniformly or specifically labeled as well. Thus the compounds designated [U-^{14}C]sucrose, [U-^{14}C-Glc]sucrose, [U-^{14}C-Fru]sucrose, and [1-^{14}C-Glc]-sucrose indicate uniformly labeled sucrose, sucrose uniformly labeled in the glucose residue, sucrose uniformly labeled in the fructose residue, and sucrose specifically labeled in C_1 of the glucose residue.

Labeled samples often contain nonlabeled carrier molecules, although labeled, carrier-free compounds can be obtained. These compounds have the highest possible specific activity (SA, see Section 6.14.1 and Equation 6-20). For example, if we had carrier-free $[^3H]NaBH_4$, we could calculate the carrier-free specific activity as shown in the accompanying box.

CALCULATION OF THE SPECIFIC ACTIVITY OF CARRIER-FREE $[^3H]$-NaBH$_4$

One mole of tritiated $NaBH_4$ has 6.02×10^{23} molecules and one millimole would have 6.02×10^{20} molecules. Since there are four ^3H atoms per molecule, there are $4 \times 6.02 \times 10^{20}$ ^3H atoms per millimole. The half-life for $^3H = 12.26$ yr and $\lambda = 0.693/t_{1/2}$ (from Equation 6-10).

$$\lambda \text{ for } {}^3H = \frac{0.693}{12.26 \text{ yr}} = 0.0565 \text{ yr}^{-1} \qquad \text{(probability for a } {}^3H \text{ nucleus to decay in 1 yr)}$$

$$SA = \frac{\text{radioactivity}}{\text{mmol}}$$

$$SA = (24.02 \times 10^{20} \, {}^3H \text{ atoms/mmol}) \times (0.0565 \text{ yr}^{-1})$$
$$= 1.36 \times 10^{20} \text{ atoms disintegrating/yr·mmol}$$

Convert to dpm:

$$= \left(1.36 \times 10^{20} \frac{\text{disintegrations}}{\text{yr·mmol}}\right)\left(\frac{1 \text{ yr}}{365 \text{ days}}\right)\left(\frac{1 \text{ day}}{24 \text{ h}}\right)\left(\frac{1 \text{ h}}{60 \text{ min}}\right)$$

$$= 2.59 \times 10^{14} \text{ dpm/mmol}$$

Convert to curies per millimole:

$$= \frac{2.59 \times 10^{14} \text{ dpm/mmol}}{2.22 \times 10^{12} \text{ dpm/Ci}} = 1.18 \times 10^2 \text{ Ci/mmol}$$

6.14 APPLICATIONS IN THE USE OF RADIOISOTOPES

Because radioisotopes can be quantitated with high precision and accuracy and their level of detection is very sensitive, they have become indispensable tools in the quantitative determination of compounds in complex mixtures. By using radioisotopes, it is possible to detect and quantitate the presence of metabolic substances present in tissues or cells at very low concentrations that are not accessible by the most sensitive chemical methods of analysis. Radioisotopes have also found qualitative uses in locating compounds in complex systems and in following the fate of a particular atom or compound in a dynamic process. This has contributed greatly to our understanding of exactly how biological processes take place in the living cell and in isolated enzyme systems: for example, citric acid cycle, photosynthesis, protein biosynthesis, and the genetic code.

6.14.1 The Quantitative Determination of Compounds by Use of Radioisotopes

The radioactive specific activity (SA) of a compound is defined as the ratio of the amount of radioactivity to the weight of the compound.

$$SA = \frac{\text{amount of radioactivity}}{\text{weight of material}} \qquad (6\text{-}20)$$

This is most often expressed as some fraction of a curie divided by some fraction of a mole ($\mu Ci/\mu mol$, $mCi/mmol$, $mCi/\mu mol$, etc.). If the molecular weight of the radioactive compound is unknown, SA also may be expressed in millicuries per milligram, microcuries per microgram and so on.

QUANTITATIVE DETERMINATION OF COMPOUNDS USING RADIOISOTOPES

Let us say, we have uniformly labeled pyruvic acid with a specific activity of 3 $\mu Ci/\mu mol$ undergoing a reaction to give ethanol and carbon dioxide.

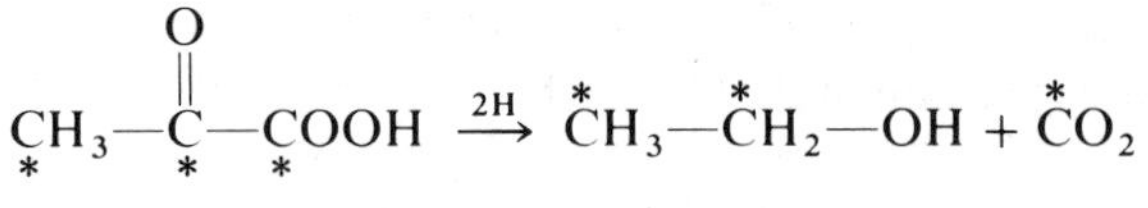

$$\overset{*}{C}H_3 - \overset{O}{\underset{*}{\overset{\|}{C}}} - \overset{*}{C}OOH \xrightarrow{2H} \overset{*}{C}H_3 - \overset{*}{C}H_2 - OH + \overset{*}{C}O_2$$

After a certain reaction time, the amount of radioactivity in CO_2 was found to be 1.00×10^4 dpm. We would like to know how many micromoles or micrograms of CO_2 and ethanol have been formed.

From the reaction, we can see that the maximum amount of radioactivity that can appear in CO_2 is one-third the amount of radioactivity in the starting material, pyruvic acid. Since we started with 3 μCi of pyruvic acid, the most radioactivity we can obtain in CO_2 is 1 μCi or 2.2×10^6 dpm. The problem is, What fraction of a microcurie does 1.00×10^4 dpm represent? The fraction is:

$$\frac{1.00 \times 10^4 \text{ dpm}}{2.2 \times 10^6 \text{ dpm}/\mu Ci} = 0.0045 \ \mu Ci$$

Since 1 μCi equals 1 μmol of CO_2, we know that 0.0045 μmol of CO_2 has been produced. Likewise, the number of micromoles of ethanol would also be 0.0045, although the amount of radioactivity in ethanol (0.009 μCi) would be twice as great as in CO_2 because there are twice as many uniformly labeled carbon atoms.

6.14.2 The Quantitative Determination of a Compound by Isotope Dilution Analysis

Many biological mixtures are complex. The concentration or amount of individual compounds in a mixture can be determined by isotope dilution analysis. Simply, isotope dilution analysis measures the changes in the specific activity of a radioactive compound when it is added to a mixture containing an unknown amount of that compound. The procedure is to add a small amount of the labeled compound having a relatively high amount of radioactivity. The sample is thoroughly mixed, an aliquot is taken, and the compound is purified, not necessarily quantitatively (by extraction, chromatography, crystallization, etc.). The radioactivity of the purified compound is then measured, the amount quantitatively determined (by spectrophotometry, fluorimetry, enzymic assay, etc.), and the specific activity computed.

The specific activity of the radioactive compound added is:

$$S_1 = \frac{A}{y} \tag{6-21}$$

where S_1 is the specific activity, A is the amount of radioactivity added, and y the weight of the compound added. The specific activity of the compound after adding the radioactive compound to the sample is:

$$S_2 = \frac{A}{x + y} \tag{6-22}$$

in which the specific activity has been decreased by x milligrams of the inactive compound being determined. From Equation 6-21 we have:

$$y = \frac{A}{S_1} \tag{6-23}$$

Substituting for y in Equation 6-22, we have:

$$S_2 = \frac{A}{x + A/S_1} \tag{6-24}$$

$$S_2 x + A \frac{S_2}{S_1} = A \tag{6-25}$$

$$S_2 x = A - A \frac{S_2}{S_1} \tag{6-26}$$

$$S_2 x = A\left(1 - \frac{S_2}{S_1}\right) \tag{6-27}$$

$$x = \frac{A/S_1(S_1 - S_2)}{S_2} \tag{6-28}$$

Substituting y for A/S_1 (Equation 6-23) and dividing each term on the right by S_2 gives the general isotope dilution equation:

$$x = y\left(\frac{S_1}{S_2} - 1\right) \tag{6-29}$$

Substituting A/y for S_1 gives:

$$x = y\left(\frac{A/y}{S_2} - 1\right) \tag{6-30}$$

then:

$$x = \frac{A}{S_2} - y \tag{6-31}$$

and in the special case when the amount of $y \lll x$, y may be neglected, and:

$$x = \frac{A}{S_2} \tag{6-32}$$

Equations 6-29 and 6-32 are the basic equations for the use of isotope dilution analysis in the quantitative determination of a compound found in a complex mixture.

QUANTITATIVE DETERMINATION OF A COMPOUND BY DIRECT ISOTOPE DILUTION

To determine how much ethanol is produced in a complex fermentation broth, for example, we can determine the amount of ethanol by isotope dilution.

We add 20,000 dpm (1 μg) of ^{14}C-labeled ethanol to a 250-mL sample of the fermentation broth. We isolate and purify the ethanol in an aliquot of the fermentation broth by HPLC and count an aliquot by liquid scintillation spectrometry to obtain the specific activity, which is found to be 800 dpm/mg.

Applying Equation 6-32, we can determine the total amount of ethanol in the fermentation broth:

$$x = \frac{A}{S_2} = \frac{20,000 \text{ dpm}}{800 \text{ dpm/mg}} = 25 \text{ mg}$$

Thus, the fermentation broth has 25 mg of ethanol/250 mL. If the density of pure ethanol is 0.8 mg/mL, there is 31.25 mL of ethanol in the 250-mL sample, giving 12.5% ethanol by volume, a fairly potent brew.

In some instances, the weight of the radioactive compound added (y) cannot be ignored because it significantly contributes to the amount of the compound being isolated. In these cases, Equation 6-29 must be used.

In inverse isotope dilution analysis, a simple variant of direct isotope dilution analysis, the change in the specific activity of a radioactive compound in a complex mixture of labeled compounds is measured after diluting with a known amount of inactive compound. The most common use of inverse isotope dilution analysis occurs when a radioactive compound has been used as a tracer in a complex system in which several compounds are labeled but the amount of only one of the compounds is wanted. The method requires the determination of two specific activities of the desired compound, one before dilution and one after dilution.

Let us say we have m milligrams of a radioactive compound, which has A radioactivity. The specific activity S_1, is given by:

$$S_1 = \frac{A}{m} \tag{6-33}$$

and we add n milligrams of the inactive compound and determine the specific activity S_2, to be:

$$S_2 = \frac{A}{m + n} \tag{6-34}$$

From Equation 6-33 we solve for A and substitute into Equation 6-34 to give:

$$S_2 = \frac{mS_1}{(m + n)} \tag{6-35}$$

$$S_2(m + n) = mS_1 \tag{6-36}$$

$$mS_1 - mS_2 = nS_2 \tag{6-37}$$

$$m(S_1 - S_2) = nS_2 \tag{6-38}$$

Solving for m in Equation 6-38 we obtain:

$$m = \frac{nS_2}{(S_1 - S_2)} \tag{6-39}$$

Equation 6-39 is the basic equation giving the amount of m in the sample using inverse isotope dilution analysis.

QUANTITATIVE DETERMINATION OF A COMPOUND BY INVERSE ISOTOPE DILUTION

A company, manufacturing radioactive DDT, has been running its wastewater into a lagoon. To find out how much radioactive DDT is present in the lagoon, an EPA analyst takes a one-liter sample from the lagoon and concentrates it to 100 μL. The specific activity S_1 of DDT is determined to be 20,000 dpm/μg. The analyst adds 1 kg of nonlabeled DDT to the lagoon and allows equilibration to occur. He takes a second one-liter sample, concentrates it to 100 μL, and determines the specific activity S_2 of DDT to be 15,000 dpm/μg. Using Equation 6-39, where

$$n = 1 \text{ kg} = 10^9 \; \mu g$$
$$S_1 = 20,000 \text{ dpm}/\mu g$$
$$S_2 = 15,000 \text{ dpm}/\mu g$$

the amount of labeled DDT in the lagoon is found to be:

$$m = \frac{(10^9 \; \mu g)(15 \times 10^3 \text{ dpm}/\mu g)}{(20 - 15) \times 10^3 \text{ dpm}/\mu g} = 3 \times 10^9 \; \mu g = 3 \text{ kg}$$

and the total amount of radioactivity from the DDT is:

$$(3 \times 10^9 \; \mu g)(2 \times 10^4 \text{ dpm}/\mu g) = 6 \times 10^{13} \text{ dpm}$$

$$= \frac{6 \times 10^{13} \text{ dpm}}{2.2 \times 10^{12} \text{ dpm}/\text{Ci}} = 27.3 \text{ Ci}$$

6.14.3 The Determination of the Mole Fractions of Several Compounds Formed from One Labeled Precursor

Suppose that an oligosaccharide is labeled exclusively in the reducing-end glucose residue $\bigcirc - \bigcirc - \bigcirc - \bigcirc - \bullet$ (where the open circles represent glucose and the solid circle is the labeled glucose), and a certain enzyme can break each of the four bonds, but at different rates, to give unequal amounts of four labeled products. We want to know what fraction of the total each of the four products represents. We run the reaction, separate the four products and the starting material by paper chromatography or TLC, and count the products by heterogeneous liquid scintillation spectrometry, obtaining the data given in Table 6-6.

We obtain the mole fraction each compound represents in the mixture by dividing the counts for each compound by the total counts. We see from Table 6-6 that 54.3% of the starting compound remains and 45.7% has undergone reaction

TABLE 6-6.   Mole fractions from a four-product reaction

Compound	cpm	% Mole fraction
Labeled product a	1,210	5.1
Labeled product b	6,890	29.1
Labeled product c	2,100	8.9
Labeled product d	620	2.6
Starting material e	12,850	54.3
	23,670	100.0

to give various percentages of the four products. One advantage of this type of analysis is that it does not depend on the size of the aliquot or on the efficiency of the counting system as long as the efficiency remains constant for all the samples. The mole fractions for analyses obtained for aliquots of different sizes may be compared, and one does not have to worry about the precision of the aliquot size.

6.14.4 Pulse and Chase Techniques with Radioisotopes

Pulse and chase techniques for identifying the presence of intermediates in a reaction sequence have been used as tools in the study of the direction of polymer biosynthesis[1-10] to answer questions such as the following: In the synthesis of proteins, are the monomers added to the carboxyl end or the amino end? In the synthesis of polysaccharides, are the monomer(s) added to the reducing end or the nonreducing end? In the synthesis of fatty acids, is acetate added to the carboxyl end or the methyl end? In the synthesis of nucleic acids, is the nucleotide monomer added to the 3′- or the 5′-ribose terminus?

In pulse and chase experiments, a labeled monomer is added in a short pulse; the reaction involves either a labeled monomer for a very short time or a very low concentration of a highly radioactive monomer for a longer time. Aliquots of the reaction mixture are taken and the product analyzed. The reaction mixture is chased by adding unlabeled monomer, and product analysis is performed by examining the amount of labeled monomer residues at the ends of the polymer chains.

As an example, let us consider the biosynthesis of a glucan polysaccharide. The enzyme responsible for biosynthesis is pulsed with ^{14}C-labeled substrate containing labeled monomer. An aliquot of the reaction is removed and the newly synthesized glucan is isolated, reduced with sodium borohydride, and acid-hydrolyzed. Then the amount of label in glucose and in glucitol (the product resulting from the reduction of the reducing glucose residue) is determined. After removal of the aliquot from the pulse reaction, a chase is performed by adding nonlabeled substrate. The reaction is allowed to proceed and the same type of analysis that was performed on the pulsed polysaccharide is performed on the chased polysaccharide. Let us say that we obtain the data in Table 6-7. The relative mole ratios of glucose to glucitol in the pulse and chase experiments, respectively, are 4:1 and 32:1. These data indicate that the biosynthesis of this polysaccharide is from the reducing end because there is much more labeled

TABLE 6-7. **Pulse–chase analysis of a glucan polysaccharide**

Product	Pulse		Chase	
	cpm	mole fraction	cpm	mole fraction
Glucose	5040	0.8	4100	0.97
Glucitol	1250	0.2	130	0.03

glucitol in the pulse experiment than in the chase experiment. This result is interpreted to mean that the labeled glucose was chased into the interior of the polymer by the addition of the nonlabeled glucose to the reducing end.

6.14.5 The Use of Radioactive Tracers in the Study of Enzyme Reaction Mechanisms and Metabolic Pathways

Specifically labeled substrates are useful in studying how an enzyme catalyzes its reaction. For example, the enzyme aldolase cleaves D-fructose-1,6-bisphosphate into two products, 3-phospho-D-glyceraldehyde and dihydroxyacetone phosphate. The use of $[1\text{-}^{14}C]$D-fructose-1,6-bisphosphate would enable us to follow the reaction to see which parts of the fructose molecule give rise to each of the two products. In this example, the following reaction (using the asterisk for the labeled C_1) would give ^{14}C-labeled dihydroxyacetone phosphate, indicating that this three-carbon product comes from carbon atoms 1, 2, and 3 of the fructose.

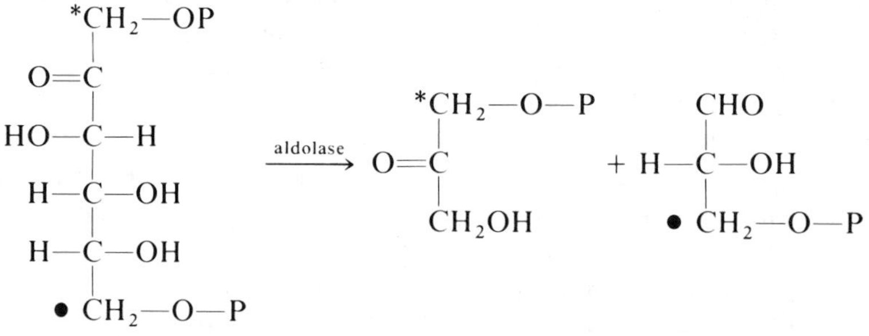

| D-Fructose-1,6- | Dihydroxyacetone | 3-Phospho-D- |
| bisphosphate | phosphate | glyceraldehyde |

Similarly if we had used $[6\text{-}^{14}C]$-D-fructose-1,6-bisphosphate as the labeled substrate, the reaction (using the dot as the labeled C_6 in the reaction above) would give ^{14}C-labeled 3-phospho-D-glyceraldehyde, indicating that this three-carbon product comes from carbon atoms 4, 5, and 6 of the fructose.

Another example is provided by the study of the series of reactions of the tricarboxylic acid cycle (citric acid cycle), which take place in intact cells. The pathway and mechanism of the cycle were established using labeled compounds, such as acetate labeled in various positions ($\overset{*}{C}H_3-COO^-$, $CH_3-\overset{*}{C}OO^-$, and $CH_3-\overset{*}{C}OO^-$), pyruvate ($\overset{*}{C}H_3-CO-COO^-$ and $CH_3-CO-\overset{*}{C}OO^-$), and bicarbonate ($H\overset{*}{C}O_3^-$). These labeled precursors gave labeled cycle intermediates in which the radioactive label was located in specific positions. The locations of the labeled isotope were determined by chemical degradation of the inter-mediates and gave information concerning the mechanisms of the reactions occurring in the cycle. Some of the label locations not only gave information about the path of carbon in the cycle but also elucidated important concepts concerning the stereochemistry involved in specific enzyme reactions (e.g., aconitase action with citrate, and the action of succinate dehydrogenase and fumarase).

6.14.6 Radioimmunoassay

Radioimmunoassay is a very sensitive method of measuring a wide variety of important biochemical substances. The development of the method is attrib-uted to Yallow and Berson.[11] The development of radioimmunoassay pro-vided methods of quantitatively determining biological substances, such as hormones, that previously were very difficult if not impossible to quantitate. The method, in essence, is the reaction of an antibody (Ab) with an antigen (Ag) to give an antibody–antigen complex (AbAg).

$$Ab + Ag \rightleftharpoons AbAg$$

Either the antibody or the antigen can be radioactively labeled. The most commonly used isotope is ^{125}I because it can be added to proteins under very mild conditions that apparently do not significantly alter the structural and chemical properties of the protein, and because it has a relatively long half-life (60 days) and is a relatively weak γ-emitter.

If we are able to label the antigen we wish to determine and have available or can produce antibodies against the antigen, the assay for the antigen will be based on competition between labeled and nonlabeled antigen for a restricted amount of antibody. In this assay, the higher the concentration of unlabeled antigen, the less labeled antigen will be complexed with a fixed amount of antibody. By separating the antigen–antibody complex from free antigen and measuring the amount of radioactivity in the complex, or measuring the amount of unbound radioactivity remaining after removal of the complex, the quantitative deter-mination of the antigen can be made. Displacement curves relating bound and free ^{125}I-labeled antigen are obtained with known amounts of unlabeled anti-gen in a series of standards. By using this type of standard curve, the amount of antigen in an unknown sample can be determined.

The first step in the development of a radioimmunoassay, as outlined above, is to obtain an antibody dilution curve in which a fixed amount of labeled antigen is incubated with different concentrations of antibody. The antibody is serially diluted by factors of 2, and each dilution is incubated with the labeled antigen. After incubation, the amount of label in the bound and the free fractions are determined. A plot of the percent of label bound versus the log of the concentration of the antibody is sigmoidal (Fig. 6-19). The optimum concen-tration of the antibody chosen for a radioimmunoassay is the amount that is necessary to bind approximately 50% of the radioactivity, which gives the

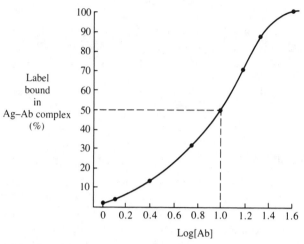

FIGURE 6-19. Antibody dilution curve. Serial dilutions of the antibody are incubated with a fixed amount of labeled antigen, and the percent of labeled antigen bound in the antigen–antibody complex is plotted against the log of the concentration of the antibody. From this plot the concentration of antibody required to incorporate 50% of the labeled antigen is determined.

maximum change in the amount of label bound to the amount of antibody added (i.e., maximum sensitivity).

The construction of a radioimmunoassay standard curve involves the incubation of fixed amounts of labeled antigen and antibody with different concentrations of unlabeled antigen. A plot of the percent of labeled antigen bound versus the log of the concentration of the antigen also gives a sigmoidal curve (Fig. 6-20). The steepest part of the curve, at which relatively small amounts of antigen produce a significant change in the amount of label bound, represents the effective range of the assay. The concentration of an unknown amount of antigen is determined by performing an assay in an identical manner to that used with the standards, and the amount of antigen in the unknown can then be obtained from the standard curve. Since the nonlabeled antigen dilutes the labeled antigen, as the concentration of the nonlabeled antigen increases, there is a decrease in the amount of labeled antigen incorporated into the antigen–antibody complex.

The radioimmunoassay procedure depends on the separation of the antibody–antigen complex from the free antigen. The separation techniques involve differences in the physicochemical properties between the antigen in its free and complexed forms. Several methods are used to separate free and bound antigen: electrophoresis on cellulose acetate or in polyacrylamide gels; gel permeation chromatography; specific adsorption of either the complex or the free antigen on charcoal, silica gel, or hydroxylapatite; fractional precipitation with ethanol, polyethylene glycol, or ammonium sulfate; or second-antibody

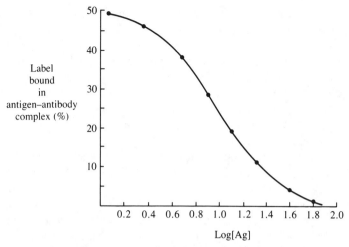

FIGURE 6-20. Determination of the amount of antigen in an unknown sample from a standard antigen–antibody curve. The unknown is incubated with a fixed concentration of antibody and labeled antigen, and the percent of the labeled antigen incorporated into the antigen–antibody complex is determined.

precipitation, in which the second antibody is either in solution or immobilized onto a solid support. For examples of radioimmunoassays see Chapter 7 (Section 7.5.2).

6.15 DESIGNING A RADIOISOTOPE EXPERIMENT

First and foremost in the design of any experiment is the formulation of the question to be studied. Next the investigator must consider what experimental techniques lend themselves to answering the question. If the use of radioisotopes is decided on, there are some special considerations to be made. It is always advisable to perform the experimental procedures as a "dry run" before the radioisotope is added. In addition, the investigator must select the type of isotope to be used. This choice, of course, depends on the experiment and the kind of question to be solved. For example, if the path of carbon in some precursor is to be followed in a metabolic pathway, ^{14}C is the obvious choice for the isotope. But possibly the fate of more than one carbon atom is desired. It might be feasible to tag one of the carbons with ^{14}C and the other with ^{3}H and perform the experiment with dual labels. Or perhaps it is desired to learn the fate of all the carbons in the precursor, in which case a uniformly labeled precursor should be used.

These considerations bring up the matter of the availability of the isotope or isotopes in the desired position(s) of the starting compound. Is the labeled compound commercially available or must it be synthesized? If it must be synthesized, are there synthetic routes (either biological or chemical, see Section

6.12) available for producing the compound incorporating the isotope(s) in the desired position(s)? If the labeled compound is commercially available, what is its cost, and how much will be required? The amount required depends on the proposed detection system (autoradiography, liquid or solid scintillation spectrometry, G–M tube) and its efficiency, and on the dilution of the isotope. An excess factor of 2-fold to 10-fold magnitude should be incorporated into the experimental design to encompass unexpected dilution factors and losses. The sampling process, the assay method, and the required precision must also be considered.

Additional factors that are sometimes important when using radioisotopes are the isotope effect and the potential disruption of the physiology and biochemistry by radiation damage. The **isotope effect** is caused by a change in the kinetics of a process when a chemical bond involving the isotope is being formed or broken. This is most pronounced for ^3H and would show up in oxidation–reduction reactions that involve the forming or breaking of a tritium bond. Although isotope effects could appear for ^{14}C (e.g., decarboxylation of a ^{14}C-labeled carboxyl group), they would be less pronounced than the isotope effect of ^3H because the ratio of the masses of ^3H to ^1H is much greater than the ratio of the masses of ^{14}C to ^{12}C.

The potential for radiation damage occurs when the radioisotope is administered to a living organism or to an actively growing culture of microorganisms. Radiation effects should be suspected when there seems to be a disruption of a physiological response or, in the case of microorganisms, when growth is retarded. In these instances, it would be advisable to repeat the experiment using 0.1 to 0.01 lower levels of radiation while maintaining the same total concentration of the administered compound.

In summary, the design of a radioisotope experiment should include the consideration of the following: (*a*) purpose and nature of the experiment, (*b*) type of radioisotope, (*c*) position and type of labeled compound, (*d*) availability of labeled compound, (*e*) specific activity and amount of radioactivity, (*f*) cost of the labeled compound, (*g*) sampling methods, (*h*) type of detection system and its efficiency, (*i*) dilution effects and the yield of the radioisotope, (*j*) controls and correction factors needed, (*k*) expression and interpretation of the results, and (*l*) possible isotope effects and possible effects of radiation damage.

As an illustrative example, let us consider the following. An enzyme is produced extracellularly by a bacterium in the amount of 5 enzyme units per milliliter. The enzyme has a well-worked-out purification procedure that yields 50% of the enzyme with an SA of 175 units per milligram of protein in a total volume of 100 mL. The molecular weight of the enzyme is 200,000, and it is suspected to be a dimer consisting of two monomers of equal weight. Furthermore, there is presumptive evidence that the enzyme has one or more covalently linked phosphate groups per monomer unit. The problem is to confirm the presence of covalently linked phosphate groups in the enzyme. A relatively simple way would be the culturing of the bacterium in the presence of [^{32}P]phosphate, purifying the enzyme, and seeing whether labeled phosphate is incorporated into the enzyme.

1. *Known and presumptive facts*
   (a) The enzyme is produced in the amount of 5 units/mL.
   (b) The enzyme can be purified in a 50% yield, giving a SA of 175 units/mg (of protein).
   (c) The enzyme is a possible dimer consisting of two monomers of equal weight, each containing one or more covalently linked phosphate groups.

2. *Considerations*
   (a) If there are 175 U/mg and the molecular weight of the enzyme is 200,000, there would be 175 U/mg × 200,000 mg/mmol or 3.5×10^7 U/mmol.
   (b) If there are two phosphate groups per molecule of enzyme (MW = 200,000)—one phosphate per monomer unit—there would be 2 mmol of phosphate per millimole of enzyme or 2 mmol of phosphate/3.5×10^7 units of enzyme.
   (c) The amount of unlabeled phosphate in the culture medium is 0.1% (w/v) or 1 g/L or 0.01 mol/L or 10 mmol/L.
   (d) Let us say that we add 10 mCi of [^{32}P]phosphate to 100 mL of culture medium. The current price and specific activity is \$120/10 mCi of carrier-free radioisotope.
   (e) The specific activity of phosphate in the culture medium would be 10 mCi/mmol.
   (f) Therefore, the hypothetical amount of ^{32}P incorporated into the enzyme would be (2 mmol phosphate/mmol of enzyme)(10 mCi/mmol of phosphate) per 3.5×10^7 units or 20 mCi ^{32}P/3.5×10^7 units of enzyme.
   (g) The amount of radioactivity per unit of enzyme would therefore be 20 mCi/(3.5×10^7 units) or 5.7×10^{-7} mCi/unit.
   (h) Since there are 2.2×10^9 dpm/mCi, there would be (5.7×10^{-7} mCi/U) (2.2×10^9 dpm/mCi) or 1.26×10^3 dpm per unit of enzyme.
   (i) In 100 mL of culture there are produced 5 U/mL for a total of 500 U. The total amount of radioactivity incorporated into the enzyme would therefore be (1.26×10^3 dpm/U)(500 U) or 6.3×10^5 dpm. This is quite sufficient for demonstrating the presence of phosphate in purified enzyme.
   (j) The counting system to be employed is heterogeneous liquid scintillation spectrometry using filter paper discs, which has an efficiency of 90% for ^{32}P.
   (k) To further demonstrate the incorporation of phosphate into the specific enzyme, polyacrylamide gel electrophoresis will be performed, followed by autoradiography of the gel to show that the specific enzyme is indeed labeled. For electrophoresis, we will need 10 μL of the purified enzyme preparation. We know that there is 6.3×10^5 dpm/mL or 6.3×10^3 dpm/10 μL (step i). If all the radioactivity occurs in protein bands of 0.1 cm², a reasonable assumption for electrophoresis gels, we would have 63,000 events/cm² per minute. The efficiency of autoradiography of ^{32}P is about 10%; thus we would have 6300 events/cm². Since we need 1×10^6 events/cm² to obtain good detection by autoradiography, we must expose the film to the electrophoresis gel containing the ^{32}P-labeled protein for

$(1 \times 10^6$ events/cm$^2)/(6300$ events cm^{-2} min$^{-1})$ or about 160 minutes. We can, thus, see that the choice of adding 10 mCi to the culture medium gives a sufficient amount of labeled phosphate to demonstrate the presence of phosphate in the enzyme. In fact, we could add 0.1 as much labeled phosphate and then expose the gel to the film for 1600 minutes and still demonstrate the presence of phosphate in the enzyme.

6.16 LITERATURE CITED

1. H. M. Dintzis, "Assembly of the peptide chains of hemoglobin," *Proc. Natl. Acad. Sci. U.S. 47*: 247 (1961).
2. R. E. Canfield and C. B. Anfinsen, "Nonuniform labeling of egg white lysozyme," *Biochemistry, 2*: 1073 (1963).
3. D. N. Luck and J. M. Barry, "The order in time in which amino acids are incorporated into pancreatic ribonuclease," *J. Mol. Biol. 9*: 186 (1964).
4. J. B. Fleischman, "Synthesis of the γG heavy chain in rabbit lymph node cells," *Biochemistry, 6*: 1311 (1967).
5. P. W. Robbins, D. Bray, M. Dankert, and A. Wright, "Direction of chain growth in polysaccharide synthesis," *Science 158*: 1536 (1967).
6. G. Eytan and I. Ohad, "Biogenesis of Chloroplast Membranes," *J. Biol. Chem. 245*: 4297 (1970).
7. P. Cozzone and G. Marchis-Mouren, "Use of pulse labeling technique in protein structure determination: ordering of the cyanogen bromide peptides from porcine pancreatic α-amylase," *Biochim. Biophys. Acta, 257*: 222 (1972).
8. J. B. Ward and H. R. Perkins, "The direction of glucan synthesis in a bacterial peptidoglycan, *Biochem. J. 135*: 721 (1973).
9. J. F. Robyt, B. K. Kimble, and T. F. Walseth, "The mechanism of dextransucrase action: direction of dextran biosynthesis," *Arch. Biochem. Biophys. 165*: 634 (1974).
10. J. F. Robyt and P. J. Martin, "Mechanism of synthesis of D-glucan by D-glucosyltransferases from *Streptococcus mutans* 6715," *Carbohydr. Res. 113*: 301 (1983).
11. R. S. Yallow and S. A. Berson, "Immunoassay of endogenous plasma insulin in man," *J. Clin. Invest. 39*: 1157 (1960).

6.17 REFERENCES FOR FURTHER STUDY

1. C. H. Wang, D. L. Willis, and W. D. Loveland, *Radiotracer Methodology in the Biological, Environmental, and Physical Sciences.* Prentice-Hall, Englewood Cliffs, NJ, (1975).
2. W. R. Hendee, *Radioisotopes in Biological Research.* Wiley, New York, (1973).
3. J. M. Chapman, and G. Ayrey, *The Uses of Radioactive Isotopes in the Life Sciences.* George Allen & Unwin, London, (1981).
4. Y. Kobayashi, and D. V. Maudsley, *Biological Applications of Liquid Scintillation Counting.* Academic Press, New York, (1974).
5. E. J. Hahn, "Autoradiography—A review of basic principles," *Am. Lab. 15*: 64–71 (July 1983).
6. T. Chard, "An introduction to radioimmunoassay and related techniques," in *Laboratory Techniques in Biochemistry and Molecular Biology*, Vol. 6, T. S. Work and E. Work, Eds. North Holland, Amsterdam, (1978).

CHAPTER 7

QUALITATIVE AND QUANTITATIVE METHODS FOR DETERMINING BIOLOGICAL MOLECULES

 Many color-producing reactions are used for the qualitative or quantitative determination of biological molecules such as carbohydrates, amino acids, proteins, nucleic acids, and lipids. These reactions are used to indicate the presence in a test solution or on a paper or thin-layer chromatogram of specific biological molecules, or to quantitatively determine the amount of the biological molecule present in a test solution by spectrophotometry. This chapter discusses the qualitative and quantitative procedures that are routinely used by biochemists.

7.1 CARBOHYDRATES

7.1.1 Qualitative Tests for Different Types of Carbohydrates

7.1.1.1 GENERAL TEST FOR CARBOHYDRATES: THE MOLISCH TEST

The Molisch test is a general test for the presence of most carbohydrates, such as monosaccharides (pentoses, hexoses, hexuronic acids, and 6-deoxyhexoses), oligosaccharides (disaccharides, trisaccharides, etc.), and polysaccharides.[1] The test is negative for sugar alcohols (alditols) and 2-deoxy-2-amino sugars.

THE MOLISCH TEST

Two reagents, concentrated sulfuric acid and 5% (w/v) α-naphthol in ethanol, are used. Two drops of the α-naphthol reagent are added to 1 mL of a solution suspected of having carbohydrate present. Then 1 mL of concentrated sulfuric acid is carefully poured down the side of the tube to form a layer on the bottom. If carbohydrate is present, a purple complex will be formed at the interface between the two solutions.

The concentrated sulfuric acid hydrolyzes any glycosidic bonds present to monosaccharides, which are then dehydrated by the concentrated acid to yield furfural or one of its derivatives (e.g., hydroxymethylfurfural). The furfural compounds condense with sulfonated α-naphthol to give a purple complex, which is characteristic of carbohydrates. The limit of detection is 10 μg/mL.

Pentoses give furfural (2-furaldehyde); hexoses give 5-hydroxymethylfurfural; hexuronic acids decarboxylate to give furfural; and 6-deoxyhexoses give 5-methylfurfural. The exact structure of the condensation product(s) with sulfonated α-naphthol (the purple complex) is unknown.

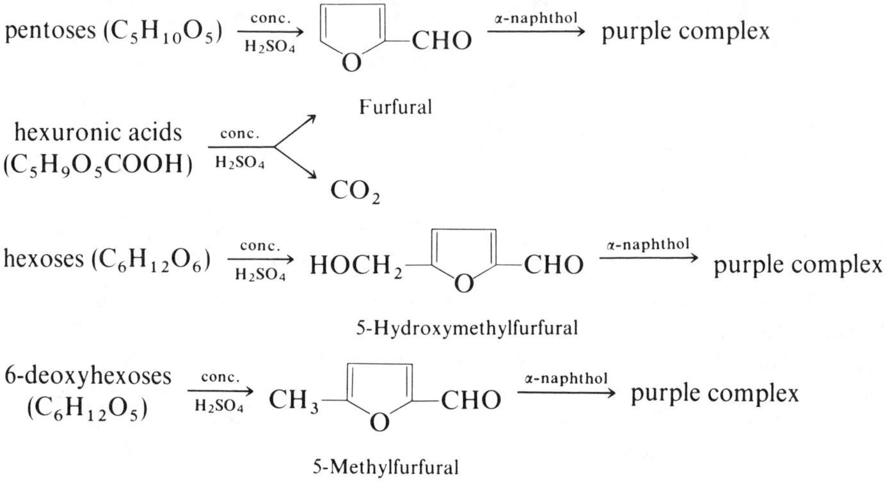

pentoses $(C_5H_{10}O_5)$ $\xrightarrow[\text{H}_2\text{SO}_4]{\text{conc.}}$
⟨structure⟩—CHO $\xrightarrow{\text{α-naphthol}}$ purple complex
O

Furfural

hexuronic acids $(C_5H_9O_5COOH)$ $\xrightarrow[\text{H}_2\text{SO}_4]{\text{conc.}}$
Furfural
CO_2

hexoses $(C_6H_{12}O_6)$ $\xrightarrow[\text{H}_2\text{SO}_4]{\text{conc.}}$ $HOCH_2$—⟨structure⟩—CHO $\xrightarrow{\text{α-naphthol}}$ purple complex
O

5-Hydroxymethylfurfural

6-deoxyhexoses $(C_6H_{12}O_5)$ $\xrightarrow[\text{H}_2\text{SO}_4]{\text{conc.}}$ CH_3—⟨structure⟩—CHO $\xrightarrow{\text{α-naphthol}}$ purple complex
O

5-Methylfurfural

7.1.1.2 THE SELIWANOFF TEST FOR KETOSES[2]

Ketoses are dehydrated more rapidly than aldoses to give furfural derivatives, which then condense with resorcinol to form a red complex.

THE SELIWANOFF TEST

Seliwanoff's reagent contains 50 mg of resorcinol in 100 mL of 3 M HCl. Four drops of the solution suspected to contain ketoses are added to 2 mL of Seliwanoff's reagent, which is placed into a boiling water bath for 1 minute. The presence of ketoses is indicated by the formation of a deep red precipitate.

The faster rate of dehydration of ketoses by HCl to furfural derivatives versus the dehydration of aldoses by HCl allows the differential detection of ketoses in the presence of aldoses. The limit of detection is 20 μg/mL.

7.1.1.3 BIAL'S TEST FOR PENTOSES[3]

When pentoses are heated with concentrated HCl, furfural is produced, which gives a blue-green color in the presence of orcinol and ferric ions.

BIAL'S TEST

Bial's reagent contains 0.3 g of orcinol dissolved in 100 mL of concentrated HCl to which 4 drops of a 10% solution of ferric chloride are added. To 100 μL of the test solution 2 mL of Bial's reagent is added. The solution is placed into a boiling water bath for 1 minute. The formation of a blue-green color indicates the presence of pentoses, and nucleotides containing pentoses.

The test is relatively specific for pentoses. Hexoses give 5-hydroxymethylfurfural, which reacts with Bial's reagent to give a yellow color that is usually masked by the blue-green color given by pentoses. Hence, hexoses interfere only slightly in the test; 2-deoxypentoses are dehydrated very slowly, hence also do not interfere with the test.

7.1.1.4 THE DIPHENYLAMINE TEST FOR 2-DEOXY SUGARS[4]

The presence of 2-deoxycarbohydrates may be tested for with diphenylamine under acidic conditions. Since 2-deoxy-D-ribose is the carbohydrate constituent of DNA, this test may also be used for DNA. The combination of this test and Bial's test can detect and differentiate RNA from DNA.

THE DIPHENYLAMINE TEST FOR 2-DEOXYPENTOSES

The diphenylamine reagent consists of 1 g of diphenylamine dissolved in 100 mL of glacial acetic acid and 2.5 mL of concentrated sulfuric acid. If DNA is being tested for, the sample is first hydrolyzed by trichloroacetic acid (final concentration 10%) at 95°C for 10 minutes.

The test solution (1 mL) is diluted with 2 mL of water, 5 mL of the diphenylamine reagent is added, and the solution is placed into a boiling water bath for 10 minutes. If 2-deoxy sugars (concentration > 10 μg/mL) are present, a blue-green color is produced. A quantitative determination can be made by measuring the absorbance at 595 nm and comparing the amount of absorbance with a standard curve obtained by using known concentrations of a 2-deoxypentose in the range of $10-200$ μg/mL.

7.1.1.5 THE TESTS FOR 2-AMINO SUGARS (ELSON–MORGAN TEST) AND N-ACETYL-2-AMINO SUGARS (MORGAN–ELSON TEST)[5]

The Elson–Morgan test for 2-amino sugars involves the acetylation of the amino sugar with 2,4-pentanedione in hot alkaline solution followed by reaction with p-dimethylaminobenzaldehyde to give an intense cherry-red color. It may be used as either a qualitative or quantitative test. For a quantitative determination, the absorbance is measured at 530 nm using $1-250$ μg of 2-amino sugars as standards.

THE ELSON–MORGAN TEST FOR 2-AMINO SUGARS

The procedure involves the addition of 0.5 mL of carbohydrate solution to 0.5 mL of acetylating agent, which consists of 0.75 mL of 2,4-pentanedione dissolved in 25 mL of 1.25 M sodium carbonate. The solution is heated in a 95°C water bath for 20 minutes and cooled, and 4 mL of Ehrlich's reagent (1.6 g of p-dimethylaminobenzaldehyde dissolved in 30 mL concentrated HCl and 30 mL of 95% ethanol) is added. The formation of a red color indicates the presence of 2-amino sugars. The limit of sensitivity is 2 μg/mL.

The Morgan–Elson test for N-acetyl-2-amino sugars involves a brief heating in alkaline solution, followed by reaction with Ehrlich's reagent.

THE MORGAN–ELSON TEST FOR N-ACETYL-2-AMINO SUGARS

The procedure involves the addition of 0.5 mL of sample to 0.5 mL of 1.25 M sodium carbonate, which is heated to 95°C for 5 minutes. The solution is cooled, and 4 mL of Ehrlich's reagent is added. A red color indicates a positive test.

The structural features required to give a positive test, are a 2-acetamido group, a free reducing group, and a free hydroxyl group at C_4. The color is not produced by free amino sugars, hence the necessity of performing an acetylation reaction to test for free 2-amino sugars. The procedure may be quantitated with standards and the measurement of the absorbance at 550 nm.

7.1.1.6 BENEDICT'S TEST FOR REDUCING SUGARS[6]

Carbohydrates may be classified as reducing sugars or nonreducing sugars. Reducing sugars have a free aldehyde group or potential aldehyde group, such as hemiacetal, or keto and hemiketal groups that are isomerized to an aldehyde under alkaline conditions. These groups may be oxidized, under alkaline conditions, by certain metal ions that become reduced. The reduction of the metal ion indicates that the sugar is a reducing carbohydrate. A commonly used ion is Cu^{2+}.

BENEDICT'S REDUCING TEST

Benedict's reagent is prepared by dissolving 173 g of sodium citrate and 100 g of sodium carbonate in 800 mL of warm distilled water and 17.3 g of copper sulfate in 200 mL of distilled water. The copper sulfate solution is slowly added to the citrate–carbonate solution with stirring.

Two hundred microliters of the test solution (containing at least 5 mg/mL of carbohydrate) is added to 1 mL of Benedict's reagent, which is placed into a boiling water bath for 5 minutes. If the carbohydrate is a reducing sugar, a brick-red precipitate of cuprous oxide (Cu_2O) results.

Another qualitative test for reducing carbohydrates involves the reduction of Ag^+ to metallic silver. This is commonly used to detect reducing sugars on paper chromatograms, as discussed in Chapter 4 (Section 4.10.3).

7.1.2 The Quantitative Determination of Carbohydrates

7.1.2.1 THE PHENOL–SULFURIC ACID METHOD[7]

An early method for quantitatively determining carbohydrates was the reaction of these compounds with an anthrone reagent, which consisted of 200 mg of anthrone dissolved in 100 mL of concentrated sulfuric acid. When the ratio of the aqueous carbohydrate solution to sulfuric acid became too high, however, investigators encountered certain problems with stability of the reagent and solubility of the anthrone. The anthrone procedure has been mainly replaced by the simpler phenol–sulfuric acid procedure because phenol is miscible with water in all proportions.

THE PHENOL–SULFURIC ACID PROCEDURE

One milliliter of 5% aqueous phenol is added to 1.0 mL of carbohydrate solution containing 10 to 100 μg of carbohydrate. The solution is mixed and 5.0 mL of concentrated sulfuric acid is added and mixed. After the reagents have been allowed to stand for 20 minutes to allow color to develop, the absorbance is measured at 470 nm. A standard curve, using the sugar being determined, is prepared to obtain the quantitative amount of carbohydrate in the unknown.

TABLE 7-1. Qualitative and quantitative tests for carbohydrates using phenolic compounds and concentrated acid

Phenolic compound	Method	Carbohydrate determined	Acid used	Test	Limit of detection
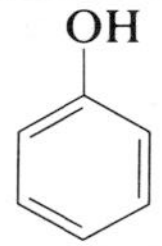 Phenol	Phenol–H_2SO_4	All carbohydrates except alditols and amino sugars	H_2SO_4	Quant.	10 µg/mL
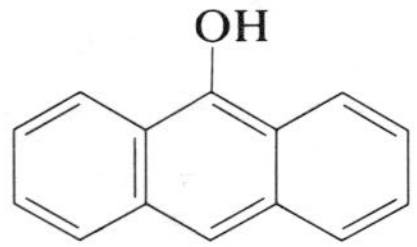 Anthrone	Anthrone	All carbohydrates except alditols and amino sugars	H_2SO_4	Quant.	50 µg/mL
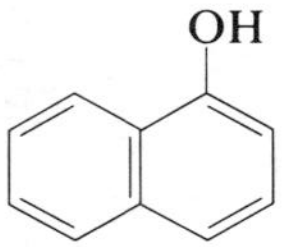 α-Naphthol	Molisch test	All carbohydrates except alditols and amino sugars	H_2SO_4	Qual.	10 µg/mL
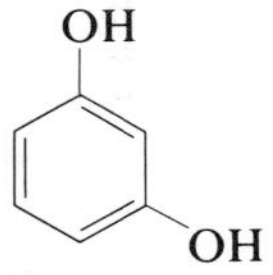 Resorcinol	Seliwanoff test	Ketoses	HCl	Qual./quant.	20 µg/mL
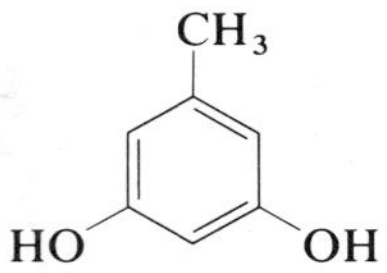 Orcinol	Bial's test	Pentoses	HCl	Qual./quant.	20 µg/mL

218

The procedure can be used to determine monosaccharides, oligosaccharides, and polysaccharides. Different monosaccharides do absorb maximally at different wavelengths and do give different molar absorbance values, but different kinds of oligosaccharide and polysaccharide containing the same monosaccharide can be used as standards for each other. For example, glucose or maltose may be used as standards for starch, glycogen, dextran, and so on. Only sugar alcohols and 2-amino sugars do not react.

Table 7-1 is a summary of the tests using phenolic compounds with strong acids to determine carbohydrates qualitatively and quantitatively.

7.1.2.2 THE QUANTITATIVE DETERMINATION OF REDUCING SUGARS

Reducing carbohydrates may be quantitated by using either a copper-reduction method, the Somogyi–Nelson procedure,[8] or a ferricyanide reduction method.[9]

THE SOMOGYI–NELSON ALKALINE COPPER METHOD

The Somogyi–Nelson procedure requires four reagents.

Reagent A: 25 g of sodium carbonate, 25 g of sodium–potassium tartrate, 20 g of sodium bicarbonate, and 200 g of anhydrous sodium sulfate, dissolved in 1 L of water.

Reagent B: 30 g of cupric sulfate pentahydrate dissolved in 200 mL of water, to which 4 drops of concentrated sulfuric acid have been added.

Reagent C: 25 g of ammonium molybdate is dissolved in 450 mL of water containing 21 mL of concentrated sulfuric acid and 3 g of disodium hydrogen arsenate heptahydrate ($Na_2HAsO_4 \cdot 7H_2O$) is dissolved in 25 mL of water. The arsenate solution is slowly added to the ammonium molybdate solution, which is diluted to 500 mL and warmed carefully at 55°C for 30 minutes or 37°C for 12–15 hours, and stored in a brown bottle.

Reagent D: a working reagent prepared fresh daily by adding 1 mL of reagent B to 25 mL of reagent A.

The procedure involves the addition of 1.0 mL of sample (containing 10–75 nmol of reducing carbohydrate) to 1.0 mL of reagent D. The solution is heated in a boiling water bath for 20 minutes and cooled in running cold water for 5 minutes; 1.0 mL of reagent C is added and the mixture is shaken until gas is no longer evolved. The solution is allowed to stand for 10 minutes and is then diluted with water to 5–25 mL (the dilution depends on the range of sensitivity required, although within any one determination the dilution must be the same as the standards); the absorbance is measured at 600 nm.

THE ALKALINE FERRICYANIDE METHOD

The ferricyanide procedure uses a single reagent composed of 0.34 g of potassium ferricyanide, 5 g of potassium cyanide, and 20 g of sodium carbonate dissolved in 1 L of water.

The procedure involves the addition of 1.0 mL of sample to 4.0 mL of the reagent. The solution is heated in a boiling water bath for 10 minutes and cooled; the absorbance is measured at 420 nm. The reducing sugar reduces the yellow ferricyanide to the colorless ferrocyanide and therefore the more reducing sugar present, the less yellow ferricyanide there is to measure. The range is 30–600 nmol of reducing carbohydrate.

A direct ferricyanide method involves the Park–Johnson modification,[10] which is approximately 10 times more sensitive than the Somoygi–Nelson and the alkaline ferricyanide procedures.

THE PARK–JOHNSON FERRICYANIDE METHOD

The Park–Johnson procedure involves the addition of 5.0 mL of a ferric ion solution (1.5 g of ferric ammonium sulfate and 1.0 g of sodium dodecyl sulfate dissolved in 1 L of 0.05 N sulfuric acid) to the cooled, ferricyanide–sample solution of the indirect ferricyanide procedure. A blue color results from the formation of ferric ferrocyanide, $Fe_4[Fe(CN)_6]_3$, which is measured at 690 nm. The method is 10 times more sensitive than the inverse ferricyanide method, having a range of 3–60 nmol of reducing carbohydrate.

7.1.2.3 THE QUANTITATIVE DETERMINATION OF GLUCOSE BY GLUCOSE OXIDASE[11]

β-D-Glucopyranose is very specifically oxidized to D-glucono-1,5-lactone by the enzyme **glucose oxidase**, which is elaborated by several microorganisms including *Penicillium notatum* and may be purchased from the biochemical supply companies. Many carbohydrates, such as D-mannose, D-galactose, D-fructose, and D-xylose are oxidized very slowly, if at all. The glucose anomer α-D-glucopyranose is oxidized only 0.007 times as fast as the β-anomer. Total glucose can be determined, however, if another enzyme, mutarotase, is also present or is added to accelerate the conversion of the α-anomer to the β-anomer. Crude preparations of glucose oxidase usually contain this second enzyme.

DETERMINATION OF GLUCOSE WITH GLUCOSE OXIDASE

The glucose oxidase reagent consists of 30 mg of glucose oxidase (Boehringer–Manheim, grade II), 3 mg of horseradish peroxidase (Boehringer–Manheim, grade II), and 10 mg of orthodianisdine hydrochloride dissolved in 100 mL of Tris–phosphate–glycerol buffer, pH 7.0, which consists of 36.3 g of Tris and 56.5 g of sodium dihydrogen phosphate dihydrate ($NaH_2PO_4 \cdot 2H_2O$) dissolved in 400 mL of water, plus 400 mL of glycerol is added, and diluted to 1 L with water. The glucose oxidase reagent is stable for one week if stored in a brown bottle at 5°C.

The procedure involves the addition of 1.0 mL of sample (containing 1–100 μg/mL of glucose) to 2.0 mL of the glucose oxidase reagent. The mixture is incubated at 37°C for 30 minutes and then 4.0 mL of 5 N hydrochloric acid is added and the absorbance measured at 525 nm. A standard curve is prepared using 1–100 μg/mL of glucose.

Glucose determination kits, based on the glucose oxidase procedure, are commercially available.

7.1.2.4 THE QUALITATIVE AND QUANTITATIVE DETERMINATION OF CARBOHYDRATES BY OPTICAL ROTATION

Many biochemical substances contain asymmetric centers and are optically active; that is, they have the ability to rotate a plane of polarized light. This is especially so for the carbohydrates, which usually have several asymmetric centers. The magnitude and the direction of the rotation of the plane-polarized light that passes through a solution of an asymmetric substance can be used for qualitative and quantitative determinations. An optically active substance may rotate the polarized light clockwise and is then said to be **dextrorotatory** (+) or counterclockwise and is then said to be **levorotatory** (−).

Quantitative optical rotation may be expressed in terms of **specific optical rotation** [α], which is defined as a function of the path length l (in decimeters) the polarized light travels through a solution of the substance, the concentration c (in grams per 100 mL of solution), and the observed angular rotation A (in degrees). The temperature, the solvent, and the wavelength of the plane-polarized light are important and should be specified.

$$[\alpha] = \frac{100A}{lc} \qquad (7\text{-}1)$$

Using the specific optical rotation $[\alpha]$, the path length l, and the measured optical rotation, the concentration c (in g/100 mL) can be obtained by the following:

$$c = \frac{100A}{l[\alpha]} \tag{7-2}$$

Optical rotation measurements can also be used to qualitatively determine the nature of a pure optically active substance of known concentration, by ascertaining the specific optical rotation and comparing it with literature values of specific optical rotations.

The value of the specific optical rotation depends on the wavelength of the plane-polarized light. The most commonly used light source is the sodium lamp; hence the sodium D-line is the wavelength most commonly used and referenced in the literature. The specific optical rotation obtained with a sodium lamp is usually indicated as $[\alpha]_D$.

Reducing carbohydrates exist primarily in a hemiacetal ring structure in solution. The hemiacetal hydroxyl group (anomeric hydroxyl) may be either above (β) or below (α) the plane of the ring. Crystallization of the carbohydrate from solution usually results in the isolation of one of the two possible anomeric forms. But, when the crystalline carbohydrate is dissolved in water, the single hemiacetal form is slowly converted into the other form and an equilibrium mixture of the α- and β-forms is produced. This is known as **mutarotation** and is accompanied by a change in the optical rotation to give a specific optical rotation $[\alpha]$ that is characteristic of the equilibrium mixture of the specific reducing carbohydrate. The mutarotation reaction is catalyzed by base (e.g., pH 10). Hence, the measurement of the specific rotation immediately after the crystals have dissolved and then after the pH has been made 10 by adding sodium carbonate will indicate the anomeric configuration of the crystalline material. If the rotation decreases after the addition of the base, the anomeric configuration of the crystalline material is α, and if the rotation increases the anomeric configuration is β. The specific optical rotation of α-anomers is always more positive (i.e., rotated clockwise) than the rotation of β-anomers. This generalization can be extended to carbohydrates containing α- and β-glycosidic linkages too. For example, an equilibrium mixture of α- and β-maltose, which contains one α-1,4 glycosidic linkage between the two glucose units, has a much higher equilibrium specific rotation $[+130.4°]$ than does an equilibrium mixture of α- and β-cellobiose, which contains one β-1,4 glycosidic linkage between the two glucose units and has a specific rotation of $[+34.6°]$. Likewise, if another (α-1,4)-linked glucose residue is added to maltose to give maltotriose, the equilibrium specific optical rotation is further increased to $+160°$; and if another (β-1,4)-linked glucose residue is added to cellobiose to give cellotriose, the equilibrium specific optical rotation is further decreased to $+22°$. Because nonreducing carbohydrates cannot undergo mutarotation, they have only one

TABLE 7-2. Specific optical rotations for selected carbohydrates

Carbohydrate	Optical rotation ($[\alpha]_D^{20°}$)[a,b]
N-Acetyl-D-muramic acid	+103°
N-Acetyl-D-neuraminic acid	−32.0°
Cellobiose	+35°
Cellotriose	+22°
6-Deoxy-D-galactose (D-fucose)	+76°
6-Deoxy-L-galactose (L-fucose)	−108°
6-Deoxy-D-glucose	+30°
6-Deoxy-L-mannose (L-rhamnose)	+8.9°
D-Fructose	−92.0°
D-Galactose	+80.2°
Gentiobiose	+8.7°
D-Glucaric acid	+20.6°
D-Glucitol (D-sorbitol)	−2.0°
D-Gluconic acid	−6.7°
δ-D-Gluconolactone	+61.7°
D-Glucose	+52.7°
α-D-Glucose-1-phosphate	+78°
D-Glucose-6-phosphate	+21°
D-Glucuronic acid	+36.3°
Glycogen	+198°
Isomaltose	+122°
Lactose	+52.3°
Maltose	+130.4°
Maltotriose	+160°
D-Mannitol	+23.0°
D-Mannose	+14.2°
Melezitose	+88°
Melibiose	+129.5°
Nigerose	+137°
Raffinose	+105.2°
D-Ribose	−21.5°
L-Sorbose	−42.7°
Starch	+220°
Sucrose	+66.5°
α,α-Trehalose	+178°
Turanose	+75.8°
D-Xylose	+18.6°

[a] The specific rotations for reducing sugars are the mutarotated equilibrium values.
[b] All rotations are for the sodium D-line at 20° in water.

specific optical rotation. The measurement of the optical rotation may, thus, be used to qualitatively determine something about the structure of the carbohydrate. Table 7-2 gives specific optical rotations for several carbohydrates.

Polarimeters are used to measure optical rotations. Although polarimeters vary, all have the following basic parts: a light source (usually a sodium lamp), a polarizer prism, an analyzer prism, a telescope with half-shaded device, a graduated circle to measure angular degrees of rotation, and a sample tube. The basic parts are illustrated in Figures 7-1 and 7-2.

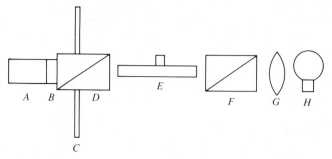

FIGURE 7-1. Basic parts of a polarimeter: A = viewing telescope, B = half-shade device, C = graduated rotating scale attached to the analyzer prism, D = Nicol analyzer prism, E = sample cell, F = polarizing Nicol prism, G = focusing lens, and H = sodium lamp.

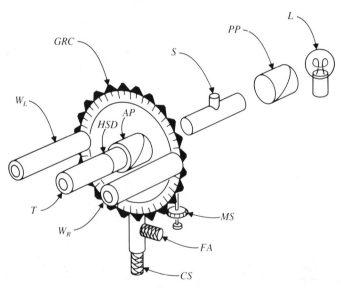

FIGURE 7-2. Schematic diagram of the Rudolph polarimeter: T = viewing telescope, S = sample cell; L = sodium lamp, W_R = right-scale viewing window, W_L = left-scale viewing window, FA = fine-adjusting knurled screw, CS = clamping screw, GRC = graduated rotating circle, HSD = half-shade device, MS = micrometer scale, AP = analyzing prism, and PP = polarizing prism.

Most polarimeters have a half-shade device that permits the optical rotation to be measured visually. With some special instruments, however, photocells or related devices are used to measure the intensity of the light coming through the two halves of the shaded device.

The rotating analyzer is at the observation end and consists of the analyzing Nicol prism, the graduated circular scale, and the telescope containing the half-shade device. In some instruments, a micrometer scale is also used for obtaining very accurate rotations.

Coarse adjustments of the analyzer are made by turning the graduated circle, which is notched for ease of handling. The scale of the graduated circle is divided into 360° and may be viewed through two illuminated windows, located on each side at the front of the polarimeter (Fig. 7-2). The zero position of the divided circle scale is normally located at the right-hand window and the 180° position is simultaneously located at the left-hand window.

OPTICAL ROTATION MEASUREMENTS WITH A RUDOLPH POLARIMETER

Turn on the sodium lamp approximately 30 minutes before making readings. Fill a polarimeter tube with the solvent, removing all the air bubbles (e.g., water if the sample is dissolved in water). Place the tube in the holder. Loosen the clamping screw at the bottom of the analyzer wheel and look into the right-scale viewing window (W_R in Fig. 7-2) and turn the large notched analyzer wheel until the hairline indicator is near zero. Then *carefully* turn the large analyzer wheel clockwise a few centimeters or so, then counterclockwise a few centimeters, and notice the half-shade changing dark from left to right (Fig. 7-3).

Next set the micrometer scale at zero. The micrometer scale is properly set at zero when the hairline bisects the arrow in the viewing window (Fig. 7-4A). Tighten the bottom clamping screw. *Carefully* turn the 'fine'-adjusting knurled screw at the side of the analyzer wheel so that one side is *slightly* darker than the other side. Darken the shaded side by turning the fine-adjusting knurled screw. Then turn the knurled screw in the other direction until the two half-fields are matched in intensity (Fig. 7-3B). Read the scale in the right-hand window.

As an example, let us say that the scale looks like Figure 7-4B, in which the hairline is between 1° and 2° or, more precisely, between 1.25° and 1.50°. Now turn the micrometer scale, located below the window, until the hairline index bisects the 1.25° line on the window scale (Fig. 7-4C). The micrometer reads 12.4 (Fig. 7-4C); the degree of rotation of the blank is then 1.250° + 0.124°, or 1.374°. This is because one full turn of the micrometer scale is equal to 0.25°, and the scale is divided into 0.002° of arc.

    (A)   *(B)*   *(C)*

FIGURE 7-3. Appearance of the half-shade device. (*A*) Rotation of analyzer counterclockwise. (*B*) Matched-fields balance point. (*C*) Rotation of analyzer clockwise.

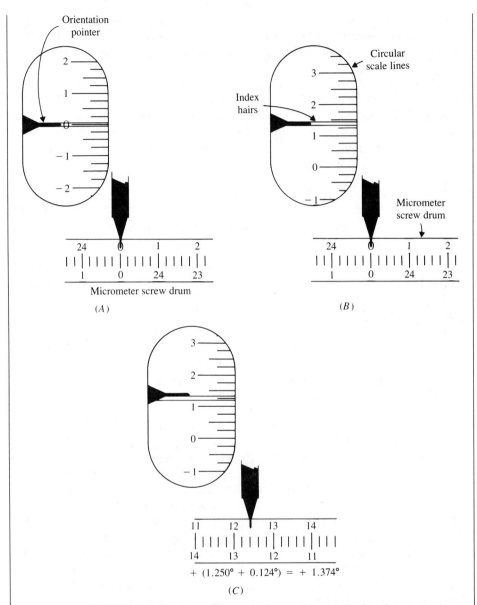

FIGURE 7-4. Reading the optical rotation scale in the Rudolph polarimeter. (*A*) Starting point, with the hairline indicator bisecting the orientation arrow and with the micrometer screw set at zero. (*B*) View of optical rotation scale after balancing the half-shade device so that both sides of the field are of equal intensity (Fig. 7-3*B*). (*C*) Setting of hairline indicator with the micrometer screw so that the hairline bisects the next lower scale value, in this example $+1.25°$. The micrometer reads 12.4, which is equivalent to $+0.124°$. The value of the optical rotation is then $(+1.250° + 0.124°) = +1.374°$.

Repeat the foregoing steps, using the fine-adjusting knurled screw, four more times to obtain five readings matching the field from the same side. Then repeat the procedure five times, matching the field from the other side. Add the 10 readings and obtain the mean for the blank and the mean deviation. Fill the polarimeter tube with the sample and obtain the rotation in the same manner as for the blank. Subtract the blank value from the sample value. You should be able to obtain polarimeter readings with a precision of $\pm 0.005°$.

7.2 AMINO ACIDS, PEPTIDES, AND PROTEINS

7.2.1 Qualitative Tests for Amino Acids and Proteins

7.2.1.1 THE NINHYDRIN TEST FOR AMINO GROUPS

Primary and secondary amines or amino-containing compounds will react with ninhydrin to give colored complexes (see Section 7.2.2.1). Free amino acids in solution react at 100°C to give a purple-blue color. The ninhydrin reaction is run by adding 0.1 mL of 0.1% ninhydrin in 95% methanol to 2 mL of the sample. The solution is heated in a boiling water bath for 10 minutes. The appearance of a blue color indicates the presence of amino groups.

7.2.1.2 NINHYDRIN AND NINHYDRIN–COLLIDINE AS QUALITATIVE TESTS FOR AMINO ACIDS ON CHROMATOGRAMS

Chromatograms can be sprayed lightly with 0.3% ninhydrin in 95% ethanol and heated at 80–100°C for 2 minutes. Most amino acids give blue spots on a colorless background; proline gives a strong yellow spot.

The addition of collidine (2,4,6-trimethylpyridine) gives distinctively colored spots for many amino acids on chromatograms and can be used to qualitatively identify them. The ninhydrin–collidine reagent is prepared by mixing 15 mL of 0.15% ninhydrin in 95% ethanol with 2 mL of collidine and 5 mL of glacial acetic acid. The chromatogram is sprayed and heated at 90°C for 2–5 minutes.[12] The resulting colors of the amino acids are given in Table 7-3.

7.2.1.3 THE ISATIN–ZINC ACETATE–PYRIDINE REAGENT AS A QUALITATIVE TEST FOR AMINO ACIDS ON CHROMATOGRAMS

The reagent is prepared by adding 1 g of isatin, 1.5 g zinc acetate, and 1 mL of pyridine to 100 mL of 2-propanol at 70–80°C. The cooled reagent is sprayed onto the chromatogram, which is heated at 90°C for 30 minutes. The resulting colors of the amino acids are given in Table 7-3.

7.2.1.4 PAULY REAGENT FOR THE DETECTION OF TYROSINE AND HISTIDINE ON CHROMATOGRAMS

The Pauly reagent is prepared by mixing 10 mL of 5% sodium nitrite with 1 mL of 0.9% sulfanilic acid in 1 M hydrochloric acid just before use. The

TABLE 7-3. Colors of amino acids on chromatograms that have been reacted with ninhydrin–collidine and isatin–zinc acetate–pyridine

Amino acid	Ninhydrin–collidine	Isatin–zinc acetate–pyridine
α-Alanine	Violet	Violet
β-Alanine	Green	Lilac
Arginine	Violet	Pink
Asparagine	Yellow	Pink
Aspartic acid	Turquoise	Violet
Cystine (Cys_2)	Brown	Pink
Glutamic acid	Violet	Pink
Glycine	Pink-purple	Pink
Histidine	Gray-green	Lilac
Isoleucine	Violet	Pink
Leucine	Violet	Pink
Lysine	Violet	Red
Methionine	Violet	Pink
Phenylalanine	Blue-gray	Lilac
Proline	Yellow	Blue
Serine	Red-purple	Orange
Threonine	Violet	Pink
Tryptophan	Gray-pink	Lilac
Tyrosine	Gray-blue	Red
Valine	Violet	Pink

Source: E. von Aux and R. Neher, *J. Chromatogr. 12*: 329 (1963).

chromatogram is sprayed with this reagent and then oversprayed with 5% sodium carbonate solution. Orange to red spots that appear immediately without heating indicate tyrosine or histidine, or peptides or proteins containing these amino acids.[13]

7.2.1.5 THE SAKAGUCHI REAGENT FOR THE DETECTION OF ARGININE ON CHROMATOGRAMS

The Sakaguchi reaction depends on the presence of the guanidino group and is positive for arginine or peptides and proteins containing arginine. Equal volumes of 0.1% butanedione and 20% potassium hydroxide are mixed and added to one volume of 25% α-naphthol in absolute ethanol. The chromatogram is sprayed, and arginine-containing spots appear bright pink within 20 minutes without heating.[14]

7.2.1.6 THE EHRLICH REAGENT FOR THE DETECTION OF TRYPTOPHAN ON CHROMATOGRAMS

Paradimethylaminobenzaldehyde in strong acid reacts specifically with the indole side chain of tryptophan to yield a purple product. The *p*-dimethyl-aminobenzaldehyde (10%) in concentrated hydrochloric acid is mixed with four volumes of acetone. After heavy spraying, tryptophan spots appear bright purple, without heating.

7.2.2 The Quantitative Determination of Amino Acids

7.2.2.1 DETERMINATION WITH NINHYDRIN[15]

When amino acids are reacted with ninhydrin hydrate (**I**) at pH 5 and 100°C for a standard period of time, a purple-blue compound, the ammonium salt of diketohydrindylidene–diketohydrindamine (**II**), is produced as the major product. Ninhydrin will also yield a similar purple-blue product with ammonia and primary amines. The absorbance of the purple-blue product is measured at 570 nm. Proline and hydroxyproline give a yellow product (**III**) that is partly transformed into enolbetaine (**IV**) by the loss of water. Extensive heating of proline and ninhydrin in acetic acid at 100°C will give a purple-blue product (**V**) when (**IV**) condenses with another molecule of ninhydrin hydrate (**I**) to give (**V**).

There are no distinct differences in the colors (maximum wavelength) of the purple-blue products for the different amino acids when the amino acids are reacted in solution. This is in contrast to the different colors obtained for the amino acids after spraying ninhydrin reagent onto thin-layer or paper chromatograms. The color yields for the different amino acids, however, are not the same and must be taken into consideration when different amino acids are being quantitatively determined (e.g., in amino acid analysis by ion-exchange chromatography or HPLC).

THE QUANTITATIVE DETERMINATION OF AMINO ACIDS WITH NINHYDRIN

The analysis is performed by the addition of 0.2 mL of ninhydrin reagent (2 mg/mL of ninhydrin in 20 mM acetic acid–acetate buffer, pH 5) to 1.0 mL of the amino acid solution. The solution is heated 10 min at 100°C and cooled, and the absorbance is measured at 570 nm for all the amino acids except proline, which is measured at 440 nm. A standard curve is prepared for the different amino acids. The limit of sensitivity is 0.1 μmol.

7.2.2.2 DETERMINATION WITH FLUORESCAMINE[16]

Fluorescamine (**VI**) reacts directly with primary amines to form fluorophors (**VII**), which are excited at 390 nm and fluoresce at 475 nm. At pH 9, the reaction with primary amines proceeds rapidly at room temperature and is complete in 100–500 ms. Excess reagent is hydrolyzed with a half-time of several seconds (5–10 s) to nonfluorescent products. In the assay procedure, primary amines are buffered to an appropriate pH (> 7) and then fluorescamine, dissolved in a nonprotic solvent such as acetone, is added. The resulting fluorescence is proportional to the amine concentration, and the fluorophors are stable over several hours. Fluorescamine does not react with proline or hydroxyproline. This disadvantage may be overcome by introducing an intermediate step that converts these amino acids into primary amines. An advantage of the fluorescamine reaction is that very little interfering fluorescence is obtained with ammonia.

Peptides will also react with fluorescamine, usually giving a higher fluorescence than their component amino acids. Peptides give maximum fluorescence at pH 7, whereas amino acids give maximum fluorescence at pH 9, with much less fluorescence at pH 7. Fluorescamine will react with the primary amino groups of proteins to give highly fluorescent derivatives that can be used for protein labeling in conformational studies or for immunochemical studies using fluorescamine-labeled antibodies. Fluorescamine can also be used for the assay of protein during a protein purification procedure.

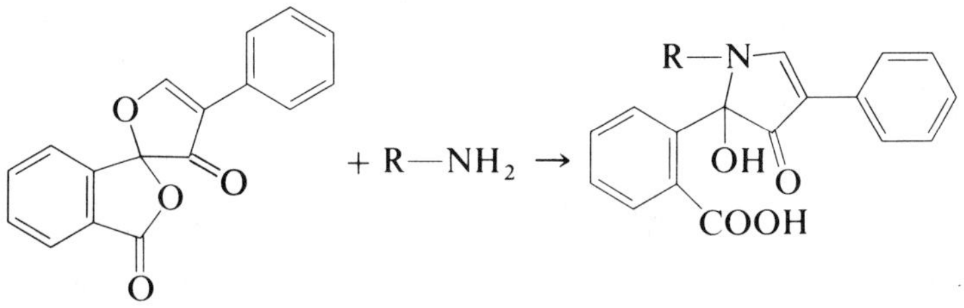

VI   VII

**THE QUANTITATIVE DETERMINATION OF AMINO GROUPS
WITH FLUORESCAMINE**

The quantitative procedure involves the addition of 0.20 mL of 0.40 M boric
acid buffer, adjusted to pH 9.7, and 0.20 mL of fluorescamine reagent (0.3 mg/
mL of fluorescamine in acetone) to 1.0 mL of amino acid solution containing
5–50 nmol of amino acid. The solution is excited at 390 nm and the fluorescence is measured at 475 nm in a fluorometer (see Section 3.4.1).

7.2.2.3 DETERMINATION WITH o-PHTHALALDEHYDE–2-MERCAPTOETHANOL

o-Phthalaldehyde, in the presence of 2-mercaptoethanol, reacts with primary amines to form highly fluorescent products. Picomole quantities of amino
acids, peptides, and proteins are detected with about 5–10 times the sensitivity
of fluorescamine. The reagent has the advantage of being soluble and stable in
aqueous buffers.

The reagent is prepared by the addition of 10 mL of 95% ethanol containing
800 mg of o-phthalaldehyde to 1.0 L of 0.4 M boric acid buffer (pH 9.7)
containing 2.0 mL of 2-mercaptoethanol and 1.0 g of BRIJ, a nonionic detergent
consisting of a series of polyoxyethylene ethers of fatty alcohols. The procedure
involves the addition of 0.20 mL of the reagent to 1.0 mL of amino acid solution
with 1–10 nmol of amino acid. The solution is excited at 340 nm and the
fluorescence is measured at 455 nm.

7.2.3 The Quantitative Determination of Proteins

Many quantitative methods for the determination of proteins have been
developed. One of the first methods was the Kjeldahl procedure for the
measurement of protein nitrogen. The Kjeldahl method, which usually requires
relatively large amounts of sample, is somewhat tedious and time-consuming.
Other methods that have emerged rely on spectrophotometric procedures
involving specific amino acid side-chain groups, the peptide linkage, or the

binding of a dye to the protein. These latter methods are of varying degrees of sensitivity and specificity, are relatively rapid, and can be performed on a semimicro or micro scale.

7.2.3.1 THE KJELDAHL METHOD

The Kjeldahl method is used when relatively large samples are available, as frequently occurs in the analysis of plant materials and animal feeds. However, the method has been used to determine the nitrogen in dried-blood samples, too. The sample or protein (1–10 g) is digested with strong (10 M) sulfuric acid (50 mL) by heating for several hours in a hood. At the end of the heating period a few drops of hydrogen peroxide may be added to assist in clarifying the solution. During the digestion, the protein nitrogen is converted into ammonium sulfate. The acid digest is made alkaline by the addition of excess sodium hydroxide, which converts the ammonium ion into ammonia. The ammonia is steam-distilled into 0.1 N hydrochloric acid, which is then back-titrated with a standard solution of 0.1 N sodium hydroxide.

Since most proteins contain about 16% nitrogen, 16 mg of nitrogen = 100 mg of protein and 1 mg of nitrogen = 100/16 or 6.25 mg of protein. The number of milligrams of nitrogen obtained by the Kjeldahl method is multiplied by 6.25 to give the number of milligrams of protein in the sample.

7.2.3.2 ULTRAVIOLET ABSORPTION

Proteins have an absorption spectrum in the ultraviolet region. Below 230 nm, the absorption of a protein rises rapidly, reaching a maximum around 190 nm due to the absorption of the peptide bond. A number of other substances, such as carboxylic acids, buffer ions, and alcohols, also absorb in this region and interfere with the measurement, making it much less specific for proteins.

Proteins also have a characteristic absorption around 275–280 nm due to the presence of tyrosine and tryptophan. Each individual type of protein has a specific extinction coefficient, $a_{1cm}^{1\%}$, which varies with the amount of tyrosine and tryptophan in the protein. The extinction coefficients for 1% solutions of protein at 280 nm vary from 6 to 60, although many proteins have values around 10. This value of 10 can be used to estimate the amount of protein in a solution if the extinction coefficient is not known or if the solution contains a mixture of proteins. A solution of 0.10 mg/mL protein with $a_{1cm}^{1\%} = 10$ will give an absorbance of 0.10. Absorbance measurements at 280 nm are frequently used to obtain an elution profile of a chromatographic separation of proteins on a column. The method is relatively fast, easily automated, nondestructive, and reasonably sensitive. Table 7-4 gives the $a_{1cm}^{1\%}$ values for 50 proteins.

A disadvantage of the method is that many other compounds will also absorb in the 280 nm region, especially nucleic acids, which have a maximum absorbance around 260 nm. Warburg and Christian[17] developed a method of correcting for the presence of nucleic acid by measuring the ratio of the absorbance at 280 nm

TABLE 7-4. Extinction coefficients ($a_{1\,cm}^{1\%}$) at 280 nm for several proteins

Protein	$a_{280\,nm}^{1\%}$
Acetyl cholinesterase (*Electrophorus electricus*)	16.1
Alcohol dehydrogenase (horse liver)	4.6
Alcohol dehydrogenase (yeast)	12.6
Aldolase (rabbit muscle)	9.4
Alkaline phosphatase (*E. coli*)	7.0
α-Amylase (porcine pancreas)	26
β-Amylase (sweet potato)	17.7
Asparaginase (*E. coli*)	7.1
Carbonic anhydrase (bovine erythrocytes)	18
Carboxypeptidase A (bovine pancreas)	19.4
Carboxypeptidase B (porcine pancreas)	21
Ceruloplasmin	14.9
Chymopapain	18.7
α-Chymotrypsin (bovine pancreas)	20.4
Collagenase (*Clostridium histolyticum*)	18
Creatine kinase (rabbit muscle)	8.8
Deoxyribonuclease (bovine pancreas)	11.1
Deoxyribonuclease II (porcine spleen)	12.1
Dextranase (*Penicillium* sp.)	20
Elastase (pig pancreas)	18.5
Fibrinogen	13.6
Fructose diphosphatase (rabbit liver)	8.9
β-Galactosidase (*E. coli*)	20.9
Glucan phosphorylase *a* (rabbit muscle)	11.9
Glucose-6-P-dehydrogenase (*Leuconostoc mesenteroides*)	12.5
β-Glucuronidase (bovine liver)	17.0
Glutamic–oxaloacetic transaminase (porcine heart)	7.2
Glyceraldehyde-3-phosphate dehydrogenase (rabbit muscle)	8.15
Glyceraldehyde-3-phosphate dehydrogenase (yeast)	8.6
Hemopexin	16.9
γA-Immunoglobulin (IgA)	13.4
γG-Immunoglobulin (IgG)	13.8
γM-Immunoglobulin (IgM)	13.3
L-Lactate dehydrogenase (baker's yeast)	31.2
Lactic dehydrogenase (beef heart)	14.9
Lactic dehydrogenase (rabbit muscle)	8.9
Lipase (porcine pancreas)	30
Lipoxidase (soybean)	17.8
Lysozyme (chicken egg white)	26.4
Malate dehydrogenase (porcine heart)	2.8
Papain	25
Polyphenol oxidase (mushroom tyrosinase)	24.9
Inorganic pyrophosphatase (yeast)	14.5
Pyruvate kinase (rabbit muscle)	5.4
Ribonuclease (bovine pancreas)	7.3
Ribonuclease T$_1$ (*Aspergillus oryzae*)	19
Serum albumin	5.8
Transcortin	7.4
Transferrin	11.2
Trypsin (bovine pancreas)	14.3
Uricase (porcine liver)	11.3
Xanthine oxidase (milk)	11.3

TABLE 7-5. Estimation of protein concentration by ultraviolet absorption at 280 and 260 nm

$R_{280/260}{}^a$	Nucleic acid (%)	Correction factor^b
1.75	0.00	1.12
1.63	0.25	1.08
1.52	0.50	1.05
1.40	0.75	1.02
1.36	1.00	0.99
1.30	1.25	0.97
1.25	1.50	0.94
1.16	2.00	0.90
1.09	2.50	0.85
1.03	3.00	0.81
0.979	3.50	0.78
0.939	4.00	0.74
0.874	5.00	0.68
0.846	5.50	0.66
0.822	6.00	0.63
0.804	6.50	0.61
0.784	7.00	0.59
0.767	7.50	0.57
0.753	8.00	0.55
0.730	9.00	0.51
0.705	10.00	0.48
0.671	12.00	0.42
0.644	14.00	0.38
0.615	17.00	0.32
0.595	20.00	0.28

^{a} Ratio of absorbance at 280 nm to absorbance at 260 nm.
^{b} Milligrams of protein per milliliter $= A_{280} \times$ correction factor.
Source: E. Layne, "Spectrophotometric and turbidimetric methods for measuring proteins," *Methods Enzymol. 3:* 447. S. P. Colowick and N. O. Kaplan, Eds. Academic Press, New York, 1957.

to that at 260 nm. This ratio may be used to estimate the amount of protein present when nucleic acids are also present (Table 7-5). Pure proteins have a ratio (A_{280}/A_{260}) greater than 1.7 and pure nucleic acids less than 0.5.

7.2.3.3 THE BIURET METHOD[18]

Polypeptides and proteins with two or more peptide bonds give a characteristic purple color when treated with dilute copper sulfate at alkaline pH values. The color is caused by the formation of a complex of a copper(II) ion with four nitrogen atoms, two from each of two peptide chains. This reaction is similar to the reaction of four ammonia molecules with copper(II) to give the deep blue complex ion, cupric tetramine. Ammonia or ammonium ions will, thus, interfere in the determination by forming a blue complex with the Cu(II) of the biuret reagent. The biuret reaction requires relatively large amounts (1–20 mg) of protein.

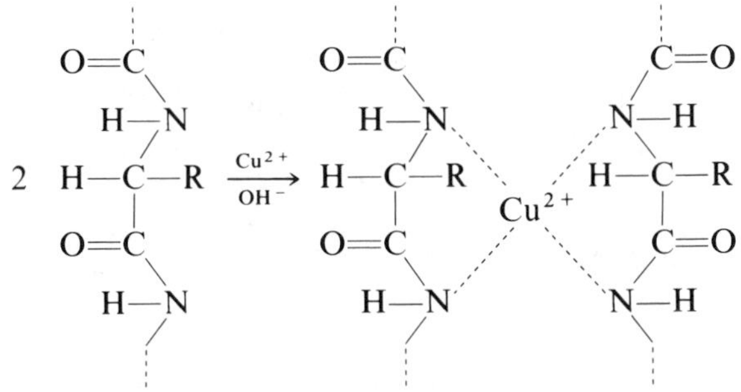

$$\text{2 } \quad \begin{array}{c} O{=}C \\ | \\ H{-}N \\ | \\ H{-}C{-}R \\ | \\ O{=}C \\ | \\ H{-}N \\ | \end{array} \quad \xrightarrow[\text{OH}^-]{\text{Cu}^{2+}} \quad \begin{array}{ccc} O{=}C & & C{=}O \\ | & & | \\ H{-}N & & N{-}H \\ & \diagdown \diagup & \\ H{-}C{-}R & \text{Cu}^{2+} & H{-}C{-}R \\ & \diagup \diagdown & \\ O{=}C & & C{=}O \\ | & & | \\ H{-}N & \text{Cu}^{2+} & N{-}H \\ | & & | \end{array}$$

Purple biuret complex

$$\text{Cu}^{2+} + 4\text{NH}_3 \rightarrow \text{Cu(NH}_3)_4^{2+}$$

Blue cupric
tetramine complex

THE BIURET METHOD

The biuret analysis is performed by adding 2.0 mL of protein solution to 3.0 mL of the biuret reagent: 3 g of copper sulfate pentahydrate ($\text{CuSO}_4 \cdot 5\text{H}_2\text{O}$) and 9 g of sodium–potassium tartrate dissolved in 500 mL of 0.2 N sodium hydroxide to which 5 g of potassium iodide is added and the whole diluted to 1 L with water. The solution of protein and biuret reagent is warmed at $37°\text{C}$ for 10 minutes and cooled to room temperature; the absorbance measured at 540 nm.

7.2.3.4 THE FOLIN–LOWRY METHOD[19]

The Folin–Lowry method is a modification of the biuret reaction to give a more sensitive determination of protein. Two color reactions are used: the biuret reaction with alkaline copper(II) and the reaction of a complex salt of phosphomolybdotungstate, called the Folin–Ciocalteu phenol reagent, which gives an intense blue-green color with the biuret complexes of tyrosine and tryptophan.

THE FOLIN–LOWRY METHOD

The copper reagent is prepared by adding 1.0 mL of 0.5% copper sulfate pentahydrate and 1% sodium–potassium tartrate to 50 mL of 0.1 M sodium hydroxide, containing 2% sodium carbonate. The protein solution (0.5 mL) is added to 5.0 mL of the alkaline copper reagent and the solution allowed to stand for 10 minutes. Then 0.50 mL of the Folin–Ciocalteu phenol reagent (which had been diluted 1:1 with water) is rapidly added and mixed. This solution is allowed to stand for 30 minutes, and the absorbance is measured at 600 nm. A standard curve is prepared using the protein under study or some other protein, such as serum albumin at concentrations of 20–400 μg/mL.

The Folin–Ciocalteu reagent may be either purchased or prepared. To prepare the reagent, add 100 g of sodium tungstate dihydrate ($Na_2WO_4 \cdot 2H_2O$) and 25 g of sodium molybdate ($Na_2MoO_4 \cdot H_2O$) to 700 mL of water to which 50 mL of 85% phosphoric acid and 100 mL of concentrated hydrochloric acid have been added. Reflux this solution gently for 10 hours and then add 150 g of lithium sulfate, 50 mL of water, and a few drops of bromine. Boil this solution for 15 minutes, cool, dilute to 1 L, and filter if necessary. The working solution is prepared by a 1:1 dilution with water. The solution is stored in a brown bottle and should be bright yellow, without a greenish tint. If the solution becomes greenish on storage, a few drops of bromine may be added and the solution boiled as before.

The Folin–Lowry method is several times more sensitive than the ninhydrin reaction, more than 10 times more sensitive than the measurement of the ultraviolet absorption at 280 nm, and 100 times more sensitive than the biuret method. The method is subject to interference by several substances, such as K^+, Mg^{2+}, NH_4^+, EDTA, Tris, carbohydrates, and reducing agents (2-mercaptoethanol, dithiothreitol, etc.).

7.2.3.5 THE BRADFORD METHOD

The method uses a dye, Coomassie brilliant blue G-250, which has a negative charge and binds with positive charges on the protein. The dye exists in a red form ($A_{max} = 465$ nm) and a blue form ($A_{max} = 595$ nm). The red form is the predominant form in solution, and when its negative charge binds to the positive charges on the protein, it is converted into the blue form. Because many proteins have nearly identical response curves, the method can be applied widely using a single set of standards. Furthermore, it is much less susceptible to interfering substances. The reaction is highly reproducible and rapid and is essentially complete after 2 minutes with color stability over 1 hour. Detergents such as sodium dodecyl sulfate (SDS) and Triton X-100 do interfere.

THE BRADFORD METHOD

The analysis is carried out by adding 0.10 mL of protein solution, containing 10–200 μg, to 5.0 mL Coomassie blue dye reagent (100 mg of Coomassie brilliant blue G-250 dissolved in 50 mL of 95% ethanol to which 100 mL of 85% phosphoric acid has been added and the whole diluted to 1 L with water). After 5 minutes, the absorbance is measured at 595 nm. Standard curves are commonly prepared by using bovine serum albumin.[20]

7.2.3.6 THE QUANTITATIVE DETERMINATION OF SULFHYDRYL GROUPS IN PROTEINS

Ellman's reagent, 5,5′-dithiobis[2-nitrobenzoic acid] (abbreviated as DTNB) is used as a standard method for the quantitative determination of reactive sulfhydryls in proteins.[21] One mole of sulfhydryl reacts with DTNB at pH 8.1 to give 1 mol of the colored thio anion (3-carboxylato-4-nitrothiophenolate [CNT]), which has a maximum absorbance at 412 nm.

SULFHYDRYL DETERMINATION WITH ELLMAN'S REAGENT

To 2.0 mL of sample containing 5–100 μM sulfhydryl is added 0.1 mL of buffer (1 M Tris and 1 M phosphate, pH 8.1) and 0.5 mL of 2 mM DTNB (buffered with 10 mM phosphate, pH 8.1). The reaction is allowed to proceed 30 min at room temperature ($\sim 22°C$). The number of sulfhydryl groups per molecule of protein is computed from the concentration of the protein (determined by absorbance measurement at 280 nm or one of the other methods for protein determination) and the concentration of the sulfhydryl groups determined by the measurement of the absorbance of CNT at 412 nm and its molar extinction coefficient, 1.14×10^4 L/mol·cm.[21]

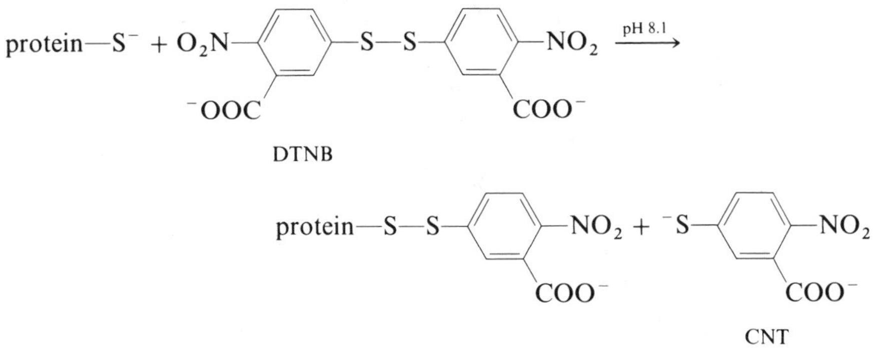

7.2.3.7 THE QUANTITATIVE DETERMINATION OF THE NUMBER OF DISULFIDE LINKAGES IN PROTEINS

a. Using Ellman's Reagent. It was observed that Ellman's reagent reacts with disulfide linkages when there is one free sulfhydryl group available for reaction.[22] Although there is only as much CNT released as there are sulfhydryl groups, the disulfide linkages are cleaved and become derivatized by the DTNB, symbolized as ESSE, as shown in the following reactions[22,23]:

$$\text{protein—S}^- + \text{ESSE} \rightarrow \text{protein—S—S—E} + \text{ES}^-$$

$$\text{protein—S—S—protein} + \text{ES}^- \rightarrow \text{protein—S—S—E} + \text{protein—S}^-$$

$$\text{protein—S}^- + \text{ESSE} \rightarrow \text{protein—S—S—E} + \text{ES}^-$$

The overall net reaction is therefore:

protein—S⁻ + protein—S—S—protein + 2 ESSE →

$$3 \text{ protein—S—S—E} + \text{ES}^-$$

THE DETERMINATION OF DISULFIDE LINKAGES WITH ELLMAN'S REAGENT

The number of disulfide linkages can be determined by making the protein solution 2 M in guanidine thiocyanate and 3 mM in EDTA and measuring the amount of CNT released as described in Section 7.2.3.4. The pH of the solution is then made 10.5, which releases all the CNT groups that are derivatized to the protein due to the CNT (ES⁻) cleavage of the disulfide linkages. The number of disulfide linkages can then be computed according to the following equation:

$$\frac{\text{number of}}{\text{disulfide linkages}} = \frac{(\text{no. CNT after pH 10.5}) - 2(\text{no. CNT before pH 10.5})}{2}$$

The release of the CNT molecules from the protein by the alkaline treatment at pH 10.5 is the result of an alkaline-catalyzed β-elimination reaction:

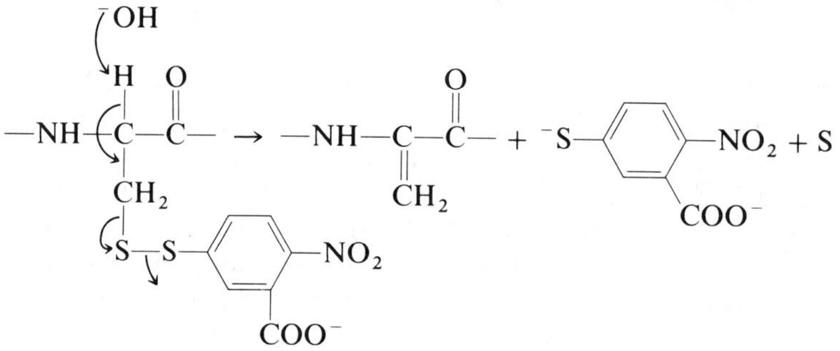

If the protein does not have an active sulfhydryl group, sulfhydryl groups can be added to initiate the reaction by adding sulfhydryl-containing compounds such as cysteine, glutathione, and dithiothreitol.

b. **Using 2-Nitro-5-Thiosulfobenzoate.** The disulfide bonds of proteins can be cleaved by reaction with excess sodium sulfite:

$$\text{protein—S—S—protein} + \text{SO}_3^{2-} \rightarrow \text{protein—S—SO}_3^- + \text{protein—S}^-$$

Since Ellman's reagent itself reacts with the sulfite, it cannot be used to determine the concentration of the sulfite-produced protein sulfhydryl. However, 2-nitro-5-thiosulfobenzoate (NTSB, ESSO_3^-) can be used because it will not react with the excess sulfite but will react with sulfite-released sulfhydryl.[24]

$$\text{protein—S}^- + \text{ESSO}_3^- \rightarrow \text{protein—S—SO}_3^- + \text{ES}^-$$

The 2-nitro-5-thiosulfobenzoate is synthesized by reacting Ellman's reagent with excess sodium sulfite. The reaction is taken to completion by oxidizing the resulting CNT back to DTNB with oxygen, which is further cleaved by the excess sulfite until all the DTNB has been converted to NTSB.

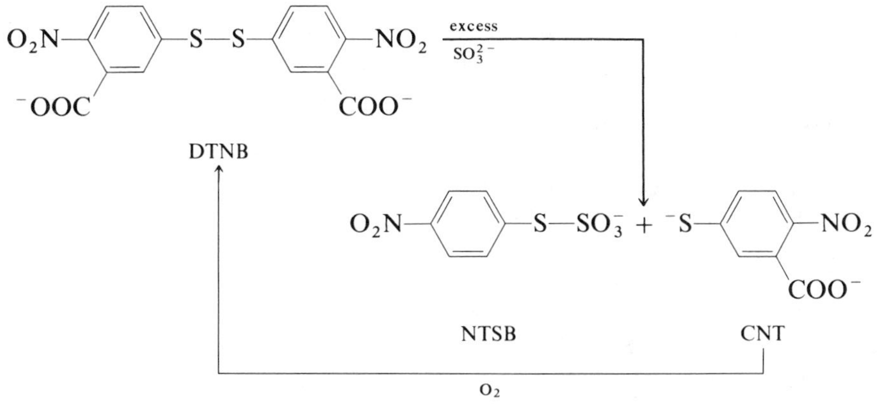

THE DETERMINATION OF DISULFIDE LINKAGES WITH 2-NITRO-THIOSULFOBENZOATE

One hundred milligrams of DTNB (2.5×10^{-4} mol) is dissolved in 10 mL of 1 M Na_2SO_3 (1×10^{-2} mol) at 38°C and the pH is adjusted to 7.5. The reaction is conducted in the dark. Oxygen is bubbled through the solution and the reaction is followed by measuring the absorbance of CNT at 412 nm. When the absorbance is near zero, the reaction is complete (~ 45 minutes).

For the assay of disulfide bonds, 0.2 mL of protein (with a disulfide bond concentration of 0.5–2 mM) is added to 3.0 mL containing 0.5 mM NTSB, 0.2 M Tris, 0.1 M sodium sulfite, 3 mM EDTA, and 2 M guanidine thiocyanate. The reaction is conducted in the dark for 25 minutes. The absorbance is then measured at 412 nm against a blank of 3.0 mL of NTSB assay solution plus 0.2 mL of water. The number of disulfide bonds is computed using the molar extinction coefficient for CNT, 1.14×10^4 L/mol · cm.

7.3 NUCLEOTIDES AND NUCLEIC ACIDS

Nucleic acids contain three distinct chemical components: purine and pyrimidine bases, D-ribose or 2-deoxy-D-ribose, and phosphorus. Consequently, methods for determining nucleic acids can be based on the UV absorption of the bases, the specific reactions of pentoses, or the determination of phosphorus. Specific color reactions for D-ribose and 2-deoxy-D-ribose can be used to distinguish between RNA and DNA.

7.3.1 The Qualitative Determination of RNA and DNA

RNA is qualitatively identified by the orcinol reaction with D-ribose in which a blue-green color is produced (Section 7.1.1.3). DNA is qualitatively identified by the diphenylamine reaction with 2-deoxy-D-ribose in which a blue-green color is produced (Section 7.1.1.4).

Specific sequences in DNA and RNA can be detected using Southern and "northern" blots. These sequences are detected by hybridizing with labeled oligonucleotide probes of known sequence. Identification of a specific gene or mRNA can be obtained with great sensitivity using radioisotope or fluorescent-labeled probes. Hybridization to DNA is termed the **Southern blot** after E. M. Southern, who developed the method to locate specific sequences in cloned DNA.[25] An analogous method for detecting specific RNA sequences is called "**northern blot.**"

DNA fragments that have been separated by electrophoresis on agarose gels are denatured and transferred to nitrocellulose. The DNA on the nitrocellulose is incubated with ^{32}P-labeled DNA or RNA probes whose sequences are known and are complementary to a DNA fragment on the nitrocellulose. The complementary sequences are located by autoradiography or fluorescence. For northern blots, the RNA is transferred to nitrocellulose in high-salt buffer solutions, but without the base denaturation step that would cleave the RNA.

SOUTHERN BLOTTING

After electrophoresis of DNA in agarose gel, the gel is stained with ethidium bromide to locate the DNA bands (Section 7.3.2.2); 0.2 μg of plasmid DNA or 10 μg of mammalian DNA is sufficient to allow detection by hybridization. The DNA in the gel is denatured by treatment with a mixture of 1.5 M NaCl and 0.5 M NaOH for 1 hour at room temperature. The gel is neutralized with 1 M Tris–HCl, pH 8 buffer, containing 1.5 M NaCl for 1 hour at 23°C. The DNA is transferred to the nitrocellulose filter (S & S BA 85 or Millipore HAHY) with 1.5 M NaCl in 0.3 M sodium citrate, pH 7. The nitrocellulose filter is moistened by floating on the buffer solution (diluted 1:5) and then submerged in the buffer. The transfer is accomplished with the system shown in Figure 7-5.

When the transfer is completed, the gel is peeled off the filter, and the filter is soaked in NaCl–Na citrate buffer for 5 minutes, dried on Whatman 3MM paper, and baked for 2 hours at 80°C under vacuum. Marker DNAs that will hybridize to the probe can be used to orient the filter and to provide size markers directly on the autoradiogram.

NORTHERN BLOTTING

RNA can be transferred to nitrocellulose filters with 0.3 M sodium citrate buffer, pH 7, containing 3 M NaCl. The technique is the same as for DNA transfer (Fig. 7-5) except that the filter is not soaked in buffer after transfer.

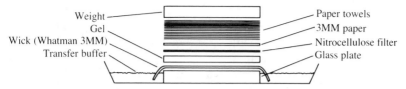

FIGURE 7-5. Method for the transfer of DNA from agarose gel to nitrocellulose filters. The following materials are stacked on a glass plate: Whatman 3MM paper wick; inverted agarose gel; wet nitrocellulose filter; Whatman 3MM paper; paper towels; and weight. The dish is filled with sodium chloride–sodium citrate buffer (1.5 M NaCl, 0.3 M sodium citrate, pH 7.0). The buffer is allowed to flow from the reservoir through the gel and the nitrocellulose filter, eluting the DNA fragments from the gel and depositing them onto the nitrocellulose, where they are adsorbed.

HYBRIDIZATION

Hybridizations are carried out in heat-sealable plastic bags, with a minimum volume of hybridization solution, approximately 40 μL/cm^2 of nitrocellulose, but with a minimum of 1.5 mL per bag. The nitrocellulose filters are wetted in 1 M NaCl in 0.1 M sodium citrate buffer, pH 7. The filters are blotted briefly and placed in the bag. For oligomer probes, filters are prehydridized with a 1:10 dilution of Denhardt's solution (1% Ficoll, 1% polyvinylpyrrolidone, and 1% bovine serum albumin), 10 mM EDTA, pH 7.5, and 0.5% SDS for 4 hours. Hybridization solutions contain: 1 M NaCl in 0.1 M sodium citate, pH 7, 1 mM EDTA, 1:10 dilution of Denhardt's solution, 0.5% SDS, 100 μg/mL heat-denatured DNA (salmon sperm DNA or tRNA), plus ^{32}P-labeled probe. Filters are incubated 18–20 hours at temperatures determined by the number of adenine–thymine and guanine–cytosine base pairs.[26]

The filters are washed in 0.3 M NaCl, 0.03 M sodium citrate, pH 7, and 0.5% SDS at 23°C. The filters are then transferred to buffer containing 0.1% SDS and washed 15 minutes at 23°C and then transferred to 0.015 M NaCl in 1.5 mM sodium citrate, pH 7, and 0.5% SDS and incubated at 68° for 2 hours. The buffer is removed and fresh buffer added, and the filters are incubated for an additional 30 minutes. The filter is dried at room temperature on Whatman 3MM paper and wrapped in plastic wrap, and an autoradiogram is made.

7.3.2 The Quantitative Determination of Nucleotides and Nucleic Acids

7.3.2.1 ABSORBANCE OF ULTRAVIOLET RADIATION AT 260 and 280 nm

Pure samples of DNA, RNA, and nucleotides absorb light in the ultraviolet region because of the presence of purine and pyrimidine bases. The concentration of the sample can be determined if the extinction coefficient is known. Double-stranded DNA has an $a_{1\,cm}^{1\%}$ at 260 nm equal to 200. Single-stranded

DNA and RNA have $a_{1\,cm}^{1\%}$ at 260 nm of 250. Pure preparations of DNA and RNA have $A_{260/280}$ values of 1.8 and 2.0 respectively. If protein or phenol is present, the absorbance at 280 nm of either substance will interfere with the accurate quantitation of the nucleic acid present. Concentrations of pure nucleic acids as low as 5 μg/mL can be accurately measured with this method.

7.3.2.2 ETHIDIUM BROMIDE FLUORESCENCE OF DNA

A rapid semiquantitative method for estimating the amount of DNA in samples containing contaminating substances is to utilize the UV-induced fluorescence emitted by ethidium bromide molecules intercalated into DNA. The quantity of DNA can be estimated by comparing the intensity of the sample fluorescence with that of a series of standards. The assay is sensitive for concentrations of 0.50 to 20 μg/mL.

A simplified procedure involves spotting equal volumes (1–5 μL) of DNA standards on a sheet of transparent plastic wrap (Saran Wrap) stretched over a UV light source. An equal volume of buffer (10 mM Tris, 1 mM EDTA, pH 7.2, containing 2 μg/mL ethidium bromide) is added to each standard and sample. The concentration of the unknown sample is estimated by comparison to the intensity of fluorescence of the standards when irridated with short-wavelength (254 nm) UV light.[27]

7.3.2.3 THE DETERMINATION OF DNA WITH DIPHENYLAMINE

Treatment of DNA with strong, hot acid causes the depurination of DNA, cleavage of the sugar phosphate backbone, and dehydration of 2-deoxy-D-ribose to ω-hydroxylevulinyl aldehyde. The aldehyde condenses with diphenylamine to yield products with an absorption maximum of 595 nm (Section 7.1.1.4).

THE DIPHENYLAMINE PROCEDURE FOR DNA

A DNA-containing sample is hydrolyzed in 10% acetic acid by heating at 95°C for 10 minutes. One part of the sample is diluted with 2 parts water; 5 parts of diphenylamine reagent (1 g of diphenylamine dissolved in 100 mL of glacial acetic acid and 2.5 mL of concentrated sulfuric acid) is added, and the solution is heated in a boiling water bath for 10 minutes. The absorbance of the cooled sample at 595 nm is measured and compared with the absorbance of standards of 2-deoxy-D-ribose in the range of 10–200 μg/mL.

7.3.2.4 THE DETERMINATION OF RNA WITH ORCINOL

The colored complex formed by D-ribose from RNA with orcinol has a maximum absorbance at 665 nm and reaches a maximum color after 15 minutes.

ORCINOL DETERMINATION OF RNA

The sample (100 μL) is added to 2 mL of orcinol reagent (0.30 g of orcinol in 100 mL of concentrated HCl to which 4 drops of a 10% solution of ferric chloride are added) and heated for 10 minutes in a boiling water bath. The sample is cooled and the absorbance is measured at 665 nm. Standards of RNA are prepared in the range of 10–200 μg/mL.

7.3.2.5 THE DETERMINATION OF NUCLEOTIDES, NUCLEOSIDES, AND BASES BY THE EFFECT OF PH ON THEIR ULTRAVIOLET SPECTRA

The ultraviolet absorbance of nucleic acids, nucleotides, and nucleosides is due to the pyrimidine and purine bases these compounds contain. The bases, nucleosides, and nucleotides, however, all have unique spectra, which are sensitive to pH. Therefore, absorbance ratios at different pH values are used to identify nucleotides and nucleosides, and they can be quantitated by the use of the corresponding extinction coefficient at the wavelength of maximum absorbance. Table 7-6 contains the spectral properties for the commonly occurring bases, nucleosides, and nucleotides. Individual nucleotides from RNA hydrolysis have distinct values of $A_{280/260}$ at pH 7: 5'-AMP = 0.16, 5'-CMP = 0.98, 5'-GMP = 0.66, 5'-UMP = 0.39. Using these absorbance ratios, one can identify the individual ribonucleotides after chromatographic or electrophoretic separation.

7.4 LIPIDS AND STEROIDS

7.4.1 Qualitative Tests for Lipids on Thin-Layer Chromatograms (also see Section 4.14)

7.4.1.1 DETECTION OF PHOSPHOLIPIDS ON TLC

Phosphate-containing compounds in the presence of molybdenum, molybdenum oxide, and strong acid yield colored complexes.

SPRAY REAGENT FOR PHOSPHOLIPIDS

Reagent 1: 8 g of MoO_3 is added to 200 mL of 25 N sulfuric acid, which is heated until the MoO_3 is dissolved.

Reagent 2: 0.39 g of powdered molybdenum is added to 100 mL of reagent 1 and the mixture is boiled gently for 15 minutes.

The spray reagent is prepared by mixing equal volumes of reagents 1 and 2 with 2 volumes of water. TLC plates are lightly sprayed. Phosphate-containing compounds appear immediately as blue spots on a gray background.[28]

TABLE 7-6.   Spectral properties of bases, nucleosides, and nucleotides

Compound	pH	λ_{max} (nm)	a_M ($\times 10^{-3}$)	Spectral ratios A_{250}/A_{260}	A_{280}/A_{260}	A_{290}/A_{260}
Adenine	1	262.5	13.2	0.76	0.38	0.04
	7	260.5	13.4	0.76	0.13	0.01
	12	269	12.3	0.57	0.60	0.03
Guanine	1	248	11.4	1.37	0.84	0.50
		276	7.35			
	7	246	10.7	1.42	1.04	0.54
		276	8.15			
	11	274	8.0	0.99	1.14	0.59
Cytosine	1	276	10.0	0.48	1.53	0.78
	7	267	6.1	0.78	0.58	0.08
	14	282	7.9	0.60	3.28	2.6
Thymine	4	264.5	7.9	0.67	0.53	0.09
	7	264.5	7.9	0.67	0.53	0.09
	12	291	5.4	0.65	1.31	1.41
Uracil	0	260	7.8	0.80	0.30	0.05
	7	259.5	8.2	0.84	0.17	0.01
	12	284	6.2	0.71	1.40	1.27
Adenosine[a]	1	257	14.6	0.84	0.22	0.03
	6	260	14.9	0.78	0.14	0.00
	11	259	15.4	0.79	0.15	—
Guanosine[a]	0.7	256	12.3	0.94	0.70	0.50
	6	253	13.6	1.15	0.67	0.28
	11.3	256–266	11.3	0.89	0.61	0.13
Cytidine[a]	1	280	13.4	0.45	2.10	1.55
	7	229.5	8.3	0.86	0.93	0.28
		271	9.1			
	14	273	9.2	—	—	—
Thymidine[a]	1	267	9.65	0.65	0.72	0.24
	7	267	9.65	0.65	0.72	0.24
	13	267	7.4	0.75	0.67	0.16
Uridine[a]	1	262	10.1	0.74	0.35	0.03
	7	262	10.1	0.74	0.35	0.03
	12	262	7.45	0.83	0.29	0.02
	14	264.5	7.5			
Adenosine-5'-	2	257	14.7	0.85	0.22	—
triphosphate	7	259	15.4	0.80	0.15	—
(ATP)	11	259	15.4	0.80	0.15	—

[a] Absorbance values of the deoxyribonucleosides and ribonucleosides and the deoxyribonucleotides and ribonucleotides are essentially the same.

7.4.1.2 DETECTION OF GLYCOLIPIDS ON TLC

In the presence of strong sulfuric acid, the carbohydrate moiety of glycolipids reacts with orcinol to yield a colored complex.

THE DETECTION OF GLYCOLIPIDS

Orcinol (200 mg) is dissolved in 100 mL of sulfuric acid–water (3:1) and stored in the dark. Plates are sprayed until moist, then heated at 100°C for 15 minutes. Glycolipids give violet spots on a white background.[29]

7.4.1.3 DETECTION OF SPHINGOLIPIDS ON TLC

Sphingolipids can be detected on TLC by the reaction of their amide groups. The amide group is first reacted with sodium hypochlorite to convert it into a chloro-substituted group that reacts with benzidene to give a blue product.

THE DETECTION OF SPHINGOLIPIDS

Reagent 1: 5 mL of Clorox, 50 mL of benzene, and 5 mL of glacial acetic acid are mixed together.

Reagent 2: 0.5 g of benzidene and one small crystal of potassium iodide are dissolved in 50 mL of 50% ethanol and the solution is filtered.

The TLC is sprayed with reagent 1 and air-dried in the hood to remove unbound chlorine gas. The TLC is then sprayed with reagent 2. Sphingolipids appear blue against a white background.[30]

7.4.2 Qualitative Tests for Steroids

7.4.2.1 DETECTION OF CHOLESTEROL AND CHOLESTEROL ESTERS ON TLC

The spray reagent consists of 50 mg of $FeCl_3 \cdot 6H_2O$ dissolved in 90 mL of water and 5 mL each of glacial acetic acid and concentrated sulfuric acid. The TLC is sprayed lightly and heated at $100°C$ for 2–3 minutes. Cholesterol and its esters appear red to violet and can be detected as low as 1 $\mu g/cm^2$. Under UV light, the compounds are detected by fluorescence with a 10-fold increase in the sensitivity.

7.4.2.2 THE LIEBERMANN–BURCHARD REACTION FOR THE DETECTION OF STEROIDS

Acetic anhydride can react with the C_3 hydroxyl group of cholesterol and related steroids in the presence of strong acids to form a blue-green complex. The reaction must be carried out in the absence of water.

THE LIEBERMANN–BURCHARD TEST

The test is performed by adding 15 drops of acetic anhydride to 1 mL of a solution of steroid in chloroform with gentle mixing. Cautiously, 1 mL of concentrated sulfuric acid is added down the side of the test tube without mixing. The appearance of a blue-green color is a positive test for cholesterol. Ergosterol quickly develops a red color.

The test can also be used quantitatively for the determination of cholesterol by measuring the absorbance at 625 nm and comparing it with a standard curve prepared with various concentrations of cholesterol.

7.4.3 The Quantitative Determination of Lipids[32]

7.4.3.1 THE DETERMINATION OF PHOSPHOLIPIDS BASED ON PHOSPHORUS

Phospholipid-containing spots can be eluted quantitatively from TLC plates and analyzed for phosphorus content. In one method, a drop of water is added to the silica gel area to be removed and the silica gel is transferred to a centrifuge tube. The phospholipids are extracted into chloroform–methanol–acetic acid–water (25:15:4:2). The solvent is removed after centrifugation, the extraction of the silica gel is repeated twice, and all the extracts are combined and the phosphorus assayed (see Section 7.5).

7.4.3.2 THE DETERMINATION OF CHOLESTEROL BY CHOLESTEROL OXIDASE

Cholesterol is oxidized to cholestenone and hydrogen peroxide by cholesterol oxidase. In the presence of peroxidase, the H_2O_2 reacts with 2-amino-antipyrine and phenol to form a colored complex that can be quantitated by measuring the absorbance at 500 nm.

$$\text{cholesterol} + O_2 \xrightarrow[\text{oxidase}]{\text{cholesterol}} \text{cholestenone} + H_2O_2$$

$$H_2O_2 + \text{phenol} + \text{4-amino-antipyrine} \xrightarrow[\text{peroxidase}]{} \text{colored complex} + H_2O$$

Since only free cholesterol is a substrate for cholesterol oxidase, the determination of total cholesterol (free cholesterol + cholesterol esters) requires that the sample first be treated with cholesterol esterase, which catalyzes the hydrolysis of cholesterol esters to free cholesterol.

CHOLESTEROL BY CHOLESTEROL OXIDASE

The reagent consists of 474 mg of sodium cholate, 19 mg of 4-amino-antipyrine, 170 mg of phenol, 110 mg of polyethylene glycol, and 12.5 mg of horseradish peroxidase (Sigma, type II) dissolved in 100 mL of 60 mM phosphate buffer, pH 7.4. Just before use, 7 IU of cholesterol esterase (Boehringer–Manheim) and 12 IU of cholesterol oxidase (Boehringer–Manheim) are added. The reagent is stable for several days at 4°C.

The procedure involves the addition of 0.10 mL of sample to 2.0 mL of cholesterol oxidase reagent and 2.0 mL of water. After careful and thorough mixing, the solution is incubated at 37°C for 25 minutes. The sample is cooled and the absorbance is measured at 500 nm. A standard curve is prepared using cholesterol in the range of 0.2–2.0 mg/mL.

7.5 MISCELLANEOUS

7.5.1 The Determination of Inorganic Phosphate: Modified Fiske–Subbarow Method

The sample volume is adjusted to 3 mL with water and the following are added in order: 1.0 mL of acid molybdate reagent, 1.0 mL of reducing reagent, and 5.0 mL of water. The solution is thoroughly mixed and allowed to stand for 20 minutes before measuring the absorbance at 660 nm.

The acid molybdate reagent is prepared by adding 13.6 mL of concentrated sulfuric acid to 36 mL of water; 2.5 g of $(NH_4)Mo_7O_{24}\cdot4H_2O$ is dissolved in 50 mL of water and added to the sulfuric acid solution. The final volume is diluted to 100 mL with water.

The reducing reagent is prepared by dissolving 15 g of sodium bisulfite $(NaHSO_3)$ and 5 g of p-methylaminophenol (Elon) in 500 mL of water and stored in a brown bottle. The reagent is stable for 10 days.

Phosphate is often found covalently combined with organic compounds. The phosphate can be released as inorganic phosphate by acid (1 M HCl) or base (1 M NaOH) hydrolysis at 95°C. Phosphomonoesters and diesters of alkyl alcohols, however, are relatively stable and require longer hydrolysis times and stronger acid.

ACID HYDROLYSIS OF COMBINED PHOSPHATE

Samples are treated with an equal volume of 2 M HCl and placed in a boiling water bath for 7 minutes to release labile phosphate (e.g., the phosphate from glucose-1-phosphate). After cooling, the pH is adjusted to 7 with 2 M NaOH, and the inorganic phosphate is determined as described above.

7.5.2 Specific Binding Assays

Specific binding assays can be used for the determination of hormones, vitamins, and other important biochemical molecules that are difficult to assay by their biological action and are normally present in cells or in body fluids at very low concentrations. The methods of assay are based on the work of Yallow and Berson.[33]

To determine the quantity of a compound in a solution, a known amount of the radiolabeled compound is added and the mixture is incubated with a limiting amount of binding substance, such as an antibody, that will specifically bind to the compound being assayed. The bound and the free fractions, containing the compound, are separated and the radioactivity of one or both of the fractions is measured (see Section 6.1.4.6 for the design of the assay). The quantity of the substance is calculated from a standard binding curve, prepared with a series of known amounts of the compound (Fig. 6-20).

7.5.2.1 THE DETERMINATION OF CYCLIC AMP

Cyclic AMP (c-AMP) and tritiated c-AMP compete for binding to c-AMP-binding protein. With increasing amounts of c-AMP, the radiolabeled phosphate is displaced from the binding protein and goes into solution. The free c-AMP in solution is adsorbed onto dextran-coated charcoal. The charcoal-bound c-AMP is removed by centrifugation, leaving the protein-bound c-AMP in the supernatant. An aliquot of the supernatant is counted in a liquid scintillation spectrometer. The amount of c-AMP in the sample is determined by comparing the amount of radioactivity to a standard curve that has been prepared with different concentrations of c-AMP in the range of 0.10–27 pmol.

SPECIFIC BINDING ASSAY FOR CYCLIC AMP

c-AMP-binding protein is obtained from Diagnostic Products Corp., Los Angeles. The dextran-coated charcoal is prepared using washed charcoal (Norit A, Sigma Chemical Co.) and dextran T70 (Sigma). The charcoal is washed by suspending 100 g in 1 L of distilled water, stirring for 5 minutes, allowing the charcoal to settle for 45 minutes, and pouring the supernatant and the nonsedimented charcoal away; the washing is repeated two more times and the washed charcoal is dried at 100°C. The dextran-coated charcoal is prepared by suspending washed charcoal (3.12% w/v) and dextran T70 (0.62% w/v) in a gelatin–phosphosaline buffer (pH 7.2) containing 0.1% w/v gelatin, 10 mM phosphate, and 150 mM sodium chloride. The suspension is stirred for 1 hour and stored overnight at 4–10°C. The dextran-coated charcoal is recovered by centrifugation.[34]

Pipet 100 μL of tritiated c-AMP (containing ~10 nCi) into a series of tubes. Add 100 μL of a sample of unknown c-AMP to two or three tubes and a series of five or six known concentrations of c-AMP in duplicate or triplicate to the remaining tubes. Add to each tube 100 μL of c-AMP-binding protein (containing an amount of protein that will bind approximately 60% of the radiolabeled c-AMP in the absence of nonlabeled c-AMP), mix thoroughly, and incubate for 90 minutes at 0°C. To each tube, add 0.5 mL of an ice-cold suspension of dextran-coated charcoal (containing an amount of dextran–charcoal that will bind all the labeled–unlabeled c-AMP in the absence of c-AMP-binding protein), mix, and incubate for 10 minutes at 0°C. Centrifuge at 5000 rpm until the charcoal is firmly packed at the bottom of the tube. Decant the supernatant into scintillation vials, effecting complete transfer of the supernatant by tapping the tube lightly or using a micropipet, but *being careful not to transfer any of the charcoal*. Add 10 mL of dioxane-based scintillation cocktail (Table 6-4) and count the samples in a liquid scintillation spectrometer. For the tubes containing the standards, calculate the percent of radioactivity bound to the c-AMP-binding protein and plot percent bound versus the amount of c-AMP (picomoles) present in the standards. Determine the amount of c-AMP in the unknown sample from the standard curve.

7.5.2.2. THE DETERMINATION OF INSULIN BY THE DOUBLE-ANTIBODY METHOD

In the assay, a limited amount of insulin-specific antibody is added to a mixture of a known amount of ^{125}I-labeled insulin and insulin from the sample. Both labeled and unlabeled insulin are bound to the antibody. The amount of radioactivity bound to the antibody is inversely proportional to the amount of unlabeled insulin in the sample. The insulin–antibody complex is removed from the solution by adding a second antibody that is specific for the insulin–antibody complex. The double insulin–antibody complex precipitates from solution and is removed by centrifugation. The supernatant, containing unbound nonlabeled insulin and ^{125}I-insulin, is decanted and counted by liquid scintillation spectrometry.

THE DETERMINATION OF INSULIN BY SPECIFIC BINDING ASSAY

Reagent A, buffer: pH 7.5, 10 mM phosphate, 10 mM EDTA, 150 mM NaCl, and 1% normal guinea pig serum (Miles Scientific, Naperville, IL).

Reagent B, anti-insulin: guinea pig anti-insulin antiserum (antibody 1), titered to bind approximately 50% of the ^{125}I-insulin in the absence of nonlabeled insulin.

Reagent C, precipitating antiserum (antibody 2): goat anti-guinea pig γ-globulins (Miles Scientific) in 10 mM phosphate, 150 mM NaCl (pH 7.5).

Reagent D, insulin standards: containing 0.1 μg/mL to 8 μg/mL (2.5–200 μU/mL).

Reagent E, ^{125}I-insulin: approximately 2 μCi of ^{125}I is required for 100 samples, including standards.

For each assay, the following mixture is prepared in duplicate or triplicate:

> 0.20 mL standard or unknown
> 0.40 mL anti-insulin (antibody 1)
> 0.40 mL ^{125}I-insulin

The solution is thoroughly mixed with a vortex mixer and incubated at 37°C for 1 hour; 0.10 mL of insulin–antibody antiserum (antibody 2) is added to all tubes and mixed. This mixture is incubated at room temperature for 1 hour or more, and centrifuged at 2000 rpm for 2 minutes in a microcentrifuge. The supernatant is decanted into a new tube. The radioactivity in either the supernatant or the precipitate, or both, is counted in a gamma scintillation spectrometer.

Blanks should be prepared for the following:

Total counts blank: ^{125}I-insulin with no antibody present.

Total binding blank: ^{125}I-insulin + antibody 1 + antibody 2 (no unlabeled insulin present).

Nonspecific binding blank: ^{125}I-insulin + antibody 2 (no anti-insulin or antibody 1 present).

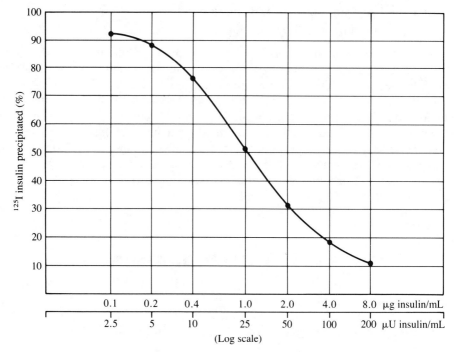

FIGURE 7-6. The standard binding curve for the radioimmunoassay of insulin. Percent [^{125}I]insulin in the double-antibody precipitate versus micrograms of insulin per milliliter or microunits of insulin per milliliter.

The data are expressed as % radioactivity bound in the double-antibody precipitate versus the concentration of the insulin standards. Figure 7-6 is an example of an insulin standard binding curve.

7.6 LITERATURE CITED

1. B. L. Oser, *Hawk's Physiological Chemistry, 14th ed.* McGraw-Hill, New York, 1965, p. 82.
2. (a) J. H. Roe, "A colorimetric method for the determination of fructose in blood and urine," *J. Biol. Chem. 107*: 15 (1934). (b) D. J. S. Gray, "Critical factors in the resorcinol reaction for the determination of fructose," *Analyst, 75*: 314 (1950).
3. (a) Z. Dische and K. Schwartz, "Mikromethode zur Bestimmung verschiedener Pentosen nebeneinander bei Gegenwart von Hexosen," *Mikrochim. Acta, 2*: 13 (1937). (b) W. Mejbaum, "Über die Bestimmung kleiner Pentosemengen, insbesondere in Derivaten der Adenylsäure," *Z. Physiol. Chem. 258*: 117 (1939). (c) Z. Dische, "Color reactions of pentoses," in *Methods in Carbohydrate Chemistry*, Vol. I, (R. L. Whistler and M. L. Wolfrom, Eds.) Academic Press, New York, 1962, pp. 484–488.
4. K. Burton, "A study of the conditions and mechanism of the diphenylamine reaction for the colorimetric estimation of deoxyribonucleic acid," *Biochem. J. 62*: 315 (1956).

5. Z. Dische, "Color reactions of hexosamines," in *Methods in Carbohydrate Chemistry*, Vol. I, (R. L. Whistler and M. L. Wolfrom, Eds.) Academic Press, New York, 1962, pp. 507–512.

6. S. R. Benedict, "The detection and estimation of glucose in urine," *J. Am. Med. Assoc. 57*: 1193 (1911).

7. M. Dubois, K. A. Gilles, J. K. Hamilton, P. A. Rebers, and F. Smith, "Colorimetric method for determination of sugars and related substances," *Anal. Chem. 28*: 350 (1956).

8. (a) N. Nelson, "A photometric adaptation of the Somogyi method for the determination of glucose," *J. Biol. Chem. 153*: 375 (1944). (b) Z. Dische, "Color reactions based on the reducing properties of sugars," in *Methods in Carbohydrate Chemistry*, Vol. I, (R. L. Whistler and M. L. Wolfrom, Eds.) Academic Press, New York, 1962, pp. 512–514.

9. (a) W. S. Hoffmann, "A rapid photoelectric method for the determination of glucose in blood and urine," *J. Biol. Chem. 120*: 51 (1937). (b) Z. Dische, "Color reactions based on the reducing properties of sugars," in *Methods in Carbohydrate Chemistry*, Vol. I, R. L. Whistler and M. L. Wolfrom, Eds., p. 513, Academic Press, New York, 1962.

10. J. T. Park and M. J. Johnson, "A submicrodetermination of glucose," *J. Biol. Chem. 181*: 149 (1949).

11. I. D. Fleming and H. F. Pegler, "The determination of glucose in the presence of maltose and isomaltose by a stable, specific enzymic reagent," *Analyst, 88*: 967 (1963).

12. E. von Arx and R. Neher, "Eine multidimensionale Technik zur chromatographischen Identifizierung von Aminosäuren," *J. Chromatogr. 12*: 329 (1963).

13. A. Chernoff and J. Liu, "Amino acid composition of hemoglobin. 2. Analytical technics," *Blood, 17*: 56 (1961).

14. I. Smith, Chromatographic and Electrophoretic Techniques, vol. 1, p. 97. W. Heinemann, Medical Books Ltd., London, and Interscience Publishers, Inc., New York, 1960.

15. R. J. F. Nivard and G. I. Tesser, in *Comprehensive Biochemistry* (M. Florkin and E. H. Stotz, eds.), vol. 6, pp. 199–200, Elsevier Publ. Co., Amsterdam, 1965.

16. S. Undenfriend, S. Stein, P. Bohlen, W. Dairman, W. Leimgruber, and M. Weigell, "Fluorescamine: a reagent for assay of amino acids, peptides, proteins, and primary amines in the picomole range," *Science, 178*: 871 (1972).

17. O. Warburg and W. Christian, "Isolierung und Kristallisation des Gärungsferments Enolase," *Biochem. Z. 310*: 384 (1941).

18. B. L. Oser, "Hawk's Physiological Chemistry," 14th ed., pp. 179–181, McGraw-Hill, New York, 1965.

19. O. H. Lowry, N. J. Rosebrough, A. L. Farr, and R. J. Randall, "Protein measurement with the Folin phenol reagent," *J. Biol. Chem. 193*: 265 (1951).

20. M. M. Bradford, "A rapid and sensitive method for the quantitation of microgram quantities of protein utilizing the principle of protein-dye binding," *Anal. Biochem. 72*: 248 (1976).

21. G. L. Ellman, "Tissue sulfhydryl groups," *Arch. Biochem. Biophys. 82*: 70 (1959).

22. J. F. Robyt, R. J. Ackerman, and C. G. Chittenden, "Reaction of protein disulfide groups with Ellman's reagent: a case study of the number of sulfhydryl and disulfide groups in *Aspergillus oryzae* α-amylase, papain, and lysozyme," *Arch. Biochem. Biophys. 147*: 262 (1971).

23. R. J. Ackerman and J. F. Robyt, "Favorable equilibria for the stoichiometry in the reaction of protein disulfide groups with Ellman reagent," *Anal. Biochem. 50*: 656 (1972).

24. T. W. Tannhauser, Y. Konishi, and H. A. Scheraga, "Sensitive quantitative analysis of disulfide bonds in polypeptides and proteins," *Anal. Biochem. 138*: 181 (1984).

25. E. Southern, "Detection of specific sequences among DNA fragments separated by gel electrophoresis," *J. Mol. Biol. 98*: 503 (1975).

26. S. Berent, M. Mahmoudi, R. M. Torczynski, P. W. Bragg, and A. P. Bollon, "Comparison of oligonucleotide and long DNA fragments as probes in DNA and RNA dot, Southern, northern, colony and plaque hybridizations," *BioTechniques, 3*: 208 (1985).

27. T. Maniatis, E. Fritsch, and J. Sambrook, *Molecular Cloning*, Cold Spring Harbor, 1982.

28. J. Dittner and R. Lester, "A simple specific spray for the detection of phospholipids on thin-layer chromatograms," *J. Lipid Res. 5*: 126 (1964).

29. O. Renkonen and A. Lunkkonen, "Thin-layer chromatography of phospholipids and glycolipids," in *Lipid Chromatographic Analysis*, Vol. 1, 2nd ed., G. Marineti, Ed. Dekker, New York, 1976, p. 1.

30. M. Bischel and J. Austin, "A modified benzidine method for the chromatographic detection of sphingolipids and acid polysaccharides," *Biochim. Biophys. Acta, 70*: 598 (1963).

31. L. Abell, "A simplified method for the estimation of total cholesterol in serum and demonstration of its specificity," *J. Biol. Chem. 195*: 357 (1952).

32. V. Skepski, R. Peterson, and M. Barclay, "Quantitative analysis of phospholipids by thin-layer chromatography," *Biochem. J. 90*: 374 (1964).

33. R. S. Yallow and S. A. Berson, "Immunoassay of endogenous plasma insulin in man," *J. Clin. Invest. 39*: 1157 (1960).

34. W. T. Schrader and B. W. O'Malley, *Laboratory Methods Manual for Hormone Action and Molecular Endocrinology*, 4th ed. Department of Cell Biology, Baylor College of Medicine, Texas Medical Center, Houston, (1980).

7.7 REFERENCES FOR FURTHER STUDY

See Appendix A for a summary of references on analytical methods for biological molecules.

CHAPTER 8

BIOLOGICAL PREPARATIONS

Biological preparations involve the isolation of a specific cell organelle or component such as a membrane, mitochondrion, chloroplast, ribosome, or Golgi apparatus, or the isolation and purification of one of the four major types of biological molecules: protein, carbohydrate, nucleic acid, and lipid. Special laboratory techniques of cell lysis, tissue homogenization, filtration, centrifugation, chromatography, salt or organic solvent precipitation, and concentration are employed. This chapter discusses these general methods and describes specific preparations to illustrate the different cell components and biochemicals that can be isolated and purified, as well as the various procedures that are used.

8.1 THE STARTING MATERIAL

The biochemist has a wide variety of starting material from microorganisms, animal tissues, and plant tissues. The choice depends on the goals of the study. For example, a certain enzyme with a specific type of action might be produced by a certain species of bacteria. The bacteria would be grown on a relatively large scale to obtain large quantities of enzyme (see Section 8.3.6 on growth of bacteria). For an extracellular enzyme secreted into the culture fluid, a cell-free culture supernatant would be required; conversely, for an intracellular enzyme, the cells themselves would be needed. In either case, it would be necessary to separate the cells by centrifugation or filtration.

The component or molecule desired might be from some specific part of an animal or plant. For example, the study might involve rabbit skeletal muscle, rat liver, bovine or porcine heart, pancreas, brain, or human blood or saliva, or the study might involve leaves, stems, flowers, roots, tubers, or seeds of specific plants. The species and the type of organ determine the kind of procedure used.

8.2 CELL LYSIS AND EXTRACTION

There are as many methods of cell lysis as there are different cell types. Animal cells generally are easy to break, whereas plant cells and bacteria are more difficult to break because of the tough cellulosic or murein cell wall. Some methods are more vigorous than others. The vigorous methods are used to obtain soluble cellular components, and the gentle methods are used to obtain cellular organelles.

The vigorous methods include the French press, in which bacteria or plant cells are forced through a small hole under very high pressure. Shearing forces disrupt the cell wall. The method is usually not practical for large amounts of material. Another vigorous method is the use of ultrasound, which produces disruption of the cell wall by shearing and cavitation. Cells also may be vigorously disrupted in a bead mill, where the cell wall is ripped from the cell when the material undergoes rapid vibration with glass beads.

Moderate methods involve homogenization with a blender. This is appropriate for most animal tissues, most plant tissues, and many bacterial species. Another moderate method, which involves grinding in a homogenizer with an abrasive such as sand or alumina, is especially used for plant tissues or bacteria. The addition of fine glass particles (100–200 μm) to the specially designed blender will enhance the breakage, especially of bacteria and bacterial or fungal spores.

Among the gentle methods are lysis by osmotic disruption of the cell membrane, lysozyme treatment of bacteria (Gram-positive bacteria are more susceptible than Gram-negative bacteria), and chemical solubilization with an organic solvent such as toluene. Animal tissues can be gently disrupted with a hand- or motor-driven homogenizer. Several types of homogenizers are available with essentially the same basic features shown in Figure 8-1. A pestle having a sintered glass or Teflon surface fits closely into a tube, which usually has a sintered glass inner surface. The tissue is placed in the tube and is homogenized by vertical strokes and rotation of the pestle.

When making a cell extract, the tissue or cells are first suspended in a suitable buffer. Plant tissues frequently pose a special problem because many plants contain phenolic compounds that are readily oxidized by endogenous phenol oxidases to form dark pigments. These pigments can adsorb proteins and react to decrease biological activity. The oxidation can be decreased or eliminated by the addition of a reducing agent such as 20–30 mM β-mercaptoethanol or 2–5 mM 1,4-dithiothreitol or by the addition of polyvinylpyrrolidone to adsorb the phenolic compounds.

Bacteria can be disrupted by vigorous treatments with sonication or the French press. These methods are useful on a small scale but usually are not convenient for big operations. For large-scale preparations the gentle method of treatment with lysozyme will generally lyse Gram-positive bacteria. The bacteria are suspended in a buffer (pH 7–8) containing 0.1–0.2 mg of lysozyme per milliliter of cells (10^8–10^{10} cells/mL) and incubated at 37°C for 15–30 minutes.

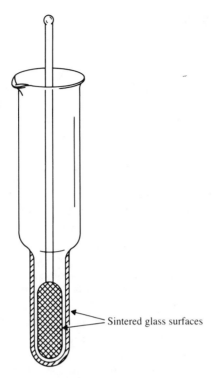

FIGURE 8-1. Potter Elvehjem-style tissue homogenizer in which tissue is homogenized by vertical strokes and rotation of the pestle. Grinding can be performed manually, or a motor-driven device can be used.

If DNA is not the component to be isolated, deoxyribonuclease can be added, or streptomycin sulfate or protamine sulfate can be added (1 mL of 10% solution per A_{260} of 1500) to remove the DNA, improving the quality of the extract by decreasing the viscosity. Gram-negative bacteria can generally be made susceptible to lysozyme action by combining a 0.1% nonionic detergent such as Triton X-100 or Tween 80 with the lysozyme or using an osmotic shock treatment in conjunction with the lysozyme. Relatively large-scale bacterial disruption can also be obtained with the combination of fine glass particles (100–200 μm) and a blender equipped with a Teflon-coated impeller. (The synthetic coating is absolutely essential to prevent the rapid disintegration of the impeller by the glass particles.)

Yeasts are conveniently lysed by treatment with toluene at 35–40°C for 1–3 hours, followed by the addition of buffer and stirring at 35–40°C for 3–4 hours. The toluene weakens the yeast cell wall by extraction of cell wall components that are soluble in organic solvents, allowing the cell wall to undergo autocatalytic degradation by yeast enzymes. After the cells have been lysed, a clear extract is obtained by centrifugation and filtration.

8.3 SPECIAL TECHNIQUES IN PREPARING BIOLOGICAL MATERIALS

8.3.1 Centrifugation

Centrifugation is the process of rapidly rotating a receptacle containing a slurry of solid particles in a fluid; the particles are sedimented by a greatly increased gravitational field. The centrifugal force or gravitational force is proportional to the square of the angular velocity (ω^2), expressed in revolutions per minute (rpm), times the radial distance r (cm) from the center of rotation. The gravitational force F is defined by:

$$F = (1.12 \times 10^{-5})(\omega^2)(r)$$

and for a rotor with a radius of 7 cm and an angular velocity of 10,000 rpm, the gravitational force is 8000 times the force of gravity or 8000g. Table 8-1 compares the g-forces that can be obtained for small, medium, and large centrifuge rotors moving at different angular velocities. Centrifuge manufacturers usually supply tables relating the speeds of their rotors to the force of gravity generated. It is better to express centrifugation in terms of the force of gravity rather than rpm because g-values provide uniform and reproducible centrifugation conditions for rotors of different types and sizes. Successful centrifugation also depends on the time of centrifugation as well as the number of g-forces developed. Table 8-2 compares the g-forces and the times of centrifugation required for sedimenting biological particles of various types. Thus, cell components can be separated by differential centrifugation using different g-forces for different periods of time.

Commercial high speed centrifuges can vary in size from micro devices taking 1-mL or smaller tubes to those taking 500-mL bottles. Many are refrigerated with variable temperature control to decrease the loss of biological activity of temperature-sensitive material. Different styles of rotors are available, such as angle rotors in which the tubes are placed at a fixed angle (e.g., 55° from the vertical) and swinging bucket rotors, primarily used for density gradient centrifugation, in which the tubes can swing out into a horizontal position during centrifugation.

Centrifugation of a relatively large volume requires a continuous flow centrifuge. An effective and reliable continuous flow centrifuge is the Sharples centrifuge, which is similar to a cream separator (Fig. 8-2). The rotor or bowl is a relatively long narrow cylinder that is turned by an electric motor or by forcing air or steam through a turbine. The slurry containing the solid particles is fed into the bottom of the rotating rotor and as the fluid flows upward, the particles are packed against the walls of the rotor; clear fluid flows out the top. An air-driven Sharples centrifuge can attain speeds of 50,000 rpm and a centrifugal force of 70,000g.

The theory of centrifugation was developed in the 1920s by Svedberg to determine the molecular weights of macromolecules. Ultracentrifuges were developed that could generate g-forces of 420,000 $\pm$ 100 with temperature control of ± 0.1°C. These analytical ultracentrifuges were equipped with optical detection systems that could determine the concentration distribution of a macromolecule at any time during a centrifugation run. The molecular weight

TABLE 8-1. Relationship of centrifuge rotor radius and angular velocity to the forces of gravity developed

Rotor radius, r (cm)	Angular velocity, ω (rpm)									
	1,000	3,000	6,000	10,000	20,000	25,000	30,000	50,000	70,000	100,000
3	35g	300g	1,200g	3,500g	14,000g	20,000g	30,000g	85,000g	160,000g	400,000g
7	100g	700g	3,000g	8,000g	32,000g	50,000g	70,000g	200,000g	400,000g	$1 \times 10^6\,g$
15	170g	1,500g	6,000g	17,000g	65,000g	100,000g	150,000g	412,000g	800,000g	—

TABLE 8-2. Comparison of the force of gravity and the time
necessary to sediment biological particles in a centrifugal field

Force	Time	Fraction sedimented
1,000g	5 min	Eukaryotic cells
4,000g	10 min	Chloroplasts, cell debris, cell nuclei
15,000g	20 min	Bacteria, mitochondria
30,000g	30 min	Lysosomes, bacterial cell debris
100,000g	3–10 h	Ribosomes

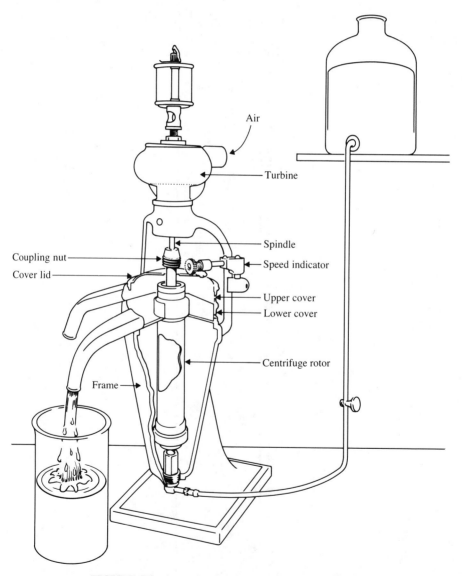

FIGURE 8-2. A continuous flow laboratory centrifuge.

of a macromolecule was found by measuring the sedimentation coefficient, the diffusion coefficient, and the partial specific volume. Because these measurements were tedious and laborious, and required very expensive instrumentation, the method has been generally replaced by the simpler procedures of polyacrylamide gel electrophoresis (Section 5.2.3.3) and gel permeation chromatography (Section 4.6).

8.3.2 Density Gradient Centrifugation

Density gradient centrifugation can be used as a means of separating macromolecules or cell components based on differences in their density. There are two basic processes: **velocity density gradient** and **equilibrium density gradient** centrifugation.

In the velocity method, a density gradient of sucrose or glycerol is prepared in a centrifuge tube by automatic pumps that mix concentrated sucrose or glycerol and water in a decreasing ratio as the tube is filled, giving a distribution of 60% w/v sucrose or glycerol in the bottom and 20% at the top. The mixture of components to be separated, dissolved or suspended in buffer, is added to the top of the gradient (Fig. 8-3A) and the tube is centrifuged in a horizontal position. This causes the components in the mixture to sediment down the density gradient at a rate determined by their density. The components of differing densities separate into bands or zones (Fig. 8-3B) that can be isolated by piercing the bottom of the centrifuge tube with a syringe needle and pumping the contents with a peristaltic pump into a fraction collector (Fig. 8-3C). Alternatively, the zones can be siphoned off manually with a syringe. The technique has been especially effective in isolating different cell components (nuclei, mitochondria, Golgi bodies, etc.) and different nucleic acids.

In the equilibrium method, a preformed gradient is not made. Instead, a 6 M solution of cesium chloride (CsCl) is prepared and the solution of macromolecules to be separated is dissolved in it. When the solution is placed in a high centrifugal field, the CsCl sediments; and at equilibrium, which can take several hours to attain, a stable gradient of CsCl is obtained. The mixture of macromolecules separates into zones according to the densities of the particles. The zones are formed when the density of the macromolecules equals the density of the gradient solution. The zones of macromolecules can be removed using the same techniques described above. The equilibrium method can give extremely high resolution and has been especially used in separating nucleic acids of different kinds (tRNA, mRNA, nuclear DNA, etc.).

8.3.3 Filtration

Filtration is sometimes used to remove solid material from a suspension. On a small scale, filtration is a good method for clarifying a solution, but the size and softness of many biological particles often leads to the rapid clogging of the filter.

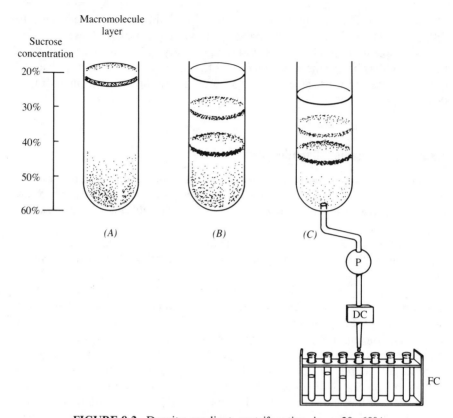

FIGURE 8-3. Density gradient centrifugation in a 20–60%
sucrose gradient. (*A*) Sample of the cellular components or
molecules to be separated layered on top of the gradient. (*B*)
Separation of components or molecules of differing densities
after centrifugation in a horizontal position. (*C*) Fractionation
of the gradient and the separated molecules by piercing the
bottom of the tube and removing the layers with a peristaltic
pump (P), counting drops with a drop counter (DC), and
collecting fractions in a fraction collector (FC).

Filtration rates can be improved by using filter aids such as Celite (a
commercial product of Johns Manville), which is a diatomaceous earth material
consisting mainly of silicon dioxide. Celite can be added to the suspension and
the slurry poured directly onto a filter in a Büchner funnel that is fitted into an
aspirator flask. Suction is applied by a water-aspirator or a vacuum pump. An
alternate method is to make a pad of Celite on the Büchner funnel by pouring a
slurry of the Celite onto the filter paper during aspiration, giving a pad 2–5 mm
thick and then carefully pouring the suspension to be filtered onto the filter pad.

Large-scale filtration has become possible with the development of the
Pellicon Cassette by Millipore (Fig. 8-4) in which the fluid of particles is continuously forced over a membrane. The liquid, under pressure, passes through
the membrane, giving a suspension of more highly concentrated particles. The

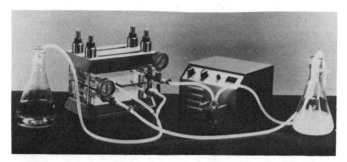

FIGURE 8-4. The Millipore Pellicon cassette for concentrating bacterial cells or solutions of large volumes. Courtesy of Millipore Corporation.

membrane has a relatively large surface area of $2 \, m^2$, giving flow rates of 1 L/min. The concentrated ($\sim 10X$) slurry can then be centrifuged in a conventional centrifuge. This technique has been particularly useful in obtaining bacteria or culture supernatant from large fermentations.

8.3.4 Concentration of Solutions

Biological solutions are frequently dilute and contain thermolabile molecules, which cannot be concentrated by heating the solution. Large volumes (10–100 L) of solution can be concentrated with the Pellicon cassette by using a membrane that does not allow the passage of the molecules of interest. Water from the solution is removed through the membrane, resulting in a concentrated solution of the molecules.

A more conventional ultrafiltration system, useful for smaller volumes (1 L–10 mL) is illustrated in Figure 8-5. The liquid in a stirred cell is forced through the membrane by pressure, leaving a concentrated solution in the cell.

Another mild method of concentration is the addition of a solid, insoluble material that has an affinity to absorb water. For example, when the highly crossed-linked Sephadexes and Bio-Gels (e.g., Sephadex G-15, Bio-Gel P6) are added to a water solution, these substances will absorb the water, leaving the biomolecules behind in a more concentrated volume. After hydration of the gel material, the slurry is filtered by aspiration, resulting in a filtrate that has one-third to one-half its original volume.

A much higher concentration (10- to 100-fold) can be obtained by filling a dialysis bag with the solution to be concentrated, placing it in a beaker or crystallizing dish, and covering it with powdered polyethylene glycol (MW > 20,000). The polyethylene glycol absorbs the water out of the dialysis bag, leaving a concentrated solution of the biomolecules in the bag. This simple, fast, gentle method can give highly concentrated solutions.

The ultimate concentration method is lyophilization or freeze-drying. The solution is quickly frozen onto the walls of a vessel in a relatively thin layer (~ 5 mm) by rapidly rolling the vessel in a dry ice–acetone bath ($\sim -70°C$).

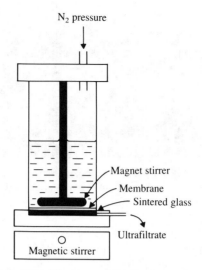

FIGURE 8-5. Ultrafiltration cell for the concentration of molecules in solution by passage through a membrane that is not permeable to the desired molecules.

The sample is judged to be completely frozen when it stops cracking. The vessel is then attached to a trap that is kept at $-70°C$ and a vacuum (≤ 100 μm) is applied (Fig. 8-6). The water in the frozen sample is sublimed and deposited in the trap. The samples are left undisturbed until all the ice has sublimed and the container, with the sample, is no longer cold. When sublimation is complete and the container has reached room temperature, air is slowly admitted to the

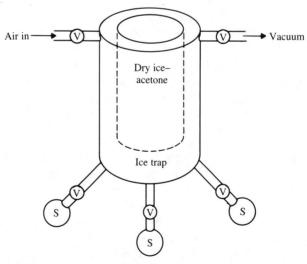

FIGURE 8-6. Freeze-drying apparatus (lyophilizer): V = valve and S = sample vessel.

system, the vacuum is turned off, and the vessels containing the solid, concentrated sample are removed. The solids still contain approximately 5–10% w/w of water. Many biological materials that rapidly deteriorate in solution or even in frozen solutions can be kept in an active or viable state for many years after lyophilization. Not all biological materials, however, are stable in lyophilized form. The stability of each type of material must be determined.

8.3.5 Dialysis

Low molecular weight materials, such as salts, and some biological materials, such as amino acids, coenzymes, and low molecular weight carbohydrates, can be removed by dialysis from macromolecular materials such as proteins, nucleic acids, and polysaccharides. **Dialysis** is the process of separating smaller molecules from larger ones in solution by use of a semipermeable membrane that permits passage of the smaller, but not the larger, molecules.

Cellophane dialysis tubing, which will not permit the passage of molecules having a molecular weight exceeding 12,000–14,000, is soaked in distilled water. The end of the dialysis tubing is closed with a tight knot and tied with a string, and the bag is filled with water to check for leaks. Enough tubing should be used to leave approximately half the bag void when the sample solution is added. (Relatively high salt solutions will take up water in the bag, and the expansion can cause the bag to explode.) After the sample solution has been poured in, the bag is closed just above the top of the solution, air is removed, and the bag is tightly knotted at the end and tied with string. Then the bag is placed in a container with 10 volumes or so of water or buffer. If necessary, the whole system is refrigerated.

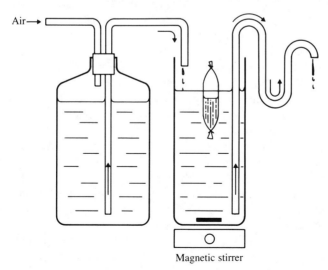

Magnetic stirrer

FIGURE 8-7. Continuous dialysis: the solution outside the dialysis bag is constantly being changed.

After 4–6 hours of dialysis, an equilibrium is established, at which point the concentration of the dialyzable material is the same outside and inside the dialysis bag. When the volume outside the bag is much larger than the volume inside, there has been a substantial reduction of the amount of dialyzable material inside the bag. For example, if the volume outside the bag is ten times that inside, the concentration of the dialyzable material in the bag at equilibrium will be one-tenth the original concentration. To reduce the concentration inside the bag further after equilibrium has been established, the outside solution can be changed and the cycle can be repeated several times, if necessary, to obtain a given concentration. It is more efficient, however, to use continuous dialysis, as illustrated in Figure 8-7.

Although dialysis is an often-used technique, the relatively long time needed to obtain equilibrium is a disadvantage. For this reason, dialysis is frequently replaced by gel filtration (Section 4.6) for the removal of low molecular weight materials from macromolecules.

8.3.6 Growth of Bacteria for the Production of Biochemical Compounds

Laboratory scaleup for the production of biochemicals from bacteria can be obtained on a scale of 1–100 L. A useful procedure involves the use of a 14-L fermenter (Fig. 8-8) or a 24-L carboy.

The bacteria are obtained from slant cultures or preserved frozen or lyophilized cultures and are transferred into 2 mL of liquid medium. Transfers are successively made using 5–10% inocula, so that for a 10-L fermentation, a 500 mL-1 L inoculum would be used.

Fermenter and medium are usually sterilized separately. The components of the medium are also frequently sterilized in separate concentrated solutions of approximately equal volume to prevent chemical interaction and modification of the components during sterilization. The carbohydrates are sterilized in 3 L, nitrogenous components of yeast extract and peptone in 3 L, and buffer and salts in 3 L. These sterilized solutions are added aseptically to the fermenter and inoculated. The fermenter is equipped with a stirrer, temperature control, and a metered valve for air or oxygen. Some fermenter designs include a pH probe and a means of controlling the pH.

The rate of aeration depends on the particular organism. A rule of thumb for most aerobic organisms is one liter of air per liter of culture per minute. Many anaerobes and facultative anaerobes can be conveniently grown in 24-L polycarbonate carboys by filling the vessels to 22 L with medium and inoculum and placing them in an incubator of the proper temperature. The organisms themselves will establish anaerobic conditions.

The fermentation is allowed to go on for 12–24 hours. A fermenter with 10 L of culture can be pumped out aseptically, leaving 1 L of inoculum for a successive fermentation by adding 9 L of fresh medium to the inoculum. As long

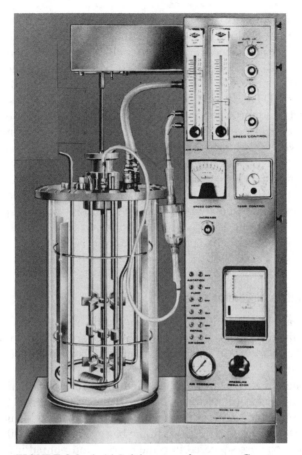

FIGURE 8-8. A 14-L laboratory fermenter. Courtesy of Virtis Corporation.

as sterile conditions are maintained, several fermentations can be performed consecutively with the same organism to obtain the desired amount of culture broth or cells required for a preparation.

In most fermentations, where aeration is used, an antifoam agent is necessary. Several types are available. For example, Dow-Corning Antifoam A will control foaming when used at $0.1-0.5$ g/10 L of culture.

8.4 SUBCELLULAR FRACTIONATION OF CELL COMPONENTS AND ORGANELLES

8.4.1 Preparation of Liver Mitochondria[1]

White quartz sand is sifted through a 500-mesh sieve and treated with concentrated HCl for 2 hours at $60°C$. The acid is removed by washing with tap water with stirring. The packed sand is mixed with three parts (w/v) of 20 mM

Na$_2$EDTA and heated for 30 minutes at 80°C (repeated two times). The sand is washed 6 times, by shaking for 30 minutes in 2 volumes of distilled water.

Fresh liver (i.e., obtained from a recently killed animal, or an organ that was frozen immediately after removal) is minced with a scissors and immediately placed into 10 volumes v/w of ice cold (4°C) isolation solution (0.33 M sucrose, 0.025 mM Na$_2$EDTA, 15 mM Tris–HCl, pH 7.4). The liver tissue (5–30 g) is gently ground in a mortar for 5 minutes with 12 times its weight of sand, which has been precooled to −20°C. The resulting pulp is filtered through glass wool and washed with 10 volumes of isolation solution. The filtrate is centrifuged at 800g for 10 minutes. The supernatant is carefully decanted and centrifuged at 8200g for 10 minutes. The supernatant is discarded and the pellet is suspended in approximately 10 mL of 0.33 M sucrose. A second centrifugation at 8200g for 10 minutes is immediately performed. The second pellet of purified mitochondria should contain 15 mg of protein per gram of liver. The mitochondria can be used to study oxidative phosphorylation.

8.4.2 Preparation of Chloroplasts[2]

Chloroplasts are isolated from the mesophyll cells of green leaves. Because of their availability from local markets, spinach leaves are a common source for the isolation. Other plant leaves, such as pea, beet, bush bean, pokeweed, and tomato can also be used.

Approximately 10 g of fresh leaves is cut into 10 cm × 10 cm squares and placed into a nylon bag (pore size ∼ 80 μm). The bag is transferred to a prechilled mortar containing the 10 mL of the isolation solution of 0.25 M sucrose, 10 mM Tris–HCl (pH 7.9) at 1°C. The leaves in the bag are ground for 10 seconds. The contents of the bag are squeezed into the mortar and poured into a prechilled 50-mL centrifuge tube that is centrifuged for 60 seconds at 1000g. The supernatant is decanted. The chloroplast pellet is resuspended by placing the centrifuge tube on a vortex mixer for 3–5 seconds. If the pellet is tightly packed, 2 mL of isolation solution can be added and the pellet resuspended by placing the tube on a vortex mixer.

A cleaner chloroplast preparation can be obtained by resuspending the pellet in isolation solution and centrifuging at 1000g for 60 seconds. The chloroplast can be used to study carbon assimilation and the associated evolution of oxygen, and photosynthetic electron transport.

8.4.3 Isolation of Liver Nuclei[3]

Liver is placed in chilled isolation solution (0.32 M sucrose and 3 mM magnesium chloride). The liver is minced with scissors, rinsed, and suspended in 3 volumes v/w of the isolation solution and 0.22 volume v/v of water; 0.8 volume of the isolation solution is carefully added with a fine long-tipped pipet to the diluted homogenate in a centrifuge tube and the contents centrifuged ($700g$) for 10 minutes.

The resulting pellet of crude nuclei is suspended in 12 mL of 2.4 M sucrose containing 1 mM magnesium chloride using 4–6 strokes of a homogenizer. The suspension is centrifuged at $50,000g$ for 1 hour. The material floating on the top after centrifugation contains whole cells, erythrocytes, and mitochondria. This debris is removed with a spatula, the dense sucrose solution is carefully siphoned off with a pipet, and the walls of the tube are wiped with Kimwipes. The pellet of nuclei is carefully suspended in 0.25 M sucrose containing 1 mM magnesium chloride (1 mL/g of pellet) using a vortex mixer for 3–5 seconds.

The procedure should give nuclei with the ability to synthesize RNA; they should have little ultrastructural damage and should be low in contamination with cytoplasmic material.

8.4.4 Preparation of *Escherichia Coli* Ribosomes

Escherichia coli cells (5 g wet weight) are weighed out into a chilled mortar, and 12.5 g of bacteriological-grade alumina (type 305, Sigma Chemical Co.) is added in three equal portions, with the thorough grinding of the cells with a pestle after each addition. Next 10 mL of Tris buffer (pH 7.4) containing 10 mM magnesium chloride and 5 U/mL deoxyribonuclease is added with mixing to suspend the paste. The suspension is transferred to centrifuge tubes and centrifuged at $10,000g$ for 15 minutes at 5°C to remove the alumina, cell debris, and unbroken cells.

The particle-free supernatant is transferred to preparative ultracentrifuge tubes and centrifuged at $105,000g$ for 2 hours. After centrifugation, the top 75–80% of the supernatant, containing soluble RNA and the components necessary for protein synthesis, is carefully removed with a pipet. The ribosomal pellet is resuspended carefully, with a glass rod, in 10 mL of 10 mM Tris–magnesium chloride buffer (pH 7.4) without deoxyribonuclease. Insoluble material is removed by centrifuging at $10,000g$ for 10 minutes. The latter pelleting step can be repeated several times to obtain a cleaner ribosome preparation. The yield of ribosomes can be determined by measuring the absorbance at 260 nm of a 1:100 dilution in Tris–Mg^{2+} buffer (pH 7.4), $a_{260\,nm}^{1\%} = 160$ for ribosomes.

8.5 PURIFICATION OF PROTEINS

A major activity of the biochemist is the purification of proteins. Enzymes are soluble, globular proteins that play a central role in the functioning of the living cell. They, therefore, are of great interest to the biochemist, and in purified form, they can also have important practical uses in medicine and in the food and pharmaceutical industries (see Chapter 9).

8.5.1 Separations Based on Solubility

8.5.1.1 ORGANIC SOLVENT PRECIPITATION

Water-soluble proteins, contained in an extracellular fluid or in a cell extract, can be precipitated by water-miscible organic solvents. Two commonly used organic solvents are ethanol and acetone. The addition of these organic solvents to an aqueous solution of proteins produces a reduction in the solvating power of water for the charged, hydrophilic protein molecule, leading to aggregation and precipitation of the protein. A feature affecting precipitation is the size of the protein. Other things being equal, the larger the protein molecule, the lower the percentage of organic solvent required to precipitate it. Precipitation also occurs at a lower organic solvent concentration around the isoelectric point (pH_I) of the protein. This supports the suggestion that protein precipitation (aggregation) by organic solvents is likely to be due to the uncovering and interaction of electrostatic and dipolar van der Waals forces contained in the protein structure.

When precipitating a protein with an organic solvent, the temperature should be maintained around $0°C$ to prevent denaturation due to the heat of mixing when organic solvent is added to water. The extract, containing the protein, and the organic solvent should be chilled before the solvent is added. The organic solvent should be added relatively slowly, with efficient stirring, to prevent the formation of locally high concentrations of organic solvent.

After the addition of the organic solvent, the precipitate is removed by centrifugation and decanting of the supernatant. Residual amounts of organic solvent can be removed from the precipitate by alternately pulling a vacuum and air over it several times until the solvent no longer can be detected. Acetone is a good precipitating solvent because of its volatility and easily detectable odor. The precipitate is then dissolved in a minimum amount of a suitable buffer.

The optimum amount of organic solvent needed to precipitate a protein is determined by adding varying amounts (5–60% v/v) of solvent to a series of aliquots of equal volumes of protein solution. The resulting precipitates are dissolved in an equal volume of buffer and various analyses are performed. With enzymes, two analyses are made: the concentration of the protein (Section 7.2.3) and the activity of the enzyme (Section 9.5). Several determinations are made from these analyses: the total amount of protein precipitated, the total enzyme activity, the yield of enzyme activity, and the *specific activity* of the enzyme (units of enzyme activity per milligram of protein). From these analyses, the amount of organic solvent needed to obtain an optimum yield and purification can be determined. Sometimes two precipitations give purification and maximum yield of the enzyme. For example, it might be found that 20% v/v of organic solvent precipitates 30% of the protein present, but only 5% of the enzyme activity, whereas 35% v/v solvent precipitates 50% of the protein and 75% of the enzyme activity. Thus the optimum purification would involve a 20% precipitation in which the precipitate is removed and discarded, followed by the addition of solvent to 35% to give a precipitate that contains 70% of the enzyme activity and only 20% of the protein present in the original solution. The double precipitation gives the maximum amount of enzyme, with an increased specific activity due to the removal of the inactive protein in the first 20% solvent precipitation. The specific activity is an index of the purification of the particular protein (enzyme) being studied. Increasing the solvent concentration above 35% in this example might have produced denaturation of the enzyme, hence loss of yield and lowering of the specific activity.

8.5.1.2 PROTEIN SALTING OUT

The salting-out technique of protein purification depends mainly on the hydrophobic character of the protein. Water solvates the added salt ions, decreasing the solvation of the protein molecule. This decrease exposes hydrophobic areas of the protein structure, which interact with each other to give aggregates that precipitate.

The procedure involves the dissolving of the salt into the solution containing the protein or the addition of a saturated solution of the salt, much like the addition of an organic solvent. The most commonly used salt is ammonium sulfate, which has a high water solubility. Saturated ammonium sulfate is 4 M or 528 g/L. Furthermore, this concentration is relatively temperature independent and varies very little between -5 and $30°C$. The optimum concentration of ammonium sulfate needed to precipitate a protein is determined using the same methods outlined for organic solvent precipitation. Appendix D contains a table for preparing different concentrations of ammonium sulfate.

A disadvantage of the salting-out technique is that relatively large amounts of salt remain with the precipitate and usually must be removed before the next step can be begun. The removal of the salt can be accomplished by dialysis (Section 8.3.5) or gel filtration (Section 4.6), but both these procedures require an additional step and usually leads to an increase in the volume of the protein solution.

8.5.1.3 SELECTIVE DENATURATION

Although proteins are generally considered to be fragile, extreme conditions can sometimes be used to selectively denature unwanted proteins. The objective of selective denaturation is the choice of a condition that will not significantly denature the protein (enzyme) of interest but will denature unwanted proteins.

The two most commonly used methods are extremes of heat and pH. For example, a crude enzyme extract may be heated at 55 or $60°C$ for a limited time. The desired enzyme is assayed before and after heating to determine how much of it has been lost and how much of the unwanted enzymes and proteins have been lost. Some enzymes initially lose their activity, but on standing at room temperature ($\sim 22–25°C$) or at refrigerator temperatures ($4–10°C$) regain nearly all or some of their original activity. Others lose their activity irreversibly. Determining specific activity before and after the heating gives an index of any purification that has occurred by removal of heat-denatured protein.

In the other method of selective denaturation, the pH is adjusted to some relatively unfavorable value (e.g., pH $2–3$ or $9–10$), and the enzyme solution is held there for some period of time; then the pH is adjusted back to the optimum range and the specific activity of the desired enzyme is determined.

Sometimes, a combination of temperature and pH treatment is used to selectively denature unwanted proteins in a crude extract.

8.5.2 Separation by Chromatography and Electrophoresis

As discussed in Chapter 4, proteins can be separated and purified by a variety of chromatographic methods: ion-exchange chromatography on diethylamino-ethyl–cellulose (DEAE–cellulose) or carboxymethyl–cellulose (Section 4.12.2), adsorption on hydroxyapatite (Section 4.12.1), gel filtration on Sephadexes or Bio-Gels of different pore sizes (Section 4.12.3), and affinity chromatography (Section 4.7).

Differences in the ionic properties of proteins can be used to effect their separation and purification by electrophoresis or isoelectric focusing (Chapter 5).

Chromatographic and electrophoretic methods frequently are used as terminal steps in a scheme of purification.

8.5.3 Crystallization of Proteins

Proteins are unique macromolecules in that many have been crystallized. For crystallization, a protein usually must be highly purified and present in relatively high concentrations (10–20 mg/mL). The crystallization process is often an arduous and fortuitous endeavor, as it is for most organic compounds. Procedures for the crystallization of two proteins are given below (Sections 8.5.4.1 and 8.5.4.4).

A new technique has been developed that greatly facilitates the crystallization process.[4] It has resulted in the crystallization of proteins that had not produced crystals by the conventional methods of purification and concentration and even has yielded crystals from a crude mixture. The procedure uses vapor diffusion in the presence of polyethylene glycol (PEG 1,000, 4,000, 6,000, and 20,000) at different concentrations between 2 to 20% w/v. PEG 20,000 is less effective than the others in producing crystals, although in the crystallization of canavalin, it was the only PEG that gave crystals. Therefore, the different sizes of PEG should all be tried as crystallizing agents.

The procedure involves the addition of various concentrations of the different PEGs to the bottom of 25 mL reservoirs. Small volumes (10–25 μL) of the protein solution are added to a container in the reservoir and an equal volume of the PEG solution from the bottom of the reservoir is added to the protein solution, giving a concentration of PEG that is half the concentration in the reservoir. The top of the reservoir is sealed with silicone grease and the reservoir is placed in the cold or at room temperature. The individual protein samples gradually lose water to the reservoir through vapor diffusion until equilibrium is attained and crystals are formed.

The individual methods of protein purification can be combined into a step-wise scheme. While there are many possible combinations of procedures, a general scheme is given.

GENERAL PURIFICATION SCHEME

1. Disruption and extraction of cells or separation of cells from a cell supernatant
2. Concentration of the extract or supernatant
   a. Ultrafiltration (Millipore Pellicon for large volumes)
   b. Organic solvent precipitation*
   c. Salt precipitation*
      i. Dialysis to remove excess salt
      ii. Gel filtration to remove excess salt
3. Chromatographic separation
   a. Ion-exchange chromatography
   b. Gel filtration chromatography
   c. Affinity chromatography
4. Concentration of active components of the chromatographic fractions
   a. Lyophilization
   b. Ultrafiltration in a stirred cell
   c. PEG-dialysis bag
5. Preparative electrophoresis or isoelectric focusing
6. Crystallization

* Both concentration and purification can be obtained with these procedures.

8.5.4 Isolation and Purification of Specific Proteins

8.5.4.1 CRYSTALLIZATION OF JACK BEAN UREASE

J. B. Sumner of Cornell University was the first to crystallize an enzyme—namely, urease;[5] and he subsequently demonstrated that urease is a globular protein. Sumner's amazingly simple purification procedure from jack bean meal can still be used to crystallize urease.

Place 100 g of dry, finely powdered jack bean meal (ICN Nutritional Biochemical Corp. Cleveland) into a 1-L beaker and slowly add, with stirring, 500 mL of 32% v/v aqueous acetone at 0°C. Stir the solution for 3–4 minutes to give a uniform suspension of the powder, and filter the suspension through a 32-cm fluted Whatman No. 1 filter paper into a 1-L cylinder. When 150 mL of filtrate has been collected, place the filtering system at 5°C (refrigerator or cold room) to allow the filtration to be completed over 15 hours. After 20–24 hours at 5°C, the urease should have crystallized from the solution. The crystals can be removed by centrifugation and dissolved in about 3 mL of distilled water; any insoluble material that remains should be removed by centrifugation.

To recrystallize the urease, add 0.03 volume of 0.5 M (pH 7) phosphate buffer containing 0.5% w/v Na_2EDTA to the clear solution. Add cold acetone (0.2 volume) dropwise with constant stirring, and place the solution at 5°C. Crystallization should begin immediately and should be complete in about an hour. The yield should be 20–30 mg of crystalline urease from 100 g of jack bean meal. The urease should be stable at 5°C over long periods when dissolved in 20 mM (pH 7) buffer containing 0.5% Na_2EDTA.

8.5.4.2 PURIFICATION OF YEAST ALCOHOL DEHYDROGENASE[6]

a. Lysis of Yeast and Extraction.   Weigh approximately 125 g of moist baker's yeast cake into a large beaker, and add 120 mL of toluene with stirring to break up the lumps. Cover the beaker and place in an incubator at 37°C for a minimum of 4 hours, with occasional stirring. Add 240 mL of cold 1 mM Na_2EDTA and gently stir for 6 hours at 4°C. Allow the phases to separate by allowing the mixture to stand for 2 hours without stirring, and then carefully remove the bottom aqueous extract. Centrifuge the aqueous extract to remove solid material, and filter through glass wool; finally filter, with gentle suction, through an analytical Celite pad 3–4 mm thick on a filter paper in a Büchner funnel. The filtrate, which should be a clear yellow, is kept at 4°C in an ice bath.

b. Heat Treatment.   Alcohol dehydrogenase is relatively stable to heat. Bring the extract rapidly to 55°C by placing it into an 80°C water bath with constant swirling and temperature monitoring. When the temperature reaches 55°C, transfer the solution to a 55°C bath for exactly 6 minutes. Cool the solution quickly in an ice bath and keep it at 0–5°C for all subsequent steps. Centrifuge at 5000*g* to remove any insoluble material that forms.

c. Acetone Fractionation.   Cool the extract in an acetone–dry ice bath to −2°C and for every 100 mL of extract, slowly add 50 mL of −10°C acetone with stirring. Remove the precipitated protein by centrifugation for 10 minutes at 25,000*g* at −6°C. Discard the precipitate and add 55 mL of −10°C acetone, with stirring, to every 100 mL of original extract. At a concentration of 51% v/v acetone, alcohol dehydrogenase should be precipitated.

Resuspend the precipitated protein in a minimum volume (<50 mL) of 1.0 mM (pH 7.5) phosphate buffer containing 1.0 mM Na_2EDTA at 4°C. If all the protein does not dissolve, centrifuge and discard the insoluble material. Transfer the solution to a dialysis tube and dialyze against the same buffer for several hours.

d. Ammonium Sulfate Precipitation.   Alcohol dehydrogenase can be precipitated with ammonium sulfate by adding 37 g of ammonium sulfate to 100 mL of the dialyzate (60% saturated ammonium sulfate). Keep the enzyme solution at 0–4°C and add the solid ammonium sulfate slowly with gentle stirring to prevent locally high concentrations of the salt. Allow the solution to stand for 30–60 minutes at 0–4°C and then centrifuge it. The precipitated alcohol dehydrogenase should be highly purified at this stage and should have a specific activity of at least 130 IU/mg.

8.5.4.3 PURIFICATION OF HUMAN SALIVARY α-AMYLASE[7]

a. Starting Material. Salivation can be stimulated by chewing pieces of pure, low-melting paraffin, collecting the saliva in a beaker. Centrifuge the collected saliva ($\sim$ 200 mL) for 30 minutes at 5000g and filter the supernatant through glass wool to remove any particles of paraffin.

b Gel Filtration. Add the saliva to the top of a column (4.5 cm × 35 cm) of Sephadex G-25 and elute with 5 mM ammonium acetate (pH 6.7) buffer containing 1 mM calcium chloride (starting buffer) at a flow rate of 500 mL/h, collecting 8-mL fractions. Two protein peaks result, with the amylase activity appearing in the first peak. Pool the fractions of the peak ($\sim$ 184 mL).

c. Concentration. Place the 184 mL of filtrate containing the amylase activity in a dialysis bag, put powdered PEG 20,000 over the bag, and concentrate the enzyme solution to approximately 20 mL. Then dialyze the bag against 2 L of starting buffer.

d. DEAE–Cellulose Chromatography. Add the dialyzed solution to the top of a DEAE–cellulose column (2.5 cm × 35 cm) that has been equilibrated with starting buffer and elute by adding 500 mL of 100 mM ammonium acetate buffer (pH 6.7) containing 1 mM calcium chloride to 500 mL of starting buffer to form a gradient. Adjust the flow rate to about 200 mL/h and collect 8-mL fractions. Increase the gradient by adding 500 mL of 200 mM ammonium acetate buffer (pH 6.7) containing 1 mM calcium chloride to the first reservoir. The amylase emerges in approximately tubes 60–70 (about 80 mL). Dialyze the pooled amylase against 2 L of starting buffer for 3–5 hours and concentrate it with PEG and a dialysis bag as before, or use lyophilization. The specific activity should be 400–500 IU/mL and there should be two to three isozymic bands of protein and amylase activity on polyacrylamide gel electrophoresis. The yield is about 60% of the original activity.

8.5.4.4 CRYSTALLIZATION OF *ESCHERICHIA COLI* ALKALINE PHOSPHATASE[8]

a. Selection of Bacterial Strain. Use *Escherichia coli* K-12, which is constitutive for the production of alkaline phosphatase. Plate out ($\sim$ 100 colonies/plate) and examine the individual colonies for alkaline phosphatase production by adding a drop of 2% *p*-nitrophenyl phosphate solution to individual colonies. Observe the colonies and select those that turn yellow in the shortest time (i.e., within at least 1–2 min).

b. Growth of Cells. Inoculate 10 L of sterile medium (composition of the medium described by Malamy and Horecker[8]) in a fermenter with a 1-L inoculum. Incubate for 16–20 hours at 37°C with stirring (200 rpm) and aeration: one liter of air per liter of culture per minute. Harvest the cells by centrifugation and wash with three 1-L portions of 10 mM Tris buffer (pH 8).

c. Preparation of Spheroplasts. Suspend the washed cells in 1 L of 20% sucrose containing 33 mM Tris buffer (pH 8). The cell suspension should have a turbidity of 3.4 at 590 nm. To a 2-L flask kept at 22–25°C, add the suspension with slow stirring by a magnetic stirrer and then add 10 mL of 10 mM Na_2EDTA and 10 mg of lysozyme. Allow the reaction to proceed for about 10 minutes until the turbidity has dropped to a constant value. Then centrifuge the solution at 5000*g* for 15 minutes at 4°C. Remove the supernatant, which should contain the majority of the alkaline phosphatase ($\sim$230,000 IU and 260 mg of protein).

d. DEAE–Cellulose Chromatography. Add the alkaline phosphatase solution to a column of DEAE–cellulose (2.5 cm $\times$ 5.5 cm) that has been equilibrated with 50 mM NaCl and adjust the flow rate to allow the enzyme solution to be adsorbed over 10 hours at 4°C. After all the enzyme solution has been adsorbed, wash the column with 20 mL of 50 mM NaCl and elute with 125 mM NaCl, collecting 4-mL fractions. The enzyme should appear in the fourth or fifth tube or after 20–36 mL of eluant has been collected.

e. Crystallization. Pool the column fractions with enzyme activity and bring it to 10 mM magnesium chloride (by adding a small volume of 1 M $MgCl_2$) and 50% ammonium sulfate saturation (by adding solid ammonium sulfate, 290 mg/mL). Adjust the pH carefully to 8 with 2 M NaOH. Add a saturated solution of ammonium sulfate (pH adjusted to 8), dropwise with mixing, until the solution becomes faintly turbid (about 61% saturated ammonium sulfate). Allow the turbid suspension to stand at 22–25°C for 1 hour, whereupon it should have increased in turbidity. Place the suspension in an ice bath for several minutes until the turbidity disappears and allow it to come to room temperature slowly. As the solution warms, the turbidity will reappear and the suspension will develop a silky sheen because of the presence of crystals of alkaline phosphatase. The crystals can be removed by settling several days or by centrifuging. The enzyme crystals should have a specific activity of 40–50 IU/mg. The crystals are stable at room temperature but should be stored at 0–4°C in 20 mM Tris (pH 8).

8.5.4.5 ISOLATION OF A RESTRICTION ENDONUCLEASE (BamI) FROM BACILLUS AMYLOLIQUEFACIENS H[9]

B. amyloliquefaciens H is grown in Penassay broth (Difco) to late log phase and the cells are harvested by centrifugation. Disrupt the cells (20 g wet weight) by grinding with alumina (3 times the wet weight of the cells) and suspend them in 10 mL of extraction buffer (25 mM Tris–HCl, pH 8.0, containing 5 mM 2-mercaptoethanol). Remove cell debris and alumina by centrifugation at 23,000*g* for 60 minutes at 4°C. Dilute the supernatant with 3 volumes of extraction buffer and precipitate the nucleic acids with streptomycin sulfate (1 mL 10% streptomycin sulfate/A_{260} of 1500). Add solid ammonium sulfate to the supernatant with constant stirring at 4°C to give 50% saturation. Remove the precipitate by centrifugation at 20,000*g* for 15 minutes at 4°C. Add solid ammonium sulfate slowly to the supernatant to give a concentration of 80%, which precipitates the endonuclease. Remove the precipitate by centrifugation at

20,000g for 20 minutes at 4°C, suspend it in 10 mL of buffer I (10 mM sodium phosphate, pH 7.4), and apply to the top of a bed of Sephadex G-25 (5 cm × 40 cm) that is eluted with buffer I, collecting 3-mL fractions. Pool the fractions containing nuclease activity* and chromatograph on DEAE–cellulose (Whatman DE52, 2.4 cm × 14 cm) by eluting with a 1-L gradient of 0 to 0.6 M NaCl in buffer I. Analyze 2-mL fractions for endonuclease activity and pool active fractions. Discard fractions eluting at high salt concentrations, which contain exonuclease activity. Pooled endonuclease activity is dialyzed against buffer I to remove salt, and is chromatographed on phosphocellulose (Whatman P11, 1.2 cm × 8.0 cm) equilibrated with 10 mM sodium phosphate, pH 7.4, containing 10 mM 2-mercaptoethanol (buffer II). Elute the enzyme from phosphocellulose with a 1-L linear gradient of 0 to 0.6 M NaCl in buffer II. The endonuclease is eluted primarily between 0.31 and 0.36 M NaCl. Pool the fractions, and store at 4°C.

8.6 PREPARATION OF CARBOHYDRATES

8.6.1 General Separation and Purification Methodology

Carbohydrates are extracted from cells with aqueous solvent and purified by removing protein and nucleic acids by various heat and pH treatments. Polysaccharides can be precipitated from solution by the addition of 2 volumes of alcohol. The polysaccharide can then be obtained as a dry powder by mixing the precipitate in a mortar with dry acetone (trituration) several times, followed by a final trituration with dry, absolute ethanol. The powdered material can then be dried *in vacuo* at 40–50°C to remove the last traces of water as the 95% ethanol azeotrope.

Charged polysaccharides can be further purified by chromatography on ECTEOLA–cellulose or DEAE–cellulose or calcium phosphate, and polysaccharides of different sizes can be separated by gel filtration.

Low molecular weight carbohydrates, such as mono- or disaccharides, are usually purified by crystallization from an extract that has had the protein removed.

8.6.2 Specific Methods for Preparing Different Carbohydrates

8.6.2.1 PREPARATION OF STARCH[10]

Starch is very widespread in the plant kingdom, occurring in algae and in the seeds, fruits, stems, leaves, bulbs, tubers, and rhizomes of higher plants. Starch granules are produced in the plastids that are embedded in the cytoplasm of plant cells.

---

* The endonuclease activity is assayed by digesting λ phage DNA (see Section for phage preparation) and analyzing the digests for fragments of DNA by electrophoresis on 0.7% agarose gel by staining with ethidium bromide.

The nonstarchy tissues are removed from the plant material and the tissue is homogenized in a blender in a cold water solution. The resulting suspension is filtered through cheese cloth, and the filtrate is centrifuged at 1000g for 20 minutes. The starch pellet is washed with cold water several times by suspension and centrifugation. Many starches contain a water-insoluble lipid wax that can be removed by refluxing with 80% aqueous methanol. The starch is washed with acetone and dried under reduced pressure at 50°C.

8.6.2.2 FRACTIONATION OF STARCH INTO AMYLOSE AND AMYLOPECTIN

Most starches are mixtures of amylose and amylopectin (with the exception of waxy-varieties, e.g., waxy-maize and waxy-rice that are 100% amylopectin). Other types of starch have varying amounts from 20–30% amylose and 80–70% amylopectin.

Ten grams of starch is stirred into 100 mL of dimethyl sulfoxide (DMSO) and the mixture is heated to 60°C for 1 hour. Stirring is continued ($\sim$10–15 hours) until all the starch is dissolved. The DMSO–starch solution is slowly diluted to 1 L with water and continuous stirring; 200 mL of 1-butanol is added and stirring is continued for 6–12 hours. The amylose–butanol complex is removed by centrifugation for 1 hour at 25,000g. The supernatant is carefully removed, 100 mL of 1-butanol is added, the solution is stirred for 6 hours, and the amylose–butanol complex is removed by centrifugation. The supernatant contains the amylopectin, which can be precipitated by adding 2 volumes of ethanol. The precipitate is collected by centrifugation at 8000g for 20 minutes.

The amylose–butanol complex and the amylopectin are further treated by tituration several times with dry acetone (10 volumes per volume of precipitate) in a mortar. The final treatment is with absolute ethanol. The ethanol is poured off and the finely powdered precipitates are dried *in vacuo* at 50°C. The yield of amylose is 1–2 g and the yield of amylopectin is 8–9 g.

8.6.2.3 ISOLATION OF GLYCOGEN FROM LIVER[11]

Animal glycogen is commonly isolated from liver or muscle tissue. The condition of the laboratory animal from which the tissue is to be taken is important because the glycogen, especially liver glycogen, will vary in properties and yield according to whether the animal is fed, fasted, ill, and so on. The tissue to be used (e.g., liver) should be removed as quickly as possible after the death of the animal and frozen by plunging thin slices into a mixture of solid carbon dioxide (dry ice) in ethanol.

The frozen tissue (30 g) is homogenized with 300 mL of 10% trichloroacetic acid in a mortar with sand at 0–4°C. The homogenate is then quickly centrifuged at 5000g at 0°C. The supernatant is poured off and 3 volumes of cold ethanol (0°C) is added to precipitate the glycogen. The glycogen is centrifuged as above and quickly dissolved in 300 mL of 5% trichloroacetic acid at 0°C, recentrifuged, and the glycogen again precipitated with 3 volumes of cold ethanol. The precipitate is centrifuged and dehydrated by several triturations with anhydrous acetone, followed by a final treatment with anhydrous, absolute ethanol and drying *in vacuo* at 50°C.

8.6.2.4 PREPARATION OF HEPARIN[12]

Heparin is a polysaccharide composed of D-glucosamine and D-glucuronic acid units with N-sulfate and O-sulfate esters. It is used as a drug for its blood anticoagulant properties and can be prepared from beef liver, lung, or small intestine.

To prepare heparin, tissue (45 kg) is finely ground through a meat grinder and suspended in 20 L of water containing 500 mL of toluene or 2 L of chloroform and is allowed to autolyze at 35°C for 24 hours. Saturated ammonium sulfate (7.5 L) is added and enough 2 M sodium hydroxide to raise the pH to 9. The solution is held for 1 hour at 60°C with stirring, and then at 70°C until complete coagulation of the proteins has occurred. The solution is filtered hot with the use of Celite filter aid. It is then cooled and acidified to pH 2–2.5 with concentrated hydrochloric acid, heated to 60°C, and filtered. The precipitate is washed with 20 L of hot 0.01 M hydrochloric acid and suspended in 6 L of 95% ethanol for 20 hours. The supernatant is decanted and the precipitate is centrifuged and dissolved in 6 L of water. The pH is adjusted to 8.0, and 25 g of trypsin and 10 mL of toluene are added. The solution is left for 36 hours at 37°C and every 12 hours the pH is checked and adjusted to 8.0; 12 L of 95% ethanol is added and the pH is adjusted to 6.0 with HCl. After 24 hours, the supernatant is decanted and the precipitate is centrifuged and suspended in 1.5 L of water to which 0.5 N sodium hydroxide is added until the precipitate completely dissolves. This solution is heated to 75°C for 20 minutes to destroy trypsin action and then centrifuged. Two volumes of acetone is added to the clear centrifugate and the solution is made slightly acidic by adding HCl. The solution is allowed to stand at room temperature, and the precipitate, which is heparin, is collected by centrifugation. It is triturated with dry acetone several times and given a final treatment with dry absolute ethanol, followed by drying in a vacuum oven at 40°C. The yield is 50–60 g from 45 kg of tissue.

Heparin subfractions can be obtained by chromatography on ECTEOLA–cellulose.[13]

8.6.2.5 ISOLATION OF LACTOSE FROM MILK

Lactose is one of three naturally occurring disaccharides, sucrose and α,α-trehalose being the other two. To isolate lactose, skimmed milk (200 mL) is warmed to 40°C on a hot plate with stirring, and dilute acetic acid (glacial acetic acid–water, 1:10) is added dropwise with continuous stirring until casein is no longer precipitated. An excess of acid should be avoided. The casein is worked into a mass and removed with a stirring rod and discarded. Five grams of finely divided calcium carbonate is immediately added, and the mixture is stirred briefly and then heated to boiling for about 10 minutes. Heating causes almost complete precipitation of the albumin. The hot mixture is filtered on a Büchner funnel, and the filtrate is concentrated to about 30 mL on a rotary evaporator at 45°C *in vacuo*; 175 mL of 95% ethanol is added with stirring, 3 g of decolorizing charcoal is added, and the mixture stirred for 10 minutes. The mixture is filtered on a Büchner funnel to which has been added a pad of Celite filter aid of 2–3 mm thick. The clear filtrate is allowed to stand for 24 hours or longer in a stoppered flask to allow crystallization of the lactose. The crystals are collected and washed with a few milliliters of cold 25% ethanol. The average yield is 5–7 g.

8.6.2.6 ISOLATION OF α,α-TREHALOSE FROM YEAST

The naturally occurring disaccharide α,α-trehalose is found in yeast cell walls and in the lymph fluid of insects. To isolate α,α-trehalose from yeast, a paste of 32 g of dried baker's yeast is made with 75 mL of water and 500 mL of 95% ethanol is added and the mixture allowed to stand for 30 minutes with occasional stirring. The mixture is filtered on a Büchner funnel and the precipitate is washed with three 150 mL portions of 70% ethanol. The washes are combined with the original extract. To the combined extract and washes, add 2 mL of 20% (w/v) zinc sulfate, 2 mL of 1% phenolphthalein, and enough saturated barium hydroxide solution to just turn the indicator pink. Add 2 g of activated charcoal, heat the mixture to 70°C on a hot plate, and filter while hot through a Büchner funnel with a pad of Celite 2–3 mm thick. The clear filtrate is adjusted to pH 7 with 0.1 M HCl and concentrated to a syrup in a rotary evaporator *in vacuo*. Ten milliliters of warm water is added to dissolve the syrup, which is transferred to a 100 mL beaker; 80 mL of 95% ethanol is slowly added with stirring. The beaker is covered with a Kimwipe held by a rubber band and set aside for the crystallization of the α,α-trehalose, which may take from 24 hours to a week. Slow crystallization often produces crystals longer than 12 mm. Crystallization can be hastened by cooling to 0°C or by adding more 95% ethanol. The crystals are filtered and dried in a vacuum oven at 40°C for 15 hours. The yield is about 5 g.

8.7 PREPARATION OF NUCLEIC ACIDS

8.7.1 General Separation and Purification Methodology

Nucleic acids are the most polar of the biopolymers and are therefore soluble in polar solvents and precipitated by nonpolar solvents. In prokaryotes, DNA is double stranded and circular and is found throughout the cytoplasm. In eukaryotes, DNA is located in the nucleus and in mitochondria or chloroplasts. The DNA in the nucleus is double stranded and linear, whereas the DNA in mitochondria and chloroplasts is like prokaryotic DNA, double stranded and circular. The DNA in prokaryotes is relatively free of associated protein, but the DNA in the nucleus of eukaryotes is associated with basic proteins, called histones. Contaminating molecules that must be removed from both prokaroytic and eukaroytic DNA are proteins and RNA. Proteins are denatured by the addition of organic solvents and detergents, and RNA is removed with a brief treatment with deoxyribonuclease-free ribonuclease. The high molecular weight DNA is soluble in the aqueous phase, from which it is obtained by cold alcohol precipitation. Care must be taken during precipitation, since high molecular weight DNA is readily sheared because of its long, fibrous tertiary structure.

As the alcohol is added, the DNA is carefully "spooled" onto a glass rod as a threadlike precipitate.

RNA has less secondary structure and is of lower molecular weight than DNA. RNA is single stranded and its secondary structure is intramolecular rather than intermolecular. There are three types of RNA: messenger RNA (mRNA), ribosomal RNA (rRNA), and transfer RNA (tRNA). Messenger RNA is found in ribosomes and free in the cytoplasm, where it is susceptible to nucleases; mRNA molecules are variable in size. Ribosomal RNA, found associated with proteins in the ribosome, is a relatively stable substance, and its molecules vary in size. Transfer RNA is the smallest of the RNAs and is distributed throughout the cell and organelles.

The N-glycosidic purine and pyrimidine linkages of DNA and RNA are stable to mild acidic and basic conditions. However, the phosphodiester linkages of RNA are cleaved at $37°C$ in 0.3 M KOH, resulting in the formation of 2'- and 3'-phosphoribonucleotides. The primary structure of DNA is stable under these conditions. In dilute acid (0.1 N TCA, HCl, or $HClO_4$) nucleic acids will precipitate. In stronger acid and at higher temperatures (1 N, $100°C$, 15 minutes) the purine bases of DNA are hydrolyzed from the 2-deoxyribose moiety. The pyrimidine bases are hydrolyzed only under extreme conditions of pH and temperature.

Many of the nucleases present in cells can digest nucleic acids. When the cell is disrupted, the nucleases can cause extensive hydrolysis. Nucleases apparently present on human fingertips are notorious for causing spurious degradation of nucleic acids during purification.

The isolation and purification of DNA consists of four steps: (a) the disruption of cells and membrane-bound structures to release DNA, (b) the inactivation of enzymes that hydrolyze the DNA, (c) the dissociation and denaturation of protein, and (d) the solvent extraction and concentration of the DNA by precipitation. Cells are broken by grinding, tissue homogenization, or treatment with lysozyme (see Section 8.2). Chelating agents are added to remove metal ions required for nuclease activity. DNA–protein interactions are disrupted with SDS, phenol, and organic solvents. The DNA remains in the aqueous phase, from which it is precipitated with cold ($0°C$) ethanol. The precipitate is usually redissolved in buffer and treated with phenol or organic solvent to remove the last traces of protein, followed by reprecipitation with cold ethanol. RNA is removed by limited treatment with deoxyribonuclease-free ribonuclease.

RNA purification begins with a solvent extraction that denatures proteins and inactivates cellular ribonucleases. Cell pastes or tissue homogenates are treated with phenol-saturated buffer at $4°C$. The aqueous and the phenol phases are separated by centrifugation. The RNA containing aqueous phase is reextracted with phenol–buffer, and the various types of RNA in the aqueous phase are precipitated with alcohol and salt. Messenger RNA is precipitated in 0.1 M NaCl and 70% ethanol, whereas rRNA is precipitated in 3 M sodium acetate and 70% ethanol in which the small RNAs (tRNA and 5S-RNA) are soluble, and tRNA is precipitated in 1 M NaCl and cold 66% ethanol.

8.7.2 Specific Methods for Preparing Different Nucleic Acids

8.7.2.1 ISOLATION OF BACTERIAL CHROMOSOMAL DNA[14,15]

Bacterial DNA is used in transformation and recombinant DNA experiments. The method described gives a relatively high yield with limited shearing of the bacterial circular DNA molecule.

Two grams of bacterial cell paste, obtained from log-phase growth, is suspended in 25 mL of 0.15 M NaCl and 0.1 M Na_2EDTA solution; 1 mL of lysozyme (10 mg/mL) is added, and the suspension is incubated at 37°C for 30 minutes with occasional shaking. Two milliliters of 25% SDS is added, and the suspension is heated at 60°C for 10 minutes. The suspension is cooled to room temperature, and sufficient 5.0 M sodium perchlorate is added to the solution to obtain a final perchlorate concentration of 1.0 M. An equal volume of chloroform–isoamyl alcohol (24:1) is added, the flask is stoppered, and the mixture is shaken by hand or a low speed shaker for 30 minutes. The resulting emulsion is separated into three layers by centrifuging for 5 minutes at 10,000g at 22–24°C. The upper, aqueous phase contains the nucleic acids and is carefully pipetted off. The nucleic acids are precipitated by gently layering 2 volumes of ethanol onto the viscous, aqueous phase. The layers are gently mixed with a stirring rod and the nucleic acids are "spooled" onto the rod as a threadlike precipitate. The precipitated DNA is drained free of excess ethanol by pressing the spooled rod against the beaker. The DNA is transferred to 10–15 mL of 15 mM sodium chloride and 1.5 mM citrate by gently moving the rod back and forth. The solution is gently shaken until dispersion of the DNA is complete. The solution is made 0.15 M in sodium chloride and 15 mM in sodium citrate (pH 7.0) by adding a pH 7.0 solution of sodium chloride (1.5 M) and sodium citrate (0.15 M).

The DNA solution is extracted by shaking with an equal volume of chloroform–isoamyl alcohol (24:1) for 15 minutes, centrifuging, removing the supernatant, and repeating the extraction several times until very little protein precipitate is seen at the interface. The DNA in the aqueous supernatant, obtained after the last deproteinization, is precipitated by adding 2 volumes of ethanol as above, and the precipitate is dispersed in 0.5–0.75 volume of 0.15 M NaCl and 15 mM Na citrate (pH 7.0). Ribonuclease is added to a final concentration of 50 μg/mL and the mixture is incubated at 37°C for 30 minutes. The digest is again subjected to a series of deproteinizations with chloroform–isoamyl alcohol as before. After the last treatment, the DNA in the supernatant is once again precipitated with 2 volumes of alcohol, and the DNA is dissolved in 9.0 mL of 15 mM NaCl and 1.5 mM Na-citrate. When all the DNA has dissolved, 1.0 mL of 3 M sodium acetate and 1 mM Na_2EDTA (pH 7) is added, and 0.54 volume of isopropyl alcohol is added dropwise with rapid stirring using a glass rod. The DNA goes through a gel phase at about 0.5 volume of isopropyl alcohol and precipitates in fibrous form. The DNA is wound around the glass stirring rod and is washed free of salt by gently stirring the DNA on the glass rod into progressively increasing concentrations of ethanol (beginning with 70% and proceeding to 95% in 5% increments). The DNA can then be dissolved in the solvent of choice.

Approximately 50% of the cellular DNA can be obtained by this procedure, or 1–2 mg of DNA from 1 g of packed, wet cells. The DNA can be stored by placing the precipitated DNA into 75% ethanol for 3–4 hours for sterilization and asceptically transferring it to a sterile solvent.

8.7.2.2 ISOLATION OF PLASMID DNA[16]

Plasmids are relatively small, double-stranded circular DNA molecules that occur naturally in bacterial cells independent of the circular chromosomal DNA. Plasmids are isolated for molecular studies, restriction enzyme mapping, and nucleotide sequencing, or for the use as vectors in the construction of new hybrid plasmids for recombinant DNA studies. A rapid method for the isolation of plasmid DNA involves alkaline extraction at pH 12.0–12.6, which denatures chromosomal DNA but not plasmid DNA.

There are 10–20 plasmids in a bacterial cell, but the number can be enhanced by adding chloramphenicol to the culture. The bacteria are grown to an absorbance of 0.8 at 600 nm, 170 μg/mL of chloramphenicol is added, and incubation is continued for 18–20 hours. Cells are harvested by centrifugation at 6000g for 10 minutes at 0°C. The bacteria are suspended in 50 mL of water and recentrifuged. The cells are suspended in 1 mL of 50 mM glucose, 10 mM cyclohexane diaminetetraacetic acid (CDTA), and 25 mM Tris–HCl (pH 8.0); 9 mL of the same solution containing 10 mg of lysozyme is added, and the cell suspension is kept at 0°C for 30 minutes. Twenty milliliters of alkaline SDS (0.2 M sodium hydroxide and 1% SDS) is added, and the mixture is stirred gently with a glass rod until it is homogeneous and clear. The solution is kept at 0°C for 10 minutes and then 15 mL of 3 M potassium acetate and 1.8 M formic acid is added. The solution is stirred for several minutes until a coarse white precipitate is formed. After 30 minutes at 0°C, the precipitate is removed by centrifugation at 12,000g for 10 minutes at 0°C.

The supernatant is transferred to another tube and 2 volumes of -10°C ethanol is added. The resulting precipitate of nucleic acid that forms on standing at -20°C for 20 minutes is collected by centrifugation and dissolved in 5 mL of 0.1 M acetate and 50 mM morpholinopropanesulfonic acid (MOPS) buffer (pH 8). The volume of the solution is measured, and an equal volume of 5 M LiCl in 50 mM MOPS (pH 8) is added. The solution is held at 0°C for 15 minutes, and the heavy precipitate that forms is removed by centrifugation at 12,000g for 10 minutes at 0°C. The clear supernatant is heated at 60°C for 10 minutes, and any precipitate that forms is removed. Plasmid DNA is precipitated by the addition of 2 volumes of -10°C ethanol to the supernatant solution. After standing at -20°C for 15 minutes, the plasmid DNA is centrifuged and dissolved in 2.5 mL of acetate–MOPS buffer (pH 8). The plasmid DNA is reprecipitated with 2 volumes of cold ethanol and dissolved in 2 mL of Tris–CDTA buffer (pH 8).

Contaminating RNA is removed by adding ribonucleases A and T1 to 10 μg and 5 units/mL, respectively, and the solution is incubated for 15 minutes at 37°C; 40 μl of 10% SDS and 2 mL of acetate–MOPS buffer are added and 4 mL of 2-propanol is added dropwise with stirring. The solution is allowed to stand for 15 minutes at room temperature, and the precipitate of plasmid DNA is centrifuged and dissolved in 2 mL of acetate–MOPS buffer and reprecipitated with 2 volumes of cold ethanol. The plasmid DNA is dissolved in 2 mL of Tris–CDTA buffer and stored frozen.

8.7.2.3 PURIFICATION OF DNA FROM BACTERIOPHAGE

a. First Method.[17] Bacteriophage-infected cultures of bacteria are centrifuged at 10,000g to pellet the bacterial cell debris. The secreted phage are precipitated from the supernatant by the addition of polyethylene glycol 6000 to 3.33% and NaCl to 0.4 M. The phage are pelleted by centrifugation at 10,000g.

Phage pellets are suspended in 100 μL Tris–EDTA buffer (10 mM Tris, 1 mM EDTA, pH 7.5); 100 μL of phenol, equilibrated with 0.5 M Tris, is added, and the solution is vigorously mixed. The suspension is centrifuged for 2 minutes at 4°C in a microcentrifuge. The aqueous layer is removed and transferred to a new tube, and 20 μL of 3 M sodium acetate (pH 6) and 400 μL of 2-propanol are added with gentle mixing. The DNA is allowed to precipitate at -20°C for 1 hour and then -70°C for 10 minutes. The DNA is pelleted for 10 minutes at 4°C in a microcentrifuge. The supernatant is carefully decanted and the pellet is rinsed with 200 μL of cold 2-propanol and centrifuged for 10 minutes at 4°C. The alcohol is decanted and the pellet is air-dried and stored at -20°C. The yield from 2 mL of culture is 15 μg of DNA.

b. Second Method. Phage pellets are suspended in 50 mM Tris–HCl, pH 8.0, containing 10 mM sodium chloride and 10 mM magnesium chloride; EDTA (0.5 M) is added to give a final concentration of 20 mM. Pronase is added to 0.5 mg/mL or proteinase K to 50 μg/mL, and SDS is added to a final concentration of 0.5%. The solution is mixed by inverting the tube several times and incubated for 1 hour at 37°C if pronase is used or 65°C if proteinase K is used. An equal volume of phenol (previously equilibrated with Tris–HCl, pH 8) is added and mixed thoroughly. The phases are separated by centrifugation (1600g) at 22°C for 5 minutes. The aqueous phase is removed and transferred to a clean tube and extracted once with 1 volume of a 1:1 mixture of phenol–chloroform, followed by extraction with 1 volume of chloroform. The aqueous phase is dialyzed against 10 mM Tris–10 mM EDTA (pH 8.0) buffer overnight at 4°C.

Extraction with phenol and chloroform takes advantage of the fact that proteins are removed more efficiently with two different solvents, and the chloroform removes the last traces of phenol.

8.7.2.4 ISOLATION OF TRANSFER RNA FROM ESCHERICHIA COLI[18]

Escherichia coli is grown to the logarithmic stage and the cells are centrifuged to obtain a cell paste. The cells can be stored frozen in thin cakes of 100–200 g wrapped in Parafilm.

One kilogram of cell paste is suspended in 1 L of standard buffer (1 mM Tris–HCl, pH 7.2, plus 10 mM magnesium chloride). A smooth suspension is obtained with a blender. The frozen cells are allowed to soften at 4°C for 2–3 hours before the addition of the buffer. The suspension is distributed equally in eight 1-L polyethylene bottles with screw caps. Equal volumes of standard buffer-saturated phenol at 5°C are added to each bottle and the mixtures are vigorously shaken for 1 hour at 5°C. The resulting emulsion is separated into three layers by centrifugation at 5000g for 30 minutes. The upper aqueous phase contains the nucleic acid and is carefully removed with a pipet. The nucleic acids are precipitated by adding 0.1 volume of cold (4°C) 20% potassium acetate (pH 5) and 2 volumes of −20°C absolute ethanol. The nucleic acid precipitate is allowed to settle overnight at −10°C. The clear supernatant is carefully siphoned or decanted and discarded. The precipitate is collected by centrifugation. The tubes are inverted to allow as much of the ethanol as possible to drain from the precipitate. The precipitate is then dispersed into 1 L of cold 1 M sodium chloride and vigorously stirred on a magnetic stirrer for 1 hour; the large clumps are broken with a spatula. The suspension is then centrifuged for 30 minutes at 5000g and the supernatant is saved. The precipitate is reextracted with 500 mL of 1 M sodium chloride and centrifuged as above. The two supernatants are combined and the tRNA is precipitated by slowly adding 2 volumes of −20°C ethanol and collected by centrifugation. Further purification can be obtained by isopropanol fractionation[19] or chromatography on DEAE–cellulose.[20,21]

Different RNA fractions can be separated according to their sizes by sucrose gradient centrifugation. The method can be used as a further purification step of the nucleic acids first isolated by phenol extraction.[22]

8.7.2.5 ISOLATION OF MESSENGER RNA FROM RIBOSOMES[23]

Messenger RNA can be isolated from ribosomes (see Section 8.4.4 for the preparation of ribosomes). A ribosome suspension is diluted with water to give an absorbance of no greater than 100 at 260 nm. Then 0.1 volume of 5% SDS and 0.1 volume of 1 M Tris buffer (pH 9) are added, followed by 1 volume of 80% aqueous phenol. The mixture is stirred vigorously for 5 minutes and centrifuged at 12,000g for 10 minutes to separate the phases. The aqueous phase is removed and saved, and the nonaqueous phase (the phenol phase plus the gel phase) is reextracted with an equal volume of 0.1 M Tris (pH 9) and 0.5% SDS by vigorous stirring as above. The aqueous phases are pooled and should contain most of the ribosomal and messenger RNA; the aqueous solution is extracted three times with an equal volume of fresh 80% phenol by brief, vigorous stirring, followed by centrifugation. The aqueous phases are pooled and contain the RNA. The RNA solution is extracted four times with diethyl ether to remove dissolved phenol. The mRNA can be precipitated by the addition of 0.1 volume of 1 M sodium chloride and 2.5 volumes of ethanol. After storage at 4°C overnight, the precipitate is collected by centrifugation and washed twice with cold 66% ethanol in 0.1 M sodium chloride. The precipitated mRNA is dissolved in water and stored at −20°C.

8.8 PREPARATION OF LIPIDS

8.8.1 General Separation and Purification Methodology

The initial purification of lipid material involves selective solvent extraction. Three types of lipid interactions are disrupted during extraction and purification: van der Waals forces in lipid–lipid and lipid–protein complexes, electrostatic and hydrogen bonding interactions between lipid and proteins, and covalent bonding between lipids, carbohydrates, and proteins. The solvent that is chosen for extraction of a particular lipid depends on the types of interaction to be disrupted. Thus, neutral lipids that are primarily bound by van der Waals, hydrophobic interactions are extracted from tissues with nonpolar solvents, whereas phospholipids that associate through electrostatic interactions are extracted by more polar solvent mixtures.

In a general method for lipid extraction, sample and solvent are homogenized in a blender or tissue homogenizer, the mixture is filtered on a Büchner funnel, and the insoluble material is reextracted. The lipid-containing solvent is washed with water or salt solution until two phases form, and the organic phase is separated and concentrated.

8.8.2 General Extraction and Purification Procedures

Fifty grams of wet tissue is homogenized with a mixture of 300 mL chloroform and 150 mL methanol. The mixture is filtered through a porous glass filter or on a Büchner funnel, and the residue is rehomogenized with the same volumes of solvent. The filtrates are combined and transferred to a graduated cylinder, and 0.2 volume of 0.6% aqueous sodium chloride is added. The solution is gently mixed to avoid the formation of an emulsion. The aqueous phase and the interface are removed. The aqueous phase may still contain some polar lipids and can be backwashed with chloroform to recover the polar lipids. Nonlipid contaminants can be removed from the organic phase by partial evaporation of the solvent in a rotary evaporator, followed by extraction with hexane–chloroform (3:1). After stirring, the mixture is allowed to settle and the supernatant is decanted from the solid, nonlipid contaminants.

Further purification of the lipid extract can be achieved by chromatography on Sephadex G-25. Water and water-soluble substances bind to the Sephadex, from which the lipids are eluted rapidly with a nonpolar solvent such as hexane. The solvent is removed by rotary evaporation at 35–40°C *in vacuo*. Final traces of solvent are removed by a stream of N_2. The purified lipids should be stored at -20°C in the absence of oxygen.

The various lipid classes can be separated by adsorption chromatography, which separates substances on the basis of differences in the relative affinity of the substances for the adsorbent. With lipids, the affinity for the adsorbent is determined primarily by the groups other than the fatty acids such as glycerolphosphate, choline, serine, and ethanolamine. Thus, lipids can be separated by classes depending on the groups present (triglycerides, choline phosphoglycerides, sphingolipids, etc.). Three adsorbents are routinely used:

silica gel (silicic acid), Florosil (magnesium silicate), and alumina (aluminum oxide), as discussed in Section 4.1. Ion exchangers such as DEAE– and TEAE–cellulose separate lipids with ionic groups or highly polar nonionic groups. Progress of the separation can be monitored by TLC (Section 4.5). TLC slides allow quick examination of column fractions and the elution process can be continuously analyzed. The lipids are visualized on the chromatogram after treatment with sulfuric acid spray and heat, or iodine vapor. Amino lipids can be detected with ninhydrin, and choline phosphatides can be detected with bismuth nitrate–iodine.

Neutral lipids and phospholipids can be separated on silica gel columns. Neutral lipids are eluted with petroleum ether–diethyl ether (1:1) or hexane–chloroform (3:1). Phospholipids can be fractionated on silica activated at 110°C and prewashed with chloroform containing 1% methanol. Phosphatidyl serines, phosphatidyl ethanolamines, and phosphatidyl cholines are eluted with an increasing gradient of methanol into chloroform. Phospholipids can be separated on aluminum oxide (grade IV, neutral). Choline phosphatides are eluted with chloroform, phosphatidyl ethanolamine with chloroform–methanol, and phosphatidyl serine with chloroform–methanol–water. Chromatography on DEAE– or TEAE–cellulose has the advantages of better separation of highly polar lipids and less oxidation or hydrolysis during chromatography. The isolation of triglycerides and cholesterol esters from liver, the isolation of lecithin from egg yolks, the purification of phosphatidyl glycerol from spinach, and the purification of isoprenyl pyrophosphate from bacteria are described in the following sections.

8.8.3 Specific Methods for Preparing Lipids

8.8.3.1 ISOLATION OF TRIGLYCERIDES AND CHOLESTEROL ESTERS FROM LIVER[24]

Fresh or frozen liver (100 g) is briefly perfused with saline (0.2% NaCl) and placed in 10-g portions in a homogenizer with 200 mL of chloroform–methanol (2:1 v/v) for 30 seconds. The homogenates are pooled and filtered through a medium sintered glass filter. An equal volume of water is added to the filtrate with mixing. The mixture is transferred to a separatory funnel and the phases separated. The water phase is discarded and the chloroform phase is evaporated to dryness in a rotory evaporator. The residue (2–3 g) is dissolved in 100–150 mL of dry benzene.

The benzene solution, containing the liver lipids, is added to a 4 cm × 50 cm column of silicic acid (∼250 g). The column is eluted with 1 L of hexane–diethyl ether (85:15) at a flow rate of 5 mL/min. The eluate is collected in 10 mL fractions, and the separation is monitored by silica gel TLC. Fractions with the same R_f are pooled and evaporated *in vacuo*. Cholesterol esters emerge in fractions 20–47, triglycerides and traces of fatty acid esters emerge in fractions 50–53, and triglycerides emerge in fractions 54–104. Free fatty acids are eluted with 250 mL of diethyl ether. The majority of the lipids are triglycerides (60%) and cholesterol esters (30%).

8.8.3.2 ISOLATION OF EGG YOLK LECITHIN[25]

The phospholipids of egg yolk are precipitated by acetone and chromatographed on aluminum oxide.

The yolks of 10 fresh eggs are blended thoroughly at room temperature with 400 mL of dry acetone. The homogenate is transferred to a 500-mL Erlenmeyer flask and kept at 5°C for 2 hours. The mixture is centrifuged at 0°C at $1000g$ for 20 minutes. The acetone is decanted and the residue is mixed with 500 mL of fresh acetone and centrifuged. The acetone is poured off and allowed to evaporate from the precipitate at room temperature. The precipitate is extracted twice with 200-mL portions of chloroform−methanol (1:1). The combined extract is evaporated in a rotary evaporator and the residue is dissolved in 50 mL of hexane. The phospholipids are precipitated by adding 10 volumes of dry acetone. The solution is kept at 0−5°C until it is clear (about an hour), and the precipitate is collected by filtration through a sintered glass filter. The precipitate on the filter, containing the phospholipids, is dissolved in benzene and again precipitated with acetone as above. The total phospholipids (~ 8 g) are dissolved in 40 mL of chloroform−methanol (19:1).

The phospholipids are chromatographed on a column (2.5 cm × 50 cm) of aluminum oxide. The column is prepared by pouring a slurry of 180 g of aluminum oxide in 200 mL of chloroform−methanol (19:1) into the column followed by washing with 400 mL of the same solvent. The chloroform−methanol−phospholipid solution is added to the column. The column is eluted with chloroform−methanol (1:1), collected in 50-mL fractions, and monitored by TLC. Dragendorff-positive spots appear in the fourth or fifth fraction. The fractions containing lecithin are pooled and evaporated. The yield of chromatographically pure phosphatidylcholine from 10 eggs is about 5 g.

8.8.3.3 ISOLATION OF SPHINGOLIPIDS AND SPHINGOMYELIN FROM BOVINE BRAIN[26]

Sphingomyelin, cerebrosides, and glycerophosphatides are extracted from bovine brain with acetone, petroleum ether, and ethanol. Sphingomyelin and glycerolphosphatides are separated from the cerebrosides by chromatography on aluminum oxide and the glycerophosphatides are removed by mild alkaline hydrolysis. Final purification of sphingomyelin is achieved by silica gel chromatography.

a. **Isolation of Crude Sphingolipids.** Fresh beef brains (500 g) are obtained from the slaughterhouse and frozen at −20°C until used. The membranes enveloping the brain are removed, the brain is sliced into thin pieces, and 100-g portions are homogenized for 3 minutes with acetone (4 mL per gram of tissue). The homogenates are pooled, the supernatant is decanted, and 2 L of fresh acetone is added to the solid. The mixture is left overnight at room temperature and then centrifuged for 30 minutes. One liter of 95% ethanol is added to the pellet, and the materials are stirred for 2 hours and filtered. The residue is then extracted with 1 L of petroleum ether (b.p. 40−60°C), stirred for 30 minutes, and allowed to stand for 30 minutes, whereupon the ether is decanted. The extraction of the residue is repeated with 500 mL of petroleum ether. The ethanol and petroleum ether extracts are combined and evaporated to a solid in a rotoevaporator at 25°C. The residue is extracted with 200 mL of diethylether (10°C) and centrifuged at 4°C. The resulting solid (~ 5 g) contains the crude sphingolipids, which are dissolved in 25 mL of chloroform–methanol (98:2) and stored at 5°C.

b. Purification of Sphingomyelin. The chloroform–methanol solution of crude sphingolipids is added to the top of a 4 cm × 30 cm column containing 250 g of aluminum oxide (grade II). The sample is washed onto the column with 200 mL of chloroform and the column is eluted with 800 mL of chloroform–methanol (1:1). The eluate is collected in 50-mL fractions, which are examined by silica gel micro-TLC plates. The sphingomyelin-containing fractions are pooled and evaporated. The residue (~ 2 g) is suspended in 5 mL of 1 N NaOH in methanol, incubated at 37°C for 8 hours, cooled to 22–24°C, and acidified with 7 mL of 0.8 N HCl. The mixture is extracted in a separatory funnel two times with 20 mL chloroform. The chloroform extracts are pooled and washed four times with 5 mL each of water-methanol (1:1) and evaporated. The residue is dissolved in 10 mL of a mixture of petroleum ether (b.p. 60–80°C) and methanol (9:1). Acetone is added to precipitate the sphingomyelin (~ 1.5 g), which is filtered, dried *in vacuo*, and dissolved in 30 mL of chloroform–methanol (10:1). The solution is added to a 3.5 cm × 60 cm column containing 150 g of silica gel. The sample is washed onto the column with 300 mL of chloroform and eluted with 300-mL portions of the following mixtures of chloroform–methanol: 95:1, 90:10, 85:15, 70:30, and 50:50, and then 600 mL of methanol. The flow of the column is adjusted to 4–5 mL/min. The 50:50 mixture and methanol eluates are collected in 50-mL fractions and analyzed by silica gel micro-TLC. Fractions containing the sphingomyelin are pooled and evaporated to give 0.9–1.1 g of pure sphingomyelin.

8.8.3.4 ISOLATION OF PHOSPHATIDIC ACID AND PHOSPHATIDYL-GLYCEROL FROM SPINACH LEAVES[27]

Phosphatidic acid and phosphatidylglycerol are obtained from spinach leaves by chloroform–methanol extraction, acetone precipitation, and chromatography on silicic acid columns.

Fresh spinach leaves (6 kg) are washed and kept at 95°C for 1 minute to inactivate phospholipases. The leaves are homogenized with 3 L of methanol, and the homogenate is filtered through cloth and washed with methanol and chloroform–methanol (1:1) until the insoluble material becomes colorless. The filtrate and washings are extracted with three portions of an equal volume of chloroform–methanol (2:1) and filtered through a coarse sintered glass filter. The chloroform layer (~ 10 L) is evaporated to dryness in a rotory evaporator (30–35°C). The residue is dissolved in 50 mL of ether and poured into 1 L of dry acetone, saturated with magnesium chloride. The mixture is stored for 16 hours at

$-15°C$ and centrifuged at $1000g$; the precipitate (~ 15 g) is dissolved in 200 mL of chloroform.

The chloroform solution is added to a 5 cm $\times$ 60 cm column of 500 g of silica gel. The column is washed with 2 L of chloroform and then eluted with 1-L portions of chloroform–methanol of the following proportions: (a) 96:4, (b) 94:6, (c) 92:8, and (d) 88:12. Eluate b contains phosphatidic acid; eluates c and d contain phosphatidylglycerol.

8.8.3.5 ISOLATION OF ^{32}P-LABELED C$_{55}$-ISOPRENYL PYROPHOSPHATE FROM *MICROCOCCUS LYSODEIKTICUS*[28]

C$_{55}$-Isoprenyl phosphate is involved in the biosynthesis of a number of bacterial polysaccharides as a coenzyme carrier of activated carbohydrate repeating units of the polysaccharide. The pyrophosphate can be isolated from the bacteria when Bacitracin, which inhibits the pyrophosphatase, is added to the culture and allows the isoprenyl pyrophosphate to accumulate. [*Note:* All glassware used in this procedure must be silanized by treating with 5% trimethylchlorosilane in benzene.] One liter of *Micrococcus lysodeikticus* ATCC 4689 is grown in a medium of 1% Bactopeptone, 0.1% yeast extract, and 0.5% NaCl adjusted to pH 7.5. At approximately one-third of maximum growth, the cells are asceptically transferred to a medium containing 25 mCi ^{32}P-labeled inorganic phosphate and 0.01% yeast extract. Bacitracin (160 μg/mL) is added after 4 hours, and the cells are harvested after an additional 3 hours and washed two or three times with 200 mL of 0.1 M Tris buffer (pH 7). The cell paste is suspended in 20 volumes of $-15°C$ acetone for 10 minutes, and the precipitate is collected on a sintered glass filter. The residue on the filter is extracted with 4 volumes of chloroform–methanol (2:1) at room temperature for 20 minutes, followed by extraction with 4 volumes of a mixture of 2 volumes of 1-butanol to 1 volume of pyridinium–acetate buffer made by adding pyridine to 6 M acetic acid until the pH is 4.2. The extraction is repeated three times with 10-second bursts of sonication in the presence of 1 mL of glass beads (5 μm diameter). The extract must be kept cold during the sonication and the subsequent extraction. The extracts are combined and kept at $0°C$ and washed four times each with 0.5 volume of cold water. The resulting aqueous phases are combined and backwashed once with 0.5 volume of 1-butanol. All butanol extracts are combined and made alkaline (pH 9) with pyridine, and evaporated *in vacuo*.

The resulting lipid is dissolved in 1 mL of carbon tetrachloride to which 9.4 mL of ethanol, 0.8 mL of water, and 0.3 mL of 1 M NaOH are added. This mixture is incubated at $37°C$ for 30 minutes. Ethyl formate (0.4 mL) is added, and the mixture incubated for an additional 5 minutes at $37°C$. The sample at pH 7 is evaporated *in vacuo* and dissolved in 6 mL of chloroform–methanol (2:1), and 4 mL of water is added. The lower organic phase is washed four times with 5 mL of water–methanol (1:1), and the aqueous washings backwashed twice with chloroform. The combined organic phase is evaporated *in vacuo*, suspended in acetone by sonication, and added to a 1-mL column of silica gel G that has been washed with acetone. The column is eluted with 3 mL of acetone followed by 3 mL of chloroform–methanol (1:1) and 3 mL of methanol; 1-mL fractions are collected and counted. The major radioactive fractions are combined and are primarily C_{55}-isoprenyl pyrophosphate and amount to about $\sim 10^8$ dpm, approximately equally distributed between the two phosphate groups. Pyrophosphatase liberates inorganic phosphate and C_{55}-isoprenyl phosphate.

8.9 LITERATURE CITED

1. F. C. Guerra, "Techniques for Mitochondria: Techniques for Rat Liver Mitochondria," in *Methods Enzymol.* S. Fleischer and L. Packer, Eds. vol. 31: 299 (1974).
2. (a) P. S. Nobel, "Rapid Isolation Techniques for Chloroplasts," in *Methods Enzymol.* S. Fleischer and L. Packer, Eds. vol. 31: 600 (1974). (b) D. A. Walker, "Chloroplasts," in *Methods Enzymol.* A. San Pietro, Ed. vol. 23: 211 (1971).
3. J. R. Tata, "Isolation of Nuclei from Liver and Other Tissue," in *Methods Enzymol.* S. Fleischer and L. Packer, Eds. vol. 31: 253 (1974).
4. A. McPherson, Jr., "Crystallization of Proteins from Polyethylene Glycol," *J. Biol. Chem. 251*: 6300 (1976).
5. J. B. Sumner, "The Isolation and Crystallization of the Enzyme Urease," *J. Biol. Chem. 69*: 435 (1926).
6. P. J. G. Butler and M. M. T. Jones, "The Preparation of Alcohol Dehydrogenase and Glyceraldehyde 3-Phosphate Dehydrogenase from Baker's Yeast," *Biochem. J. 118*: 375 (1970).
7. R. Shainkin and Y. Birk, "Isolation of Pure α-Amylase from Human Saliva," *Biochim. Biophys. Acta, 122*: 153 (1966).
8. M. Malamy and B. L. Horecker, "Purification and Crystallization of the Alkaline Phosphatase of *Escherichia coli*," *Biochemistry, 3*: 1893 (1964).
9. G. A. Wilson and F. E. Young, "Isolation of a Sequence-specific Endonuclease (Bam I) from *Bacillus amyloliquefaciens* H," *J. Mol. Biol. 97*: 123 (1975).
10. N. P. Badenhuizen, "General Method for Starch Isolation," in *Methods Carbohydr. Chem.* R. L. Whistler, R. J. Smith, and J. N. BeMiller, Eds, vol. 4: 14 (1964).
11. J. N. BeMiller, "Glycogen," in *Methods Carbohydr. Chem.* R. L. Whistler, R. J. Smith, and J. N. BeMiller, Eds. vol. 5: 138 (1965).
12. R. W. Jeanloz, "Heparin," in *Methods Carbohydr. Chem.* R. L. Whistler, R. J. Smith, and J. N. BeMiller, Eds. vol. 5: 150 (1965).
13. C. T. Laurent, "Studies on Fractionated Heparin," *Arch. Biochem. Biophys. 92*: 224 (1961).
14. J. Marmur, "Procedure for the Isolation of Deoxyribonucleic Acid from Microorganisms," *J. Mol. Biol. 3*: 208 (1961).
15. J. Marmur, "A Procedure for the Isolation of Deoxyribonucleic Acid from Microorganisms," in *Methods Enzymol.* S. P. Colowick and N. O. Kaplan, Eds. vol. 6: 726 (1963).
16. H. C. Birnboim, "A Rapid Alkaline Extraction Method for the Isolation of Plasmid DNA," in *Methods Eznymol.* R. Wu, L. Grossman, and K. Moldave, Eds. vol. 100: 247 (1983).
17. S. A. Williams, B. E. Slatko, L. S. Manan, and S. M. DiSimone, "Isolation and Purification of Bacteriophage DNA," *Biotechniques, 4*: 138 (1986).
18. G. von Ehrenstein, "Isolation of sRNA from intact *Escherichia coli* Cells," in *Methods Enzymol.* L. Grossman and K. Moldave, Eds. vol. 12, part A: 588 (1967).

19. G. Zubay, "The Isolation and Fractionation of Soluble Ribonucleic Acid," *J. Mol. Biol. 4*: 347 (1962).

20. R. W Holley, J. Apgar, B. P. Doctor, J. Farrow, M. A. Marini, and S. H. Merrill, "A Simplified Procedure for the Preparation of Tyrosine- and Valine-acceptor Fractions of Yeast Soluble Ribonucleic Acid," *J. Biol. Chem. 236*: 200 (1961).

21. D. Bell, R. V. Tomlinson, and G. M. Tener, "Chemical Studies on Mixed Soluble Ribonucleic Acids from Yeasts," *Biochemistry, 3* 317 (1964).

22. M. Girard, "Isolation of Ribonucleic Acids from Mammalian Cells and Animal Viruses," in *Methods Enzymol.* L. Grossman and K. Moldave, Eds. vol. 12, part A: 581 (1967).

23. G. Brawerman, "The Isolation of Messenger RNA from Mammalian Cells," in *Methods Enzymol.* K. Moldave and L. Grossman, Eds. vol. 30: 605 (1974).

24. E. V. Dyatlovilskaya, "Isolation of Triglycerides from Rat Liver," in *Lipid Biochemical Preparations*, L. D. Bergelson, Ed. Elseiver/North Holland, New York/Amsterdam, 1980, pp. 122–124.

25. L. I. Barsukov, V. I. Volkova, and L. D. Bergelson in *Lipid Biochemical Preparations*, L. D. Bergelson, Ed. Elseiver/North Holland, New York/Amsterdam, 1980 pp. 128–129.

26. L. D. Bergelson, in *Lipid Biochemical Preparations*, L. D. Bergelson, Ed. Elseiver/North Holland, New York/Amsterdam, 1980 pp. 155–156.

27. F. Haverkate and L. L. M. van Deenen, "Isolation and Chemical Characterization of Phosphatidyl Glycerol from Spinach Leaves," *Biochim. Biophys. Acta, 106*: 78 (1965).

28. K. J. Stone and J. L. Strominger, in *Methods Enzymol.* V. Ginsburg, Ed. vol. 28: 306 (1972).

8.10 REFERENCES FOR FURTHER STUDY

1. "Enzyme Purification and Related Techniques," vol. 22 in *Methods in Enzymology.* W. B. Jakoby, Ed., Academic Press, New York, 1971.

2. R. K. Scopes, *Protein Purification: Principles and Practice.* Springer-Verlog, New York, 1982.

3. *Lipid Biochemical Preparations*, L. D. Bergelson, Ed., Elsevier/North Holland, Amsterdam/New York, 1980.

4. See Appendix A for a summary of other references on biological preparations.

CHAPTER 9

ENZYMOLOGY

Enzymes, the proteins that catalyze reactions in living cells, make up one of the important classes of biological molecules. Enzymes increase the velocity of a reaction between 10^6- and 10^8-fold over the velocity of the uncatalyzed reaction. Besides greatly increasing the velocity, enzymes have the special property of specificity. An individual enzyme will catalyze a specific type of reaction with a unique substrate or group of related substrates. Currently there are approximately 2200 recognized enzymes, although it is estimated that a single bacterial cell contains 10^4 different kinds of enzymes. Many of the recognized enzymes have been highly purified and even crystallized.

In 1835 J. J. Berzelius,[1] the Swedish chemist, developed the theory of chemical catalysis. Berzelius gave a number of examples of enzyme-catalyzed reactions and cited the tremendous potency of the enzyme "diastase" to hydrolyze starch in comparison with the hydrolysis by acid. The specificity of enzymes impressed Emil Fischer,[2] who in 1894 proposed the famous lock-and-key model to explain enzyme action and specificity. According to this model, the molecule undergoing catalysis fits exactly into a site on the enzyme at which the catalysis occurs. The first recognized cell-free enzyme preparation was made from yeast by E. Büchner[3] in 1897. The preparation, which catalyzed the conversion of sugar into alcohol, was a mixture of enzymes called zymase.

The kinetic and catalytic properties of enzymes were studied in the early part of the twentieth century. In 1913 L. Michaelis and M. L. Menten[4] published their classic study of the substrate concentration dependence of enzyme reactions and postulated the formation of an enzyme–substrate complex, the Michaelis complex or intermediate, to explain their kinetic data.

J. B. Sumner[5] crystallized the first enzyme, jack bean urease, in 1926. Shortly thereafter (1930–1936) J. H. Northrup and M. Kunitz[6] reported the crystallization of pepsin, trypsin, and chymotrypsin. Analysis of these crystalline compounds, which were shown to be proteins with enzymatic activity, established the protein nature of enzymes.

291

The study of the structures of enzymes was aided by the development of a number of techniques:

1. The use of the ultracentrifuge to determine the molecular weight of macro-molecules (T. Svedberg,[7] 1930–1940)
2. The determination of amino acid composition by ion-exchange chroma-tography (W. Stein and S. Moore,[8] 1940–1950)
3. The determination of the amino acid sequence (F. Sanger,[9] 1950–1960)
4. The determination of purity and molecular weight by disc gel electrophoresis (B. J. Davis,[10] 1964; K. Weber and M. J. Osborn,[11] 1969)
5. The use of X-ray crystallographic techniques to determine three-dimensional structure (M. F. Perutz[12] and J. C. Kendrew,[13] 1960–1970). In 1965 L. N. Johnson and D. C. Phillips[14] demonstrated the actual formation of an enzyme complex by taking a X-ray "picture" of the binding of an inhibitor at the active site of lysozyme.

Research on the action of enzymes continues to contribute to the general understanding of chemical catalysis and the mechanism of chemical reactions. Enzymes are finding increasing use as "reagents" in the specific synthesis of compounds that are difficult to synthesize by the usual chemical methods (Section 6.12). Enzymes, immobilized on solid supports, are used over and over in chemical reactors for the production of compounds in the chemical industry. Enzymes also have found use as chemotherapeutic agents, especially in the treatment of hereditary diseases, and as diagnostic indicators in the detection of disease. Furthermore an understanding of the specificity, regulation, structure, and mechanism of action of enzymes provides a major contribution to the understanding of the functioning of the living cell.

Enzymologists use special terms in discussing enzymes. The molecule undergoing reaction is called the **substrate** and the complete reaction mixture containing substrate, buffer, and enzyme is called a **digest**. The **active site** of an enzyme is the part of the enzyme molecule in which the substrate binds and catalysis occurs to give products. Other terms will be defined as they occur in the sections that follow.

9.1 KINETICS AND THEORY OF ENZYME ACTION

Michaelis and Menten found that the rate or velocity v of catalysis by enzymes varies with the substrate concentration. The velocity increases with the increase in substrate concentration up to a certain point and then becomes constant and reaches a **maximum velocity** V_m as shown in Figure 9-1. Michaelis and Menten proposed a simple model to accommodate the kinetics they observed. In their model, the enzyme E combines with the substrate S to form an enzyme–substrate complex ES. The ES complex can either dissociate back to E and S or proceed to

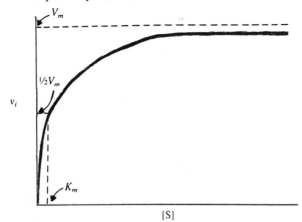

FIGURE 9-1. Initial velocity v_i as a function of substrate concentration [S] for an enzyme-catalyzed reaction following Michaelis–Menten kinetics. The units of v_i are in some amount (μmol, mmol, etc.) of product produced or substrate converted per minute. The units of [S] are in concentration (mmol/L, mol/L, etc.).

form a product, P. This was expressed in the following reactions:

$$E + S \underset{k_{-1}}{\overset{k_1}{\rightleftharpoons}} ES \overset{k_2}{\rightarrow} E + P \tag{9-1}$$

It was assumed that none of the product reverts back to the substrate. Michaelis and Menten developed a mathematical treatment, given below, for this process and obtained their now-famous equation.

The velocity of forming ES is given by:

$$v_f = k_1(E_f)(S) \tag{9-2}$$

where E_f, the concentration of free enzyme, equals the total amount of enzyme added minus the amount of enzyme complexed with substrate, $(E_t) - (ES)$. Substituting for E_f in Equation 9-2, we have:

$$v_f = k_1[(E_t) - (ES)](S) \tag{9-3}$$

The velocity of disappearance of ES is given by:

$$v_d = k_{-1}(ES) + k_2(ES) \tag{9-4}$$

At the steady state, when the velocities of formation and disappearance of ES are equal ($v_f = v_d$), we have:

$$k_1[(E_t) - (ES)](S) = k_{-1}(ES) + k_2(ES) \tag{9-5}$$

Rearrangement of the terms in Equation 9-5 to put all the rate constants on one side of the equation gives:

$$\frac{(S)[(E_t) - (ES)]}{(ES)} = \frac{k_{-1} + k_2}{k_1} \tag{9-6}$$

$$\frac{k_{-1} + k_2}{k_1} = K_m \qquad \text{the \textbf{Michaelis constant}} \tag{9-7}$$

Further rearrangement of Equation 9-6 gives:

$$(ES) = \frac{(E_t)(S)}{K_m + (S)} \tag{9-8}$$

The **initial velocity** v_i of the enzyme-catalyzed reaction is expressed:

$$v_i = k_2(ES) \tag{9-9}$$

where k_2 is the rate constant for the conversion of the ES complex into products (Equation 9-1) and is sometimes called K_{cat} or the *turnover* number.

The **maximum velocity** V_m, is obtained when all the enzyme is complexed with substrate:

$$V_m = k_2(ES) = k_2(E_t) \tag{9-10}$$

Multiplying both sides of Equation 9-8 by k_2 gives:

$$k_2(ES) = \frac{k_2(E_t)(S)}{K_m + (S)} \tag{9-11}$$

and substituting Equations 9-9 and 9-10 into 9-11 gives:

$$v_i = \frac{V_m(S)}{K_m + (S)} \qquad \text{the \textbf{Michaelis–Menten equation}} \tag{9-12}$$

This equation explains the kinetic plot of the velocity dependence on substrate concentration shown in Figure 9-1. When (S) is much less than K_m, the initial velocity v_i is directly proportional to the concentration of S, and we write:

$$v_i = \frac{V_m(S)}{K_m} \tag{9-13}$$

At high substrate concentrations, when (S) is much greater than K_m,

$$v_i = \frac{V_m(S)}{(S)} \tag{9-14}$$

$$v_i = V_m \tag{9-15}$$

When (S) = K_m, it can be shown from Equation 9-12 that:

$$v_i = \tfrac{1}{2} V_m \tag{9-16}$$

and it can be further shown that when $(S) = nK_m$

$$v_i = \frac{n}{n+1} V_m \qquad (9\text{-}17)$$

Thus, when $[S] = 10$ times K_m, $v_i = 10/11 \, V_m$ or 91% of the maximum velocity.

9.2 DETERMINING THE INITIAL VELOCITY v_i

The initial velocity is equal to the amount of product formed per unit time or to the amount of substrate lost per unit time:

$$v_i = \frac{d(P)}{dt} = \frac{-d(S)}{dt} \qquad (9\text{-}18)$$

The v_i is determined by quantitatively measuring the amount of one of the products at various times or by quantitatively determining the amount of unused substrate at various times. A plot of (P) versus time t or of (S) versus time is prepared (see Fig. 9-2). The slope of the linear part of the curve is the initial velocity.

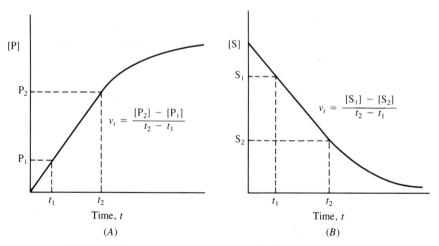

FIGURE 9-2. Determination of v_i. (A) Plot of the concentration of product versus time. (B) Plot of the concentration of substrate versus time. The slopes of the linear part of the two lines are equal to the v_i of the reaction.

The v_i, thus, can be obtained by taking samples of the reaction at various times after adding enzyme. The enzyme reaction is stopped by some appropriate means of denaturing the enzyme such as by heating in a boiling water bath, lowering the pH by the addition of acid, or the addition of alcohol. The concentration of the product or the substrate in each aliquot is determined by some chemical or physical analysis (Table 9-1, p. 301, gives methods of assaying different enzymes). The concentration of product or substrate is then plotted against time. The plot that results should be linear over a minimum of three time points, and the slope of the linear part of the line is then the initial velocity of the reaction (Fig. 9-2). Continuous, automated methods of product or substrate analysis give a continuous measure of the reaction over the entire time course.

9.3 ENZYME REACTIONS AS A FUNCTION OF pH

The velocity of enzyme-catalyzed reactions depends on pH. Enzymes have pH optima and often give bell-shaped curves of velocity versus pH (Fig. 9-3), although other shapes have been observed. The pH optimum for different enzymes varies depending on the nature of the catalytic groups. The stability of the tertiary and/or quaternary structures of the enzyme may also be pH dependent and may affect the velocity of the enzyme reaction, especially at extreme alkaline or acidic pH values.

The pH of the enzyme digest is fixed by the use of buffers (Section 2.8). Because most enzymes do not consume or produce hydrogen ions during the reaction, the buffering capacity need not be high. Buffer concentrations of 20–50 mM usually are sufficient.

To determine the optimum pH of an enzyme reaction, the reaction mixture is buffered at different pH values and the activity (Section 9.5) of the enzyme is determined. To determine the pH-stability profile of an enzyme, various aliquots of the enzyme are buffered at different pH values and held for a given period (one

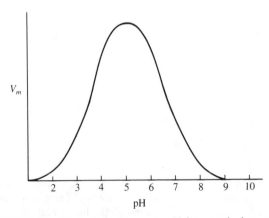

FIGURE 9-3. Plot of the V_m versus pH for a typical enzyme.

hour, one day, etc.). The pH of the aliquots is then adjusted to the optimum pH value and each aliquot is assayed. The effect of pH on stability over the period of time chosen is, thus, obtained.

9.4 ENZYME REACTIONS AS A FUNCTION OF TEMPERATURE

The velocity of an enzyme-catalyzed reaction increases with an increase in the temperature. The velocity approximately doubles for every 10°C rise in temperature. In contrast with ordinary chemical reactions, enzyme reactions have optimum temperatures and then rapidly decrease with further temperature increases (Fig. 9-4). The loss of activity at the higher temperatures is due to thermal conformational (denaturation) changes of the enzyme. Most enzymes are inactivated at temperatures above 55–60°C, but the thermostable enzymes have increased activities between 60 and 90°C. These enzymes are obtained primarily from thermophilic bacteria.

The optimum temperature for an enzyme-catalyzed reaction is obtained by determining the activity of the enzyme at different temperatures. The V_m is determined at different temperatures by using the same amount of enzyme at each temperature and a saturating concentration of substrate (Section 9.5). The plot of the V_m versus temperature shown in Figure 9-4 represents a typical enzyme.

The effect of temperature on the rate of an enzyme reaction is described by the **Arrhenius equation**:

$$\log_{10}k = -2.303\frac{E}{R}\left(\frac{1}{T}\right) + \log_{10}C \qquad (9\text{-}19)$$

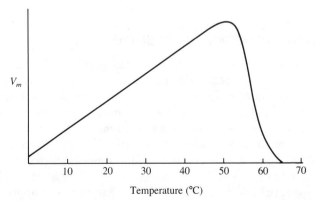

FIGURE 9-4. Plot of the V_m versus temperature for a typical enzyme.

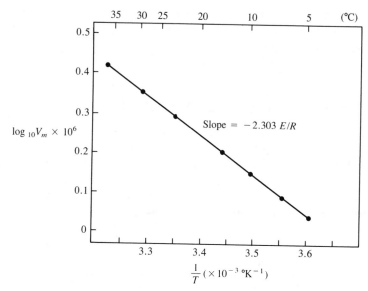

FIGURE 9-5. Arrhenius plot of the effect of temperature on the maximum velocity of an enzyme reaction $\log_{10}V_m$ versus the reciprocal of the temperature in degrees kelvin.

where k is the rate constant of the reaction and the $\log_{10}k$ is inversely proportional to the absolute temperature in degrees kelvin ($^{\circ}C + 273$) multiplied by R, the gas constant (1.98 cal/mol-$^{\circ}K$), and E is the **activation energy**. Sometimes also called the **transition-state energy**, E is the amount of energy necessary to place the substrate into a reactive state; C is an integration constant.

In Michaelis–Menten kinetics, when the concentration of S is high, $v_i = V_m = k_2[E_t]$ and k_2 is the rate constant for the enzyme reaction. Thus, a plot of $\log_{10}V_m$ versus $1/T$ gives a straight line (Fig. 9-5), and the activation energy E can be obtained from the slope of the line.

9.5 DETERMINING THE ACTIVITY OF AN ENZYME

The activity of an enzyme is the amount of reaction that a certain amount of enzyme will produce in a specified period of time. The activity is determined by measuring the amount of product produced or the amount of substrate used up per unit of time under high concentrations or saturating conditions of substrate. The activity is essentially the measurement of the initial velocity under conditions that make it the maximum velocity. The activity should be proportional to the amount of enzyme added. That is, if two or three times as much enzyme is added, the activity should be two or three times as great. Enzymes are usually assayed at their optimum pH and temperature. But to determine the optimum pH and temperature (Sections 9.3 and 9.4), the enzyme is assayed at various pH values and temperatures.

In an enzyme assay it is critical that the substrate concentration be saturating during the entire period the reaction is sampled (i.e., [S] is at least $20K_m$) and that the amount of product or substrate measured be linear over the period during which the reaction is sampled. The activity of the enzyme is then obtained as the slope of the linear part of the line of a plot of [P] or [S] versus time (Fig. 9-2). If there are a large number of samples to assay, a single aliquot is frequently taken at a specified time. This can be risky and will give valid results only if the time at which the sample is taken falls on the linear part of a plot of [P] or [S] versus time. These plots will become nonlinear if the substrate concentration falls below the concentration necessary to saturate the enzyme, if the increase in the concentration of the product produces a significant amount of the reverse reaction, or if denaturation of the enzyme causes a loss of activity.

The activity of an enzyme is determined by measuring the amount of product formed or the amount of substrate that has disappeared. This is usually accomplished by performing a chemical analysis for the product or substrate. For example, β-galactosidase will catalyze the hydrolysis of o-nitrophenyl-β-D-galactopyranoside to give o-nitrophenol, which absorbs maximally at 405 nm and can, thus, be determined spectrophotometrically. Polysaccharide hydrolysis by various hydrolytic enzymes can be assayed by determining the increase in the reducing value (with alkaline copper or ferricyanide reagents; see Section 7.1.2.2) that results when low molecular weight saccharides are released from the polysaccharides. Sometimes the amount of radioactivity that is incorporated into a product from a radioactive substrate is measured, as for example, the assay of DNA polymerase in which ^{3}H incorporation into DNA from tritiated dTTP is determined by separating the polymeric DNA from the monomer dTTP and counting the amount of ^{3}H incorporated into DNA by liquid scintillation spectrometry.

Unfortunately, an easily measured product is not always produced. It may still be possible, however, to assay the enzyme by coupling the reaction with another enzyme that can convert one of the products into another substance that can be easily measured. Many coupled assays have been developed using dehydrogenases that use NAD^+ or NADH as coenzymes. NADH can be easily measured spectrophotometrically at 340 nm. For example, the reaction of aldolase can be assayed by reacting one of the products of the aldolase reaction, 3-phospho-D-glyceraldehyde, with α-glycerophosphate dehydrogenase and NADH. The disappearance of NADH is measured by the loss of absorbance at 340 nm.

$$\text{D-fructose-1,6-bisphosphate} \xrightarrow{\text{aldolase}} \text{dihydroxyacetonephosphate}$$
$$+$$
$$\alpha\text{-glycerophosphate} \xleftarrow{\text{dehydrogenase}} \text{3-phospho-D-glyceraldehyde}$$
$$\text{NAD}^+ \qquad \text{NADH} + \text{H}^+$$

Another type of coupled assay is the assay of oxidases (e.g., D-glucose oxidase and L-amino acid oxidase) that produce hydrogen peroxide. Peroxidase and a dye that is colorless in the reduced state are added to the digest. The peroxidase converts the hydrogen peroxide to oxygen that oxidizes the dye (o-dianisidine) to a colored product that absorbs at 460 nm.

D-glucose $\xrightarrow{\text{glucose oxidase}}$ δ-gluconolactone + hydrogen peroxide

oxidized dye $\longleftarrow$ reduced dye + oxygen + water $\longleftarrow$ peroxidase

In developing coupled assays, the concentration of the second enzyme should be fixed and nonlimiting. Likewise, if for the second enzyme there are any coenzymes or cosubstrates (e.g., NADH is a coenzyme and the reduced dye is a cosubstrate in the examples above), their concentration too must be fixed and nonlimiting. An additional factor to be considered is the pH and temperature profiles of the enzymes. It is essential that all the component enzymes be active at the pH and temperature employed in the assay. Usually the pH and temperature optimum of the enzyme being assayed (i.e., the first enzyme) is used, and if the second enzyme is active, although maybe not optimally, enough of the second enzyme is added to ensure that its activity is not limiting. Table 9-1 lists enzymes, their substrates, and methods of assay.

In 1964, the International Union of Biochemistry established a standard unit for the expression of enzyme activity. One **international unit** (IU) of enzyme activity was defined as "the amount of enzyme necessary to produce one micromole of product (or the loss of one micromole of substrate) per minute, under specified conditions of substrate concentration, pH, and temperature." Prior to this, there were almost as many units as there were enzymes or investigators. The standardization of the enzyme unit permits a comparison of the activities of different kinds of enzymes and different preparations.

A TYPICAL ENZYMATIC ASSAY

The reaction is initiated by adding 5.0 mL of enzyme, which has been diluted 1:20 with 25 mM buffer, to 5 mL of substrate buffered with 25 mM buffer. 1.0 mL aliquots are removed at 3, 6, 9, 12, and 15 minutes and the amount of product measured by a color-producing chemical reaction. A graph is prepared of the amount of product versus time. The slope of the linear part of the line is found to be 0.90 μmol/min. Therefore:

units of enzyme/mL = (0.90 μmol/mL) $\times$ 10 $\times$ 2 $\times$ 20 = 360 U/mL

The factor 10 gives the total amount of micromoles produced in the digest because only 1.0 mL is assayed and there are 10 mL per assay digest. The factor 2 corrects for the dilution of the enzyme (5 mL of enzyme added to 5 mL of substrate gives a 1:2 dilution of the enzyme). The factor 20 corrects for the initial dilution (1:20) of the enzyme before the addition of the substrate.

The factors would be changed, depending on the design of the assay. For example, the foregoing assay might be run by adding 0.5 mL of 1:50 diluted enzyme to 9.5 mL of substrate and 0.1 mL aliquots taken. In this example, the correction factors would be 100 $\times$ 20 $\times$ 50 to obtain the total number of micromoles produced, the dilution of the enzyme in the digest, and the initial dilution, respectively.

TABLE 9-1. Assay methods for enzymes

Enzyme	pH	Temp. (°C)	Substrate	Assay Principle	Ref.
acetylcholine esterase	7	25°	acetylcholine	titration of acetic acid produced	(18)
acid phosphatase	5	25°	o-carboxyphenyl phosphate	increase in absorbance at 300 nm due to formation of salicylic acid	(19)
yeast alcohol dehydrogenase	7.5	25°	ethanol	increase in absorbance at 340 nm due to formation of NADH	(20)
L-amino acid oxidase	7.6	25°	L-leucine	increase in absorbance at 436 nm due to oxidation of o-dianisidine coupled through peroxidase	(21)
α-amylase	7	37°	starch	increase in reducing value by copper or ferricyanide reagent	(22)
β-amylase	4.8	25°	starch	increase in reducing value by copper or ferricyanide reagent	(22)
aspartate amino-transferase (GOT)	7.4	25°	L-aspartate	loss of absorbance at 340 nm due to loss of NADH in coupled reaction with malate dehydrogenase	(23)
aldolase	7.5	25°	D-fructose-1,6-bisphosphate	loss of absorbance at 340 nm due to loss of NADH, coupled through α-glycerophosphate dehydrogenase	(24)
alkaline phos-phatase	8	25°	p-nitrophenyl phosphate	increase in absorbance at 410 nm due to release of p-nitrophenol	(25)
carboxypeptidase	7.5	25°	hippuryl-L-phenylalanine	increase in absorbance at 254 nm due to release of phenylalanine	(26)
catalase	7	25°	hydrogen peroxide	decrease in absorbance at 240 nm due to disappearance of peroxide	(27)
cellulase	5	37°	microcrystalline cellulose	increase in reducing value by copper or ferricyanide reagent	(22)
chymotrypsin	7.8	25°	N-benzoyl-L-ethyl tyrosine	increase in absorbance at 256 nm	(28)
creatine kinase	8.9	25°	creatine + ATP	decrease in absorbance at 340 nm due to loss of NADH, coupled through pyruvate kinase and lactate dehydrogenase	(29)

(continued)

TABLE 9-1. (*continued*)

Enzyme	pH	Temp. (°C)	Substrate	Assay Principle	Ref.
deoxyribonuclease	5	25°	DNA	increase in absorbance at 260 nm due to hydrolysis of DNA	(30)
dextranase	6	37°	dextran	increase in reducing value by copper or ferricyanide reagent	(22)
DNA ligase	8.0	30°	linear [³H]DNA	acid precipitation of exonuclease-stable (cyclic) DNA and scintillation counting	(31)
DNA polymerase	6.8	37°	activated DNA, dATP, dGTP, dCTP, [³H]dTTP	incorporation of ³H into DNA by precipitation with TCA and liquid scintillation counting	(32)
elastase	8.8	37°	orcein-elastin	increase in absorbance at 590 nm due to release of orcein dye	(33)
β-galactosidase	7.5	25°	p-nitrophenyl β-D-galactopyranoside	increase in absorbance at 405 nm due to release of p-nitrophenol	(34)
glucose oxidase	6	37°	glucose	increase in absorbance at 420 nm due to oxidation of o-dianisidine coupled through peroxidase	(35)
glucose-6-phosphate dehydrogenase	7.8	30°	glucose-6-phosphate	increase in absorbance at 340 nm due to formation of NADPH	(36)
β-glucosidase	5	37°	p-nitrophenyl-β-D-glucopyranoside	absorbance at 420 nm due to released p-nitrophenol	(37)
glutamic acid decarboxylase	5	37°	[1-¹⁴C]glutamic acid	trapping of released ¹⁴C-carbon dioxide in alkali and scintillation counting	(38)
glyceraldehyde 3-phosphate dehydrogenase	8.5	25°	glyceraldehyde-3-phosphate	increase in absorbance at 340 nm due to formation of NADH	(39)
glycerol kinase	9	25°	glycerol + ATP	increase in absorbance at 340 nm due to formation of NADH, coupled through glycerol-3-phosphate dehydrogenase	(40)
glycogen phosphorylase	6.5	30°	glycogen + P_i	increase in absorbance at 340 nm due to formation of NADPH, coupled through phosphoglucomutase and glucose-6-phosphate dehydrogenase	(41)
hexokinase	7.5	30°	glucose + ATP	increase in absorbance at 340 nm due to formation of NADH, coupled through glucose-6-phosphate dehydrogenase	(42)
inorganic pyrophosphatase	7.2	25°	sodium pyrophosphate	increase in orthophosphate by molybdateaminonaphthol-sulfonic acid reagent	(43)

	pH	temp.	substrate	assay	
lactate dehydrogenase (beef)	7.3	25°	lactic acid	increase in absorbance at 340 nm due to formation of NADH	(44)
lactate dehydrogenase (yeast)	8.4	25°	lactic acid	decrease in absorbance at 420 nm due to reduction of ferricyanide	(45)
lactoperoxidase	7.0	25°	iodide + hydrogen peroxide	increase in absorbance at 350 nm due to formation of triiodide	(46)
lipase, pancreatic	8	25°	[U-^{14}C]triolein	isolation of released oleic acid by ion-exchange, and scintillation counting	(47)
luciferase	6.8	25°	flavin mononucleotide	production of light (495 nm) upon oxidation of FMN	(48)
lysozyme	7	25°	bacterial cells	decrease in turbidity measured as absorbance at 450 nm	(49)
neuraminidase	5	37°	mucin	colorimetric determination of periodate oxidation of released sialic acid	(50)
nitrogenase	7.2	30°	acetylene	gas-liquid chromatographic analysis of ethylene produced by acetylene reduction	(51)
papain	6.2	25°	benzoyl L-arginine ethyl ester	titration of released benzoic acid	(52)
peroxidase	7	25°	guaiacol, hydrogen peroxide	increase in absorbance at 470 nm due to oxidation of guaiacol	(53)
restriction endonuclease	7.6	37°	cyclic [^{3}H]DNA	liquid scintillation counting of acid-soluble products of exonuclease action	(54)
ribulose 1,5-bisphosphate carboxylase	7.9	30°	ribulose 1,5-bisphosphate, [^{14}C]-KHCO$_3$	incorporation of ^{14}C into products not volatile from acid solution, and scintillation counting	(55)
RNA polymerase	7.9	25°	ATP, GTP, CTP, [^{3}H]UTP	incorporation of ^{3}H into acid-insoluble products by scintillation counting	(56)
ribonuclease	5	37°	RNA	absorbance at 260 nm of ribonucleotides following acid precipitation of RNA	(57)
trypsin	8.1	25°	p-toluenesulfonyl L-arginine methyl ester	increase in absorbance at 247 nm due to release of p-toluenesulfonyl L-arginine	(58)
urease	7.0	38°	urea	titration of ammonia produced	(59)

Enzymes are increasingly being used as catalysts in the preparation of specific compounds. In these preparations, the amount of time theoretically necessary to convert all the substrate to product by the amount of enzyme added is called the **conversion period**. A 100% conversion of substrate to product is almost never attained because of reverse reaction and the establishment of equilibrium.

DETERMINING THE CONVERSION PERIOD

If a reaction digest has 10 mol of substrate and 10,000 IU of enzyme is added, what would be the conversion period?

$$10 \text{ mol} = 10 \times 10^6 \text{ } \mu\text{mol} = 10^7 \text{ } \mu\text{mol}$$

$$10,000 \text{ units} = 10^4 \text{ } \mu\text{mol/min}$$

$$\frac{10^7 \text{ } \mu\text{mol}}{10^4 \text{ } \mu\text{mol/min}} = 10^3 \text{ minutes}$$

$$\frac{10^3 \text{ min}}{60 \text{ min}} = 16.67 \text{ hours}$$

9.6 DETERMINING THE MICHAELIS CONSTANT K_m, THE MAXIMUM VELOCITY V_m, AND THE TURNOVER NUMBER k_2 OF AN ENZYME

The Michaelis constant is an indication of the relative affinity of the enzyme for the substrate; the lower the Michaelis constant, the higher the affinity.

Lineweaver and Burk[15] converted the Michaelis–Menten equation (9-12) into straight-line form by taking the reciprocal of each side of the equation and rearranging the terms to obtain:

$$\frac{1}{v_i} = \frac{K_m}{V_m}\left(\frac{1}{[\text{S}]}\right) + \frac{1}{V_m} \tag{9-20}$$

A plot of $1/v_i$ versus $1/[\text{S}]$ gives a straight line with a slope of K_m/V_m and a y-intercept of $1/V_m$ (Fig. 9-6). If the line is extrapolated to intercept the x-axis, the intercept is $-1/K_m$. Thus, V_m and K_m can be obtained from the plot. This double reciprocal plot is probably the most used method of plotting enzyme kinetic data, although it has the disadvantage of compressing the data points at high substrate concentrations into a small region near the y-axis and over-emphasizing the data points at low substrate concentrations.

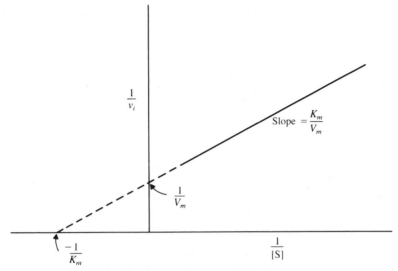

FIGURE 9-6. Lineweaver–Burk double reciprocal plot of $1/v_i$ versus $1/[S]$ for the dependence of enzyme-catalyzed reaction velocity on substrate concentration.

Another transformation of the Michaelis–Menten equation can be obtained by dividing both sides of Equation 9-12 by [S] and rearranging the terms to obtain:

$$v_i = -K_m\left(\frac{v_i}{[S]}\right) + V_m \qquad (9\text{-}21)$$

A plot of v_i versus $v_i/[S]$ also will give a straight line and is called an **Eadie–Hofstee plot**[16,17] (Fig. 9-7). The y-intercept is V_m, the slope is $-K_m$, and the x-intercept is V_m/K_m. The Eadie–Hofstee plot yields K_m and V_m directly and spreads the data points relatively evenly, even for high values of [S].

If the concentration and the molecular weight of the enzyme are known, k_2, the **catalytic constant** or **turnover number** can be obtained from V_m as

$$k_2 = \frac{V_m}{[E_t]} \qquad (9\text{-}22)$$

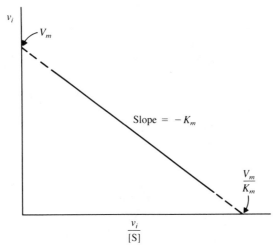

FIGURE 9-7. Eadie–Hofstee plot of v_i versus $v_i/[S]$ for the dependence of enzyme-catalyzed reaction velocity on substrate concentration.

COMPUTATION OF AN ENZYME TURNOVER NUMBER

If 0.5 μg of enzyme having a molecular weight of 50,000 is added to 1.0 mL of substrate and a value of V_m of 20 μmol of product per liter per minute is obtained, the k_2 would be calculated as follows:

$$[E_t] = \frac{0.50 \ \mu\text{g of enzyme/mL}}{50,000 \ \mu\text{g}/\mu\text{mol}} = 1 \times 10^{-5} \ \mu\text{mol/mL}$$

$$[E_t] = 1 \times 10^{-2} \ \mu\text{mol/L}$$

$$k_2 = \frac{V_m}{[E_t]} = \frac{20 \ \mu\text{mol of product L}^{-1} \ \text{min}^{-1}}{1 \times 10^{-2} \ \mu\text{mol of enzyme/L}}$$

$$k_2 = 2000 \ \mu\text{mol of product per micromole of enzyme per minute}$$

Thus, the turnover number k_2 is the number of substrate molecules converted to product molecules per molecule of enzyme per unit of time. Turnover numbers range widely from 36,000,000 min^{-1} for carbonic anhydrase, 12,500 min^{-1} for β-galactosidase, and 30 min^{-1} for lysozyme.

9.7 INHIBITION OF ENZYMES

The activity of an enzyme can be reversibly decreased by the noncovalent binding of inhibitors. The study of enzyme inhibitors is important because many drugs function as enzyme inhibitors. There are four major kinds of reversible inhibition: competitive, noncompetitive, uncompetitive, and mixed.

9.7.1 Competitive Inhibition

In competitive inhibition, the inhibitor I competes with the substrate S for binding to the active site of the enzyme:

$$E + S \underset{k_{-1}}{\overset{k_1}{\rightleftharpoons}} ES \overset{k_2}{\rightarrow} E + P$$

The dissociation constant of the enzyme–inhibitor complex is:

$$K_I = \frac{k_{-i}}{k_i} = \frac{[E][I]}{[EI]} \tag{9-23}$$

The Lineweaver–Burk equation for competitive inhibition is:

$$\frac{1}{v_i} = \frac{K_m}{V_m}\left(\frac{1}{[S]}\right)\left(1 + \frac{[I]}{K_I}\right) + \frac{1}{V_m} \tag{9-24}$$

and the Eadie–Hofstee equation for competitive inhibition is:

$$v_i = -K_m\left(\frac{v_i}{[S]}\right)\left(1 + \frac{[I]}{K_I}\right) + V_m \tag{9-25}$$

Lineweaver–Burk and Eadie–Hofstee plots for competitive inhibition are shown in Figure 9-8. The values of the intercepts and slopes, from which K_I can be determined, appear in the figure and in Table 9-2 (p. 309). Competitive inhibition is most easily recognized from these plots. When a competitive inhibitor is present, the apparent K_m is increased by the value $(1 + [I]/K_i)$, but the V_m is not changed.

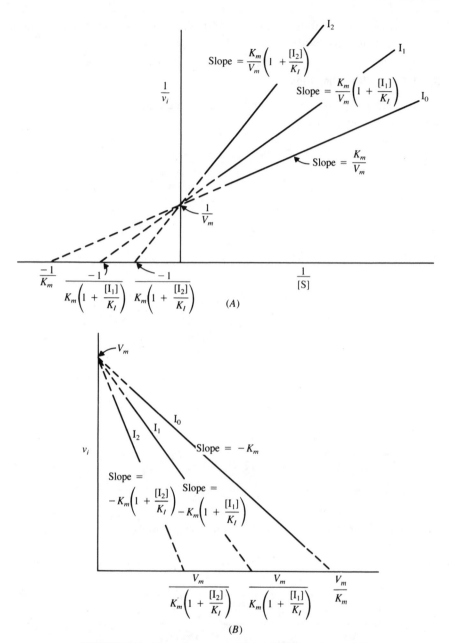

FIGURE 9-8. Plots for competitive inhibition: I_0 = reaction with no inhibitor present; I_1 and I_2 = reactions with inhibitor present at two different concentrations, where $[I_2] > [I_1]$. (A) Lineweaver–Burk plot. (B) Eadie–Hofstee plot.

TABLE 9-2. Intercept and slope functions for the four types of enzyme inhibition and the two methods of plotting

Type of inhibition	Lineweaver–Burk			Eadie–Hofstee		
	x-Intercept	y-Intercept	Slope	x-Intercept	y-Intercept	Slope
None	$-\dfrac{1}{K_m}$	$\dfrac{1}{V_m}$	$\dfrac{K_m}{V_m}$	$\dfrac{V_m}{K_m}$	V_m	$-K_m$
Competitive	$-\dfrac{1}{K_m(1+[I]/K_I)}$	$\dfrac{1}{V_m}$	$\dfrac{K_m}{V_m}\left(1+\dfrac{[I]}{K_I}\right)$	$\dfrac{V_m}{K_m(1+[I]/K_I)}$	V_m	$-K_m\left(1+\dfrac{[I]}{K_I}\right)$
Noncompetitive	$-\dfrac{1}{K_m}$	$\dfrac{1}{V_m}\left(1+\dfrac{[I]}{K_I}\right)$	$\dfrac{K_m}{V_m}\left(1+\dfrac{[I]}{K_I}\right)$	$\dfrac{V_m}{K_m(1+[I]/K_I)}$	$\dfrac{V_m}{(1+[I]/K_I)}$	$-K_m$
Uncompetitive	$-\dfrac{1}{K_m}\left(1+\dfrac{[I]}{K_I}\right)$	$\dfrac{1}{V_m}\left(1+\dfrac{[I]}{K_I}\right)$	$\dfrac{K_m}{V_m}$	$\dfrac{K_m}{V_m}$	$\dfrac{V_m}{(1+[I]/K_I)}$	$\dfrac{-K_m}{(1+[I]/K_I)}$
Mixed	$-\dfrac{1}{K_m}\dfrac{(1+[I]/K_{II})}{(1+[I]/K_I)}$	$\dfrac{1}{V_m}\left(1+\dfrac{[I]}{K_{II}}\right)$	$\dfrac{K_m}{V_m}\left(1+\dfrac{[I]}{K_I}\right)$	$\dfrac{V_m}{K_m(1+[I]/K_I)(1+[I]/K_{II})}$	$\dfrac{V_m}{(1+[I]/K_{II})}$	$-K_m(1+[I]/K_I)$

9.7.2 Noncompetitive Inhibition

In noncompetitive inhibition, I does not compete with S. Both inhibitor and substrate can bind simultaneously to the enzyme to give a ternary complex, EIS, as shown in the following reaction scheme. When the dissociation constant of S from EIS is the same as that from ES and EIS does not react to form product, noncompetitive inhibition occurs.

$$
\begin{array}{ccc}
E + S & \underset{k_{-1}}{\overset{k_1}{\rightleftharpoons}} \ ES & \overset{k_2}{\longrightarrow} \ E + P \\
+ & + & \\
I & I & \\
k_{-i} \Big\Vert k_i & k'_{-i} \Big\Vert k'_i & \\
EI + S & \underset{k'_{-1}}{\overset{k'_1}{\rightleftharpoons}} \ EIS &
\end{array}
$$

The Lineweaver–Burk equation for noncompetitive inhibition is:

$$
\frac{1}{v_i} = \left(\frac{K_m}{V_m} \left(\frac{1}{[S]} \right) + \frac{1}{V_m} \right) \left(1 + \frac{[I]}{K_I} \right) \tag{9-26}
$$

and the Eadie–Hofstee equation for noncompetitive inhibition is:

$$
v_i = -K_m \left(\frac{v_i}{[S]} \right) + \frac{V_m}{(1 + [I]/K_I)} \tag{9-27}
$$

Lineweaver–Burk and Eadie–Hofstee plots for noncompetitive inhibition are shown in Figure 9-9. A noncompetitive inhibitor binds to the enzyme in a way that is independent of the concentration of the substrate. The EIS complex is inactive and the V_m is lowered by the factor $1 + [I]/K_I$, but the K_m is not changed.

9.7.3 Mixed Inhibition

Mixed inhibition is a special type of noncompetitive inhibition in which the lines in a Lineweaver–Burk plot intersect in the second quadrant (Fig. 9-10A). This type of inhibition occurs when S can bind with the EI complex and I can bind with the ES complex, both giving the nonproductive ESI complex; and EI has a lower affinity for S than E has for S and the inhibition affects both V_m and K_m giving two inhibition constants K_I and K_{II}. The K_m is changed by the factor $(1 + [I]/K_I)$ and V_m is changed by the factor $(1 + [I]/K_{II})$, where K_I is the dissociation constant of EI and K_{II} is the dissociation constant of I from ESI.

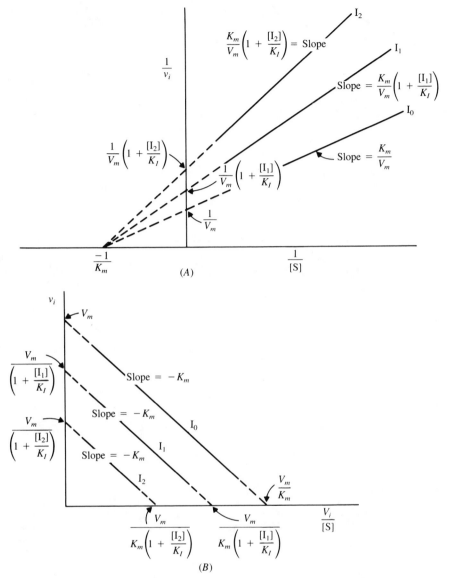

FIGURE 9-9. Plots for noncompetitive inhibition: (A) Lineweaver–Burk and (B) Eadie–Hofstee.

The Lineweaver–Burk equation for mixed inhibition is:

$$\frac{1}{v_i} = \left(\frac{K_m}{V_m} \left(\frac{1}{[S]} \right) \right) \left(1 + \frac{[I]}{K_I} \right) + \frac{1}{V_m} \left(1 + \frac{[I]}{K_{II}} \right) \tag{9-28}$$

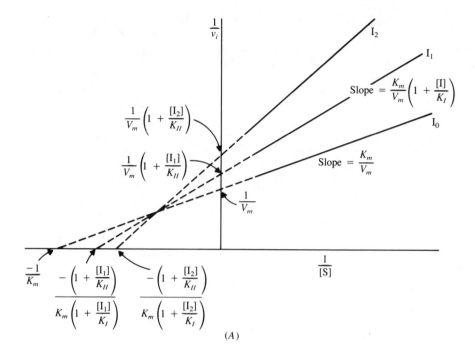

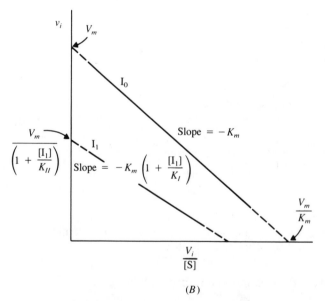

FIGURE 9-10. Plots for mixed inhibition: (*A*) Lineweaver–Burk and (*B*) Eadie–Hofstee.

and the Eadie–Hofstee equation for mixed inhibition is:

$$v_i = -K_m\left(\frac{v_i}{[S]}\right)\left(1 + \frac{[I]}{K_I}\right) + \frac{V_m}{(1 + [I]/K_{II})} \qquad (9\text{-}29)$$

Mixed inhibition will give the Lineweaver–Burk and Eadie–Hofstee plots shown in Figure 9-10.

9.7.4 Uncompetitive Inhibition

Uncompetitive inhibition occurs when I binds to ES but not to E:

$$E + S \underset{k_{-1}}{\overset{k_1}{\rightleftharpoons}} ES \overset{k_2}{\rightarrow} E + P$$
$$+$$
$$I$$
$$k_{-i}\Big\Vert k_i$$
$$EIS$$

The Lineweaver–Burk equation for uncompetitive inhibition is:

$$\frac{1}{v_i} = \frac{K_m}{V_m}\left(\frac{1}{[S]}\right) + \frac{1}{V_m}\left(1 + \frac{[I]}{K_I}\right) \qquad (9\text{-}30)$$

and the Eadie–Hofstee equation for uncompetitive inhibition is:

$$v_i = \frac{-K_m}{(1 + [I]/K_I)}\left(\frac{v_i}{[S]}\right) + \frac{V_m}{(1 + [I]/K_I)} \qquad (9\text{-}31)$$

The Lineweaver–Burk and Eadie–Hofstee plots for uncompetitive inhibition are given in Figure 9-11.

Enzyme inhibition is studied by using at least two concentrations of the inhibitor, one approximately twice the concentration of the other. The inhibitor is added to a series of different substrate concentrations. The reactions are started by adding the same amount of enzyme to each of the substrate concentrations

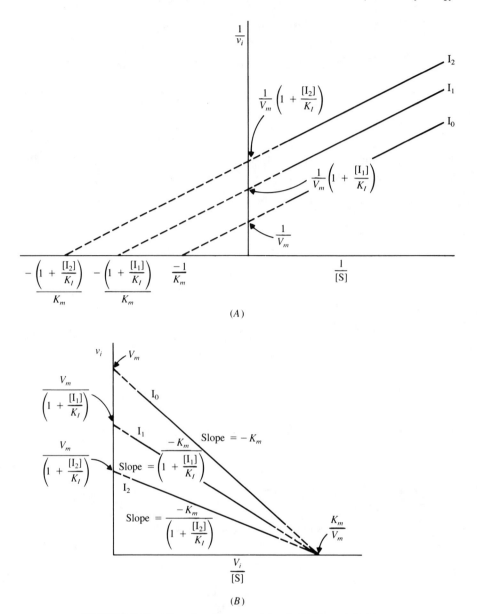

FIGURE 9-11. Plots for uncompetitive inhibition: (*A*) Lineweaver–Burk and (*B*) Eadie–Hofstee.

containing the inhibitor. The initial velocity is determined for each digest by measuring the amount of product formed (or substrate that has disappeared) as a function of time. The initial velocity data is then plotted in either the Lineweaver–Burk or the Eadie–Hofstee format. The type of inhibition can then be determined by inspection of the inhibition plot as shown in Figures 9-8, 9-9, 9-10, and 9-11. Table 9-2 summarizes the x- and y-intercept and the slope functions for the four types of inhibition and the two methods of plotting.

DETERMINATION OF THE TYPE OF INHIBITION AND THE VALUE OF K_I

Let us consider the Lineweaver–Burk plot in Figure A. Inspection shows that the noninhibited line and the inhibitor lines intersect on the y-axis, indicative of competitive inhibition. In the absence of inhibitor, the x-intercept equals $-1/K_m = -5 \times 10^{-2}$ mM^{-1} and $K_m = 20$ mM. In the presence of 25 mM inhibitor, the x-intercept is -3×10^{-2} mM^{-1}. The value for the x-intercept for a competitive inhibitor (Table 9-2 and Fig. 9-8) is:

$$\frac{-1}{K_m(1 + [I]/K_I)}$$

Substituting 20 mM for K_m and 25 mM for $[I]$, we have:

$$\frac{-1}{20(1 + 25/K_I)} = -3 \times 10^{-2}$$

$$1 = 0.6 + \frac{15}{K_I}$$

$$K_I = 0.6K_I + 15$$

$$0.4K_I = 15$$

$$K_I = 37.5 \text{ mM}$$

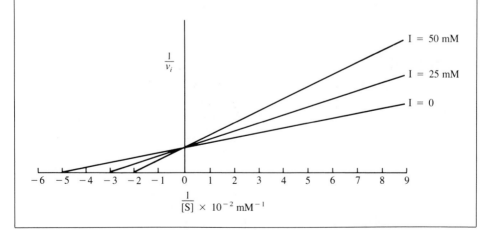

9.8 ALLOSTERIC ENZYMES THAT DO NOT FOLLOW MICHAELIS–MENTEN KINETICS

Allosteric enzymes are frequently involved in metabolic control and often do not follow Michaelis-Menten kinetics. These enzymes give sigmoidal plots (Fig. 9-12) of the initial velocity v_i versus substrate concentration [S], rather than the usual hyperbolic plot of Figure 9-1. At low concentrations of S, there is a relatively low velocity, but as the concentration increases, there is a point at which the velocity rapidly increases. Enzymes of these types have at least two binding sites for substrate. As the first substrate is bound, the second binding site is changed from a site with low affinity for S to one with a higher affinity. It is postulated that the binding of the first substrate molecule changes the conformation of the second binding site, altering its affinity for S.

A second type of allosteric activation occurs when the product P activates the enzyme. A plot of [P] versus time (Fig. 9-13) indicates a low initial velocity that is increased as the concentration of the product is increased. The product binds at a second site that is not an active site but is exclusively a binding site that when occupied by the product alters the conformation of the active site to give a more efficient enzyme, which either binds the substrate with higher affinity at the active site or increases the maximum velocity of the reaction, giving a higher turnover number.

Enzymes also can be allosterically activated or inhibited by molecules other than substrate or product. These molecules, called **effectors**, bind at sites other than the active site but affect the active site by altering its conformation to give activation or inhibition. For the action of different allosteric effectors, see Figure 9-14. Figure 9-14A shows one effector that decreases the K_m (activating effect), whereas the other increases the K_M (inhibiting effect). In Figure 9-14B, one effector increases the V_m (activating effect) and the other decreases the V_m (inhibiting effect).

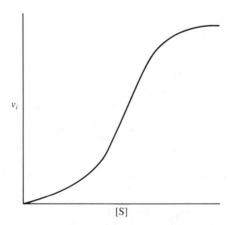

v_i

[S]

FIGURE 9-12. Activation of an enzyme-catalyzed reaction by substrate.

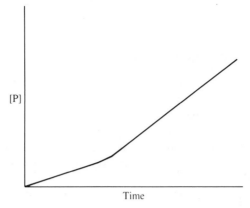

FIGURE 9-13. Activation of an enzyme-catalyzed reaction by product.

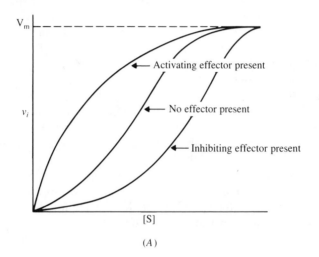

(A)

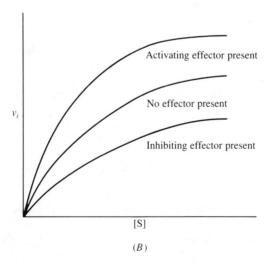

(B)

FIGURE 9-14. Plots of activating and inhibiting allosteric effectors on (A) substrate binding (K_m) and (B) maximum velocity. Different effectors will affect either substrate binding or maximum velocity, but both are usually not affected simultaneously by a single effector.

9.9 SPECIFICITY OF ENZYMES

As has been indicated, one of the hallmarks of enzyme action is specificity. In some cases, an enzyme may be so specific that it will interact with only a single substrate. A good example of this is glucose oxidase, which catalyzes the oxidation of β-D-glucopyranose only. It will not catalyze the oxidation of the α-anomer, or any other carbohydrate such as L-glucose, D-fructose, or D-mannose. Other enzymes, however, will interact with a series of different but related substances, usually with different affinities, and will catalyze the reaction at different rates. An example of this type of enzyme is alcohol dehydrogenase, which readily catalyzes the dehydrogenation or oxidation of ethanol but also will oxidize methanol, 1-propanol, 1-butanol, and ethylene glycol to varying degrees. Enzymes catalyzing reactions with polymers have two types of specificities: *exoenzymes* catalyze reactions at the ends of the polymer, and *endo-enzymes* catalyze reactions at the inner parts of the polymer chain. Examples of the former are carboxypeptidase, leucine amino peptidase, β-amylase, and glycogen phosphorylase and examples of the latter are trypsin, chymotrypsin, α-amylase, and restriction endonucleases.

To quantitatively measure the scope of the specificity, the maximum velocities of the different substrates are compared by measuring the initial velocities under conditions of saturating substrate and constant amount of enzyme. An additional and possibly more accurate quantitative measurement of specificity is the comparison of the ratios of k_2 to K_m. This type of comparison considers both the velocity of the reaction and the affinity of the enzyme for the substrate. For example, if the k_2/K_m ratios for two substrates are 1000 and 100 min^{-1}, respectively, the specificity of the enzyme for the first substrate is greater than the second substrate by a factor of 10.

9.10 LITERATURE CITED*

1. J. J. Berzelius, "Quelques Idées sur une nouvelle Force agissant dans les Combinaisons des Corps Organiques," *Ann. chim. phys. 61* [2]: 146 (1836).
2. (a) E. Fischer, "Einfluss der Configuration auf die Wirkung der Enzyme," *Ber. 27* [3]: 2985 (1894). (b) E. Fischer, "Bedeutung der Stereochemie für die Physiologie," *Z. physiol. Chem. 26*: 60 (1898).
3. E. Büchner, "Alkoholische Gährung ohne Hefezellen," *Ber. 30*: 117 (1897).
4. L. Michaelis and M. Menten, "Die Kinetik der Invertinwirkung," *Biochem. Z. 49*: 333 (1913).
5. J. B. Sumner, "The Isolation and Crystallization of the Enzyme Urease," *J. Biol. Chem. 69*: 435 (1926).
6. (a) J. H. Northrup and M. Kunitz, "Crystalline Trypsin," *J. Gen. Physiol. 16*: 267 and 295 (1932). (b) J. H. Northrup and M. Kunitz, *Crystalline Enzymes*, Columbia Univ. Press 1948.
7. T. Svedberg and K. O. Pedersen, *The Ultracentrifuge*, Oxford Univ. Press 1940.

* Listings 18–59 refer to Table 9-1.

8. S. Moore and W. H. Stein, "Chromatography of Amino Acids on Sulfonated Polystyrene Resins," *J. Biol. Chem. 192*: 663 (1951).

9. F. Sanger, "Chemistry of Insulin: Determination of the Structure of Insulin," *Science, 129*: 1340 (1959).

10. B. Davis, "Disk Electrophoresis. II. Method and Application to Human Serum Proteins," *Ann. N. Y. Acad. Sci. 121*: 404 (1964).

11. K. Weber and M. Osborn, "The Reliability of Molecular Weight Determination by Dodecyl sulfate-polyacrylamide Gel Electrophoresis," *J. Biol. Chem. 244*: 4406 (1969).

12. M. F. Perutz, "The Hemoglobin Molecule," *Sci. Amer. 211* [5]: 64 (1964).

13. J. C. Kendrew, "The Three-dimensional Structure of a Protein Molecule," *Sci. Amer. 205* [6]: 96 (1961).

14. (a) L. N. Johnson and D. C. Phillips, "Structure of Some Crystalline Lysozyme-inhibitor Complex Determined by X-ray Analysis at 6 Å Resolution," *Nature, 206*: 761 (1965). (b) D. C. Phillips, "The Three-dimensional Structure of an Enzyme Molecule," *Sci. Amer. 215* [5]: 78 (1966).

15. H. Lineweaver and D. Burk, "The Determination of Enzyme Dissociation Constants," *J. Am. Chem. Soc. 56*: 658 (1934).

16. G. S. Eadie, "The Inhibition of Cholinesterase by Physostigmine and Prostigmine," *J. Biol. Chem. 146*: 85 (1942).

17. B. H. J. Hofstee, "Non-inverted Versus Inverted Plots in Enzyme Kinetics," *Nature, 184*: 1296 (1959).

18. S. Hestrin, *J. Biol. Chem. 180*: 249 (1949).

19. B. H. J. Hofstee, *Arch. Biochem. Biophys. 51*: 139 (1954).

20. B. L. Vallee and F. L. Hoch, *Proc. Nat. Acad. Sci. USA 41*: 327 (1955).

21. D. Wellner and A. Meister, *J. Biol. Chem. 235*: 2013 (1960).

22. J. F. Robyt, p. 90 in *Starch: Chemistry and Technology* (R. L. Whistler, J. N. BeMiller, and E. F. Paschall, eds.). Academic Press (1984).

23. E. Amador and W. E. C. Wacker, *Clin. Chem. 8*: 343 (1962).

24. O. C. Richards and W. J. Rutter, *J. Biol. Chem. 236*: 3177 (1961).

25. A. Garen and C. Levinthal, *Biochim. Biophys. Acta 38*: 470 (1960).

26. J. E. Folk and E. W. Schirmer, *J. Biol. Chem. 238*: 3884 (1963).

27. R. F. Beers and I. W. Sizer, *J. Biol. Chem. 195*: 133 (1952).

28. B. C. W. Hummel, *Can. J. Biochem. Physiol. 37*: 1393 (1959).

29. C. E. Dinovo, D. S. Miyada, and R. M. Nakamura, *Clin. Chem. 19*: 994 (1973).

30. M. Kunitz, *J. Gen. Physiol. 33*: 349, 363 (1950).

31. P. Modrich and I. R. Lehman, *J. Biol. Chem. 245*: 3626 (1970).

32. P. Setlow, *Methods in Enzymol. 29*: 3 (1974).

33. L. A. Sacher, K. K. Winter, and S. Frankel, *Proc. Soc. Exp. Biol. Med. 90*: 323 (1955).

34. G. R. Craven, E. Steers, and C. B. Anfinsen, *J. Biol. Chem. 240*: 2468 (1965).

35. I. D. Fleming and H. F. Pegler, *Analyst 88*: 967 (1963).

36. T. Lessie and F. C. Neidhardt, *J. Bacteriol. 93*: 1337 (1967).

37. S. Gatt and M. M. Rapport, *Biochim. Biophys. Acta 113*: 567 (1966).

38. C. Ressler and T. Koga, *Biochim. Biophys. Acta 242*: 473 (1971).

39. E. J. Hill, T. Chou, M. C. Shih, and J. H. Park, *J. Biol. Chem. 250*: 1734 (1975).

40. S.-I. Hayashi and E. C. C. Lin, *J. Biol. Chem. 242*: 1030 (1967).

41. O. H. Lowry, D. W. Schulz, and J. V. Passonneau, *J. Biol. Chem. 239*: 1947 (1964).

42. M. D. Joshi and V. Jagannathan, *Methods in Enzymol. 9*: 371 (1966).

43. S. A. Heppel, *Methods in Enzymol. 2*: 570 (1955).

44. W. J. Reeves, Jr. and G. M. Fimognari, *Methods in Enzymol. 9*: 288 (1966).

45. C. A. Appleby and R. K. Morton, *Biochem. J. 71*: 492 (1959).

46. M. Morrison, *Methods in Enzymol. 17A*: 653 (1970).
47. J. C. Khoo and D. Steinberg, *Methods in Enzymol. 35*: 181 (1975).
48. C. Balny and J. W. Hastings, *Biochem. 14*: 4719 (1975).
49. D. Shugar, *Biochim. Biophys. Acta 8*: 302 (1952).
50. D. Aminoff, *Biochem. J. 81*: 384 (1961).
51. R. C. Burns and R. W. F. Hardy, *Methods in Enzymol. 24*: 480 (1972).
52. R. Arnon, *Methods in Enzymol. 19*: 226 (1970).
53. P. George, *J. Biol. Chem. 201*: 413 (1953).
54. P. Modrich and D. Zabel, *J. Biol. Chem. 251*: 5866 (1976).
55. M. Wishnick and M. D. Lane, *Methods in Enzymol. 23*: 590 (1971).
56. J. J. Jendrisak and R. R. Burgess, *Biochem. 14*: 4639 (1975).
57. G. Kalnitsky, J. P. Hummel, and C. Dierks, *J. Biol. Chem. 234*: 1512 (1959).
58. K. A. Walsh, *Methods in Enzymol. 19*: 41 (1970).
59. R. L. Blakeley, E. C. Webb, and B. Zerner, *Biochem. 8*: 1984 (1969).

9.11 REFERENCES FOR FURTHER STUDY

1. K. M. Plowman, *Enzyme Kinetics*. McGraw-Hill, New York, 1972.
2. H. J. Fromm, *Initial Rate Enzyme Kinetics*. Springer-Verlag, Berlin/New York, 1975.
3. I. H. Segel, *Enzyme Kinetics*. Wiley, New York, 1975.
4. A. Fersht, *Enzyme Structure and Mechanism*, Freeman, New York, 1987.
5. L. B. Spector, *Covalent Catalysis by Enzymes*. Springer-Verlag, Berlin/New York, 1982.
6. D. L. Purich, Ed., *Contemporary Enzyme Kinetics and Mechanism*, in *Selected Methods in Enzymology Series*. Academic Press, New York, 1983.

CHAPTER 10

STRUCTURAL ANALYSIS
OF BIOLOGICAL MOLECULES

The knowledge of structures, in molecular terms, of biomolecules is indispensable for an understanding of the function of these molecules in a living cell. The chemistry of the various functional groups tells much about their roles in determining the properties of any given molecule. Biochemistry, however, is set apart from other disciplines of chemistry by the subtle roles played by the location and stereochemistry of the functional groups in the macromolecules found in cells.

As an example of the importance of the differences in chemical properties of functional groups and their stereochemistry, let us consider D-glucopyranose, which has four alcohol hydroxyl groups and one hemiacetal hydroxyl group. The hemiacetal hydroxyl group can react with any one of the four alcohol hydroxyls of another D-glucopyranose, giving two possible stereochemical products, α and β, at each of the four positions. This gives eight different disaccharides. When multiple α-1,4-glycosidic linkages are formed with many glucose residues, starch- or glycogen-type polymers result, and when multiple β-1,4-glycosidic linkages are formed with many glucose residues, cellulose-type polymers result. The subtle difference in the stereochemistry of these two types of linkage results in molecules of vastly different structures, functions, and metabolism.

The presence or absence of a single hydroxyl group can greatly change the structure and function of a molecule. For example, the presence of a hydroxyl group at the 2' position of D-ribose in a polynucleotide gives RNA, whereas the substitution of a hydrogen for a hydroxyl group at the 2' position in a polynucleotide gives DNA. This difference further influences the secondary and tertiary structures of polynucleotides. The absence of the hydroxyl group at the 2' position permits the formation of double-helical structures in DNA, whereas the presence of the 2'-hydroxyl on D-ribose does not readily give rise to double-helical structures but permits the other types of intramolecular secondary structures found in RNA.

The nucleotides in DNA and RNA have a precise order that gives explicit instructions for the assembly of specific sequences of amino acids in proteins. A single change in the DNA sequence can give a modified protein with a single amino acid change, which can affect the primary, secondary, and tertiary structures. For example, the substitution of glutamic acid for valine in the β-chains of hemoglobin changes the hemoglobin tertiary structure, giving different physical and chemical properties that greatly change the morphology of the red blood cells. Changes in the polynucleotide sequence further provide an important molecular mechanism for producing biological diversity by evolution.

In this chapter we present the techniques for investigating the primary structures of biological molecules necessary for the functioning of the cell. These studies bring together many of the techniques that have been presented earlier and illustrate their use in structure determinations. We have tried to present general methods that can be used to begin the investigation of higher levels of structure found in membranes, chromatin, ribosomes, endoplasmic reticulum, and multienzyme complexes. After the primary structures of the molecules found in these substances have been determined, the higher structural levels can be studied by X-ray diffraction, electron microscopy, Raman spectroscopy, electron–spin resonance spectroscopy, light scattering, and other advanced techniques not covered in this text.

10.1 DETERMINING THE STRUCTURE OF OLIGOSACCHARIDES AND POLYSACCHARIDES

Carbohydrates are ubiquitous in the biological world and represent a wide diversity of structure and function. They are comprised of monosaccharides, monosaccharide derivatives (sugar acids, sugar alcohols, amino sugars, deoxy-sugars, etc.), oligosaccharides, which are composed of two to several monosac-charides covalently combined by acetal or ketal linkages, and polysaccharides, which are composed of several hundred to thousands of monosaccharides covalently combined by acetal or ketal linkages, giving carbohydrate polymers.

D-Glucose and its derivatives are the most abundant of the carbohydrates found on the earth. D-Glucose is produced from the photosynthetic process combining CO_2 and H_2O and represents the conversion of the energy of the sun into chemical energy. D-Glucose is then converted or combined into other forms such as starch, which is the storage form of glucose in plants and serves as a major food source and link to the energy of the sun. Starch can be broken down in the digestive process and converted into D-glucose, which in turn provides the energy for the synthesis of many other substances, including ATP, the universal source of chemical energy used by all living organisms.

D-Glucose is converted into D-ribose and 2-deoxy-D-ribose, which are the major carbohydrate components in RNA and DNA, respectively. D-Glucose also is converted into naturally occurring disaccharides such as lactose (the primary nutrient carbohydrate in mammalian milk), sucrose (found in plants and used by man as a food and confection since early times), and α,α-trehalose (the carbohydrate component of insect lymph fluid, which serves the same function in insects that D-glucose serves in the blood of mammals).

D-Glucose is polymerized into cellulose, which serves as the major structural component of plant cell walls, and when modified to 2-amino-2-deoxy-D-glucose is polymerized into murein (the major structural component of all bacterial cell walls) and chitin (the major structural component in the exoskeletons of insects and crustaceans).

Oligosaccharides containing different kinds of monosaccharides such as D-glucosamine, D-galactosamine, D-mannose, D-neuraminic acid, L-rhamnose, and L-fucose are found covalently linked to proteins and lipids to form glyco-proteins and glycolipids. These carbohydrates function in a wide variety of ways, including the following: secreting and transporting of proteins through membranes, acting as signals in the control of protein turnover, and forming recognition sites whereby cells distinguish each other and control processes of differentiation; they also play significant roles in fertilization and in organo-genesis. Changes in cell surface carbohydrates have been observed in cancerous growths and may be involved in the undifferentiated and uncontrolled growth of tumors.

These are but a few of the types and functions of carbohydrates. A knowledge of the structure of carbohydrates is, thus, important for an understanding of the diverse functions of these molecules.

The isolation and purification of carbohydrates was covered in Chapter 8 (Section 8.6).

10.1.1 Determining the Monosaccharide Composition

The monosaccharide composition can be determined by a variety of the chromatographic procedures discussed earlier (Section 4.10): thin-layer chroma-tography, gas chromatography, paper chromatography, and high performance liquid chromatography. In the TLC analysis of a carbohydrate sample, 5 μL of a 10 mg/mL solution is applied to a silica gel TLC plate that can be irrigated four times with 85:15 (v/v) acetonitrile–water (solvent A, Section 4.10.4) to separate the major monosaccharides and disaccharides. Amino sugars do not migrate using this solvent system but can be separated by silica gel TLC using three as-cents of acetonitrile–acetic acid–ethanol–water (65:5:10:20, v/v).[1]

If the unknown carbohydrate does not move from the origin or only moves about one-third of the way up the plate using these solvent systems, it is most likely a polysaccharide (which will not move from the origin) or an oligosac-charide (which will move only about one-third of the way or less up the plate). If this is the case, the sample must first be acid-hydrolyzed before the TLC analy-sis is made to determine the monosaccharide composition.

The conditions for acid hydrolysis must be controlled to ensure that complete hydrolysis is obtained with little or no degradation of the monosaccharide units. Because of differences in the susceptibility of different kinds of linkages and the presence of different kinds of monosaccharide units, more than one set of hydrolysis conditions must be used. Three parameters can be varied: the length of time of hydrolysis, the acid concentration, and the temperature. The usual starting conditions for acid hydrolysis are 1 M acid (HCl or trifluoroacetic acid are used because of their volatility) at 100°C for various periods of time. Partial acid hydrolysis, especially from polysaccharides, can give a series of oligosaccharides that can be isolated by preparative paper chromatography or HPLC (Sections 4.4 and 4.9) and the oligosaccharide structures subsequently determined to help elucidate the structure of the polysaccharide.

**ACID HYDROLYSIS OF OLIGOSACCHARIDES
AND POLYSACCHARIDES**

A solution of the carbohydrate (10 mg/mL) is acid-hydrolyzed by adding 10 μL of concentrated HCl or trifluoroacetic acid to each of several 100-μL samples of carbohydrate solution. The samples are poured into ampules, sealed, and placed into a boiling water bath for various periods of time (30–240 minutes). The TLC analysis is then performed on the hydrolyzates to determine the monosaccharide composition.

10.1.2 Determining the Structure of Oligosaccharides

Oligosaccharides can be composed of a single monosaccharide or several different types of monosaccharides. These molecules can be determined by TLC analysis after acid hydrolysis as described in Section 10.1.1. The monosaccharide units can be covalently linked in a variety of ways to give different oligosaccharide structures. For example, if the oligosaccharide is composed of a single D-aldopyranose unit, 11 types of disaccharides are possible: 8 reducing disaccharides (4 α-linked and 4 β-linked to positions 2, 3, 4, and 6) and 3 nonreducing disaccharides, in which position 1 is linked to 1′ in α,α-, β,β-, and α,β-configurations. If the oligosaccharide is composed of two different kinds of D-aldopyranose units, the number of possible disaccharides is increased. There are 16 possible reducing disaccharides: (8 with unit A linked to unit B and eight with unit B linked to unit A) and 4 nonreducing disaccharides (α-A linked to α-B, β-A linked to β-B, α-A linked to β-B, and β-A linked to α-B). With higher saccharides there can be an even greater complexity, giving an increased number of possible structures.

Although not all the possible combinations occur naturally, all the D-glucopyranose disaccharides have been found to occur naturally or have been obtained from naturally occurring substances (see Table 10-1 for the structures and occurrence of some di- and trisaccharides containing glucose and other monosaccharides).

TABLE 10-1. Structure and properties of some naturally occurring oligosaccharides

Oligosaccharide	Structure	Source	Type
Maltose (Mal)	α-Glc-p(1→4)Glc-p	Hydrolysis product of starch	Reducing
Cellobiose (Cel)	β-Glc-p(1→4)Glc-p	Hydrolysis product of cellulose	Reducing
Lactose (Lac)	β-Gal-p(1→4)Glc-p	Mammalian milk	Reducing

(continued)

TABLE 10-1 (*continued*)

Oligosaccharide	Structure	Source	Type
Melibiose (Mel)		Produced from raffinose by top yeast fermentation	Reducing
Turanose (Tur)		Mild acid hydrolysis of melezitose	Reducing

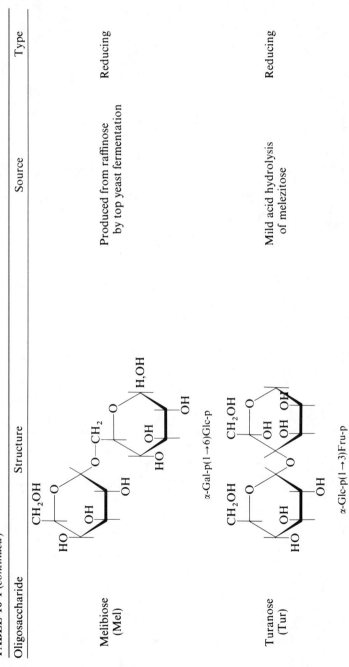

α-Gal-p(1→6)Glc-p

α-Glc-p(1→3)Fru-p

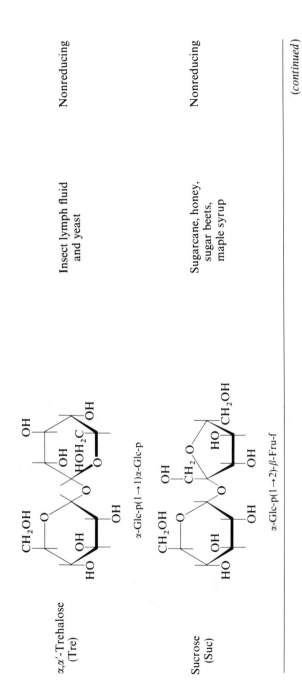

α,α′-Trehalose
(Tre)

α-Glc-p(1→1)α-Glc-p

Insect lymph fluid
and yeast

Nonreducing

Sucrose
(Suc)

α-Glc-p(1→2)-β-Fru-f

Sugarcane, honey,
sugar beets,
maple syrup

Nonreducing

(*continued*)

327

TABLE 10-1 (*continued*)

Oligosaccharide	Structure	Source	Type
Raffinose (Raf)	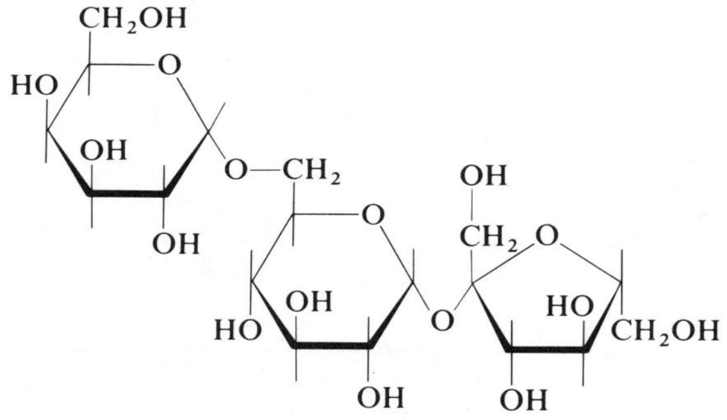 α-Gal-p(1→6)-α-Glc-p(1→2)-β-Fru-f	Beet molasses cottonseed meal	Nonreducing
Melezitose (Mez)	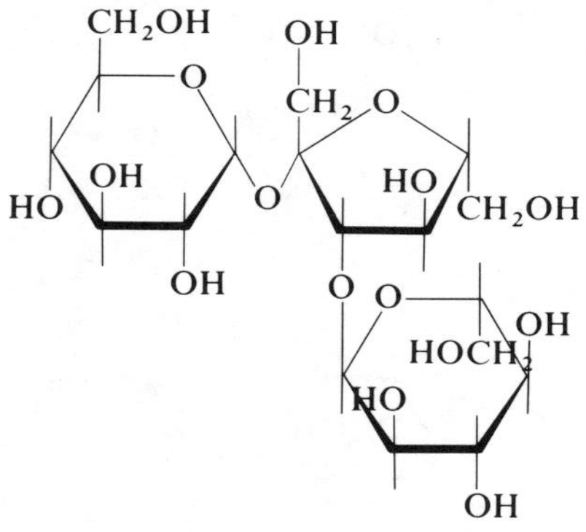 α-Glc-p(1→2)-β-Fru-1-(3←1)-α-Glc-p	Sweet exudate of fir, pine, and larch trees	Nonreducing

10.1.2.1 DETERMINATION OF THE PRESENCE OF A REDUCING GROUP IN OLIGOSACCHARIDE

One of the first tests performed to obtain information about the structure of an oligosaccharide is the determination of whether it is reducing or nonreducing. This can be determined using Benedict's test (Section 7.1.1.6).

10.1.2.2 PARTIAL ACID HYDROLYSIS AS A MEANS TO DETERMINE THE STRUCTURE OF OLIGOSACCHARIDES

Because different types of linkages are susceptible to acid hydrolysis to different degrees, partial acid hydrolysis can give important information about the types of linkages present in an oligosaccharide. The following generalities apply to linkage susceptibility to acid hydrolysis: (*a*) β-linkages are more stable than α-linkages; (*b*) linkages to primary alcohols are more stable than linkages to secondary alcohols; (*c*) the α,α-acetal–acetal linkage of α,α-trehalose is very stable, but the β,β-acetal–acetal linkage of β,β-trehalose would be the most stable of all, and the α,β-acetal–acetal linkage of α,β-trehalose would be intermediate in stability; (*d*) the α-acetal–β-ketal linkage of sucrose is very labile. Thus, the order of decreasing stability of glycosidic linkages to acid hydrolysis is β,β-trehalose $> \alpha,\beta$-trehalose $> \alpha,\alpha$-trehalose $> \beta\text{-}1\rightarrow6 > (\beta\text{-}1\rightarrow2 = \beta\text{-}1\rightarrow3 = \beta\text{-}1\rightarrow4) > \alpha\text{-}1\rightarrow6 > (\alpha\text{-}1\rightarrow2 = \alpha\text{-}1\rightarrow3 = \alpha\text{-}1\rightarrow4) > \alpha,\beta$-linkage of sucrose.

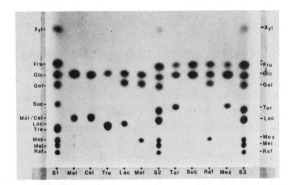

FIGURE 10-1. TLC analysis of the partial acid hydrolysis of nine oligosaccharides. Hydrolysis was performed under identical conditions (1 M trifluoroacetic acid, 100°C, 30 minutes) for each sugar, showing differences in linkage susceptibility. S1 is a standard series containing xylose, fructose, glucose, galactose, sucrose, trehalose, melezitose, melibiose, and raffinose; S2 contains xylose, fructose, glucose, galactose, turanose, lactose, melezitose, melibiose, and raffinose. The carbohydrates were separated on Whatman K5 silica gel plates with four multiple ascents of acetonitrile–water (85:15, v/v) and visualized by spraying with sulfuric acid–methanol (1:3, v/v) followed by heating at 110°C for 5 minutes.

In the comparison of the susceptibility of the acid hydrolysis of two 1,4-linked saccharides, cellobiose would be more stable than maltose. Figure 10-1 shows a TLC analysis of the partial acid hydrolysis (obtained by a 30-minute hydrolysis as described above) of several oligosaccharides.

10.1.2.3 THE USE OF ENZYMES TO DETERMINE THE STRUCTURE OF OLIGOSACCHARIDES

Enzymes have specificity that can be used in the analysis of different types of structures. For example, α-glucosidases can be used to determine whether α-linkages are present, β-glucosidases can be used to determine whether β-linkages are present, and invertase can be used to determine whether the sucrose linkage is present. The products that are formed from the use of enzymes can also tell something about the structure of the original oligosaccharide. If no reaction occurs, however, caution must be exercised in concluding that the supposedly susceptible linkage is not present, as certain structural features can prevent the action of the enzyme. For example, α,α-trehalose is not susceptible to α-glucosidase, even though there are two α-linkages present, and melezitose is not susceptible to invertase even though it contains a sucrose linkage. The presence of the two α-linkages in a "head-to-head" arrangement in trehalose prevents the hydrolysis by α-glucosidase, and the presence of the (α-1,3)-linked glucose attached to fructose in the melezitose structure produces a steric effect that prevents the action of invertase on the sucrose linkage of melezitose. Figure 10-2 shows a TLC analysis of the specificity of action of α-glucosidase, β-glucosidase, and invertase on several oligosaccharides.

THE ENZYME HYDROLYSIS OF OLIGOSACCHARIDES

Enzyme hydrolysis is conducted by adding 20 μL of the enzyme solution, buffered at its optimum pH and containing 50 IU/mL, to 100 μL of saccharide solution (10 mg/mL) containing 0.02% sodium azide. Hydrolysis is allowed to proceed for 12 hours at room temperature; 4-μL aliquots are then spotted onto Whatman K5 silica gel plates along with 4 μL of standards containing 2 mg/mL of each carbohydrate. The TLC plates are irrigated four times with acetonitrile–water (85:15, v/v), drying thoroughly after each irrigation. Each TLC plate is dried and sprayed with sulfuric acid–methanol (1:3, v/v) and heated in an oven at 110–120°C for 5 minutes.

10.1.2.4 DETERMINING THE REDUCING-END RESIDUE OF A REDUCING OLIGOSACCHARIDE

The reducing-end residue can be identified by reduction with sodium borohydride, followed by acid hydrolysis and paper chromatographic analysis. The reduction of the reducing-end residue results in a sugar alcohol that can be

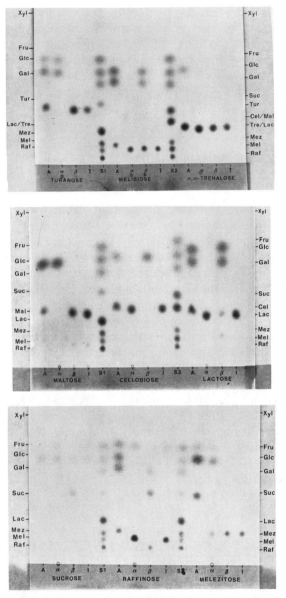

FIGURE 10-2. TLC analysis of enzymic hydrolysis and acid hydrolysis of nine oligosaccharides. Hydrolysis was performed with A, acid (same conditions as Fig. 10-1), $\alpha = \alpha$-glucosidase, $\beta = \beta$-glucosidase, and I = invertase. Enzymic hydrolysis was performed with 1.0 unit/100 μL for each sugar at 22°C for 48 hours. The standards and method of chromatography were the same as in Figure 10-1.

identified in a mixture of the other residues from the saccharide. For example, the reduction of lactose, followed by acid hydrolysis gives galactose and glucitol, indicating that glucose was at the reducing end; the reduction of turanose gives glucose and two sugar alcohols, glucitol and mannitol. The formation of these two sugar alcohols indicates that fructose is at the reducing end, since fructose is a ketose that on reduction gives the two epimeric sugar alcohols.

THE DETERMINATION OF THE REDUCING-END RESIDUE BY REDUCTION, ACID HYDROLYSIS, AND PAPER CHROMATOGRAPHY[2]

A solution of the carbohydrate (10 mg/mL) is reduced by adding 10 μL of sodium borohydride (10 mg/mL) to 100 μL of carbohydrate in an ampule. The sample is heated in a boiling water bath for 15 minutes; 12 μL of concentrated trifluoroacetic acid is added, and the ampule is sealed and placed in an oven at 120°C for 1 hour. The ampule is opened and 25 μL is placed onto a sheet of Whatman 3MM paper along with 25 μL of glucose, fructose, galactose, glucitol, and mannitol (each 10 mg/mL). The chromatogram is irrigated in a descending chromatography chamber (Section 4.4) with nitromethane–ethanol–acetic acid–boric acid saturated water (8:1:1:1, v/v) for 12 hours and dried. The chromatogram is developed by dipping into reagent A, which is 4 parts methanol and 1 part 0.5 M sodium hydroxide and 4% pentaerythritol, dried and dipped into reagent B, which is 1 L of methanol and 100 mL of 0.1 M sodium periodate, dried and dipped into reagent C, which is 1 L of acetone, 10 mL of water, and 5 mL of saturated silver nitrate, dried and dipped into reagent A again. Brown spots on a tan background appear for the carbohydrates. When the darkness of the spots is maximum, the chromatogram is quickly dipped into a solution of 240 g of sodium thiosulfate in 1 L of water, and immediately washed in water for 10 minutes and dried.

10.1.2.5 OPTICAL ROTATION AS A MEANS OF DETERMINING THE STRUCTURE OF AN OLIGOSACCHARIDE

As described in Section 7.1.2.4, structures with α-linkages have relatively high positive rotations and structures with β-linkages have relatively low, sometimes even negative optical rotations.

After a structure has been determined by the methods described above, the structure can be confirmed by measuring the specific optical rotation and comparing it with the specific optical rotation in the literature for the presumptive structure. If the saccharide is new and has not been reported in the literature, the specific optical rotation is a physical constant that can be used to identify the new compound in the future.

AN EXAMPLE OF THE DETERMINATION OF THE STRUCTURE OF AN UNKNOWN OLIGOSACCHARIDE STRUCTURE

Let us say that we have just isolated a new saccharide from avocado seeds. It gives a positive Benedict's test. The TLC of it (A in Fig. 10-3) shows that it is pure and migrates about 20 mm from the origin. Partial acid hydrolysis gives B, α-glucosidase hydrolysis gives C, β-glucosidase gives D, and invertase gives no reaction (E). When the saccharide is reduced with $NaBH_4$ and acid-hydrolyzed, and the products separated by descending paper chromatography, the products are D-galactose, D-glucose, D-glucitol, and D-mannitol.

Because the saccharide migrated 20 mm from the origin, it is either a di- or a trisaccharide. Partial acid hydrolysis produced three monosaccharides (D-fructose, D-glucose, and D-galactose), indicating that it is probably a trisaccharide. α-Glucosidase gave D-fructose and a disaccharide migrating where the disaccharide, lactose, migrates; β-glucosidase gave D-galactose and a disaccharide migrating where the disaccharide, turanose, migrates. The products from reduction and acid hydrolysis indicate that the reducing-end residue is D-fructose.

The structure of the saccharide (*a*) contains three monosaccharides (fructose, glucose, and galactose) and (*b*) is most probably a trisaccharide because of its migration on the TLC. The structure contains both an α- and a β-linkage, as indicated by enzyme hydrolysis in which lactose is probably linked α- to D-fructose and D-galactose is probably linked β- to turanose. Thus, the structure has both the elements of lactose (β-D-galactopyranosyl-(1→4)-D-glucopyranose)

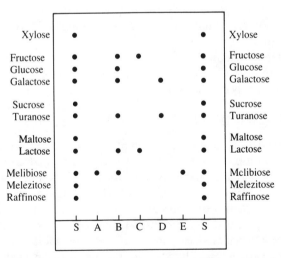

FIGURE 10-3. TLC analysis of the saccharide isolated from avocado seeds: S = standards, A = isolated saccharide, B = partial acid hydrolysis, C = α-glucosidase hydrolysis, D = β-glucosidase hydrolysis, and E = invertase hydrolysis.

and turanose (α-D-glucopyranosyl-(1→3)-D-fructose) and has D-fructose at the reducing terminus.

The structure of the saccharide, therefore, is:

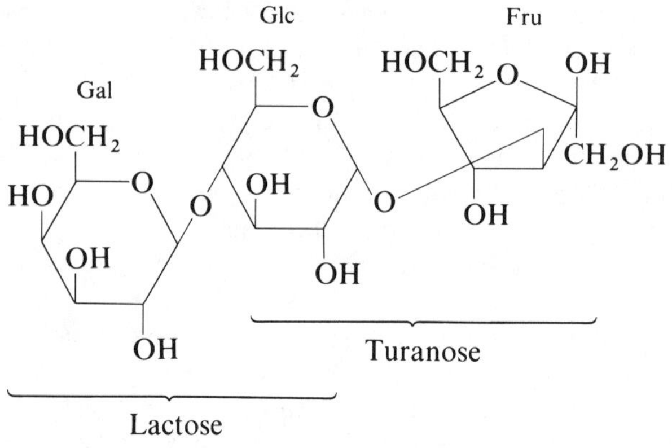

β-D-Galactopyranosyl-(1→4)-α-D-glucopyranosyl-(1→3)-D-fructose

10.1.3 Determining the Structure of Polysaccharides

After isolation (Sections 8.6.2.3, 8.6.2.5, 8.6.2.6), the polysaccharide is dissolved in hot water or an aprotic solvent such as dimethyl sulfoxide (DMSO). Many polysaccharides initially insoluble in water can first be dissolved in DMSO or 1 M NaOH and then carefully diluted with water to give an aqueous solution and neutralized if NaOH is used. For polysaccharides not soluble in DMSO or NaOH, 4-methylmorpholine N-oxide monohydrate (MMNO) can be used to effect solubilization.

The first step in determining the structure is the identification of the mono-saccharide composition, which can be obtained by TLC analysis of an acid hydrolyzate as described in Section 10.1.1.

10.1.3.1 ANALYSIS OF THE TYPES OF LINKAGE PRESENT

We begin by determining the position of the linkage (i.e., attachment of the acetal group to positions 2, 3, 4, or 6 for hexopyranoses) and the configuration (α or β). The position of the linkage is most directly determined by methylation analysis or ^{13}C NMR (Section 3.6.2). Methylation analysis will also provide information about whether the polysaccharide is branched and, if so, the amount of branching. The free hydroxyl groups in the polysaccharide are methylated by reaction with methylsulfinyl carbanion and methyl iodide. The methylated polysaccharide is then hydrolyzed, and the resulting, partially methylated monosaccharides are reduced, acetylated, separated by gas–liquid chroma-tography (Section 4.10.7) or HPLC and identified by their retention times.

For example, if the polysaccharide is a branched glucan with D-glucose linked 1→4 in the main chains and (1→6)-linked branch linkages, methylation and acid hydrolysis will give three types of partially methylated glucose residues: 2,3,6-tri-O-methyl glucose (I), 2,3-di-O-methyl glucose (II), and 2,3,4,6-tetra-O-methyl glucose (III) (see Fig. 10-4). The first methylated glucose (I) comes from the main chain residues, the second methylated glucose (II) comes directly from the branched glucose, and the third methylated glucose (III) comes from the non-reducing-end residues. The latter two (II and III) are the result of branching, and a quantitative determination of the three types of methylated glucoses will give the degree of branching. For example, if a polysaccharide gave 78% 2,3,4-tri-O-methyl glucose, 11% 2,4-di-O-methyl glucose, and 11% 2,3,4,6-tetra-O-methyl glucose, we would have $(11/[78 + 11]) \times 100 = 12\%$ branch linkages.

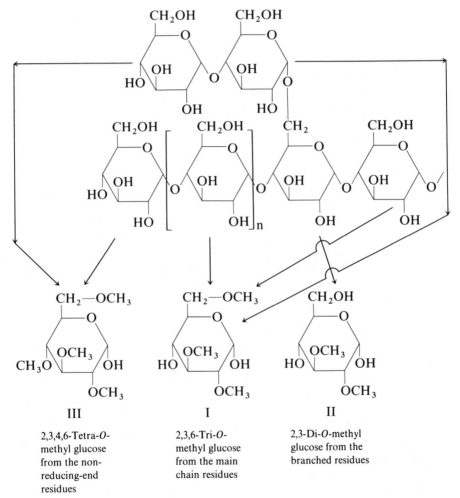

III

2,3,4,6-Tetra-O-methyl glucose from the non-reducing-end residues

I

2,3,6-Tri-O-methyl glucose from the main chain residues

II

2,3-Di-O-methyl glucose from the branched residues

FIGURE 10-4. Structures of the products formed from the methylation analysis of an (α-1,4)-linked glucan with α-1,6 branch linkages. In methylation analysis the oligosaccharide or polysaccharide is methylated and acid-hydrolyzed, and the products are analyzed by gas–liquid chromatography.

METHYLATION ANALYSIS[3]

The methylating reagent consists of methyl iodide and potassium methylsulfinyl carbanion. The latter is prepared by reacting potassium hydride with DMSO. About 1 g of potassium hydride in 5 g of mineral oil is washed four times with 20 mL of hexane by centrifugation and suspension. The potassium hydride is suspended in 20 mL of hexane, transferred to a 250-mL filter flask in which the potassium hydride is allowed to sediment, and the hexane is removed by aspiration. The flask is placed into an ice bath. (*Caution: After removal of the hexane, the potassium hydride is very reactive and should be handled only in small quantities, with extreme care*). DMSO (13 mL) is added and the solution is mixed under argon as it is warmed to room temperature. After 30 minutes of reaction at room temperature, the solution is transferred to a 20-mL borosilicate glass centrifuge tube fitted with a Teflon-lined screw cap and centrifuged for 3 minutes at $1500g$ to produce a clear, gray-green solution. The supernatant, containing methylsulfinyl carbanion, is decanted and stored under argon at $-20°C$.

The methylation is performed by adding 200 μL of potassium methylsulfinyl carbanion to 1–5 mg of polysaccharide dissolved in 50–200 μL of DMSO and placed under argon. After 10 minutes the sample is placed on ice and 150 μL of ice cold methyl iodide is added. The mixture is allowed to warm to room temperature and is stirred for 10 minutes; 3 mL of chloroform–methanol (2:1) is added, followed by 2 mL of water with mixing. The phases are separated by centrifugation ($200g$, 30 s) and the upper phase is removed by aspiration, taking care not to disturb the interface. The organic phase (containing the methylated sugars) is washed four times with 2 mL of water. After the final wash, 2 mL of 2,2-dimethoxypropane, 20 μL of 18 M acetic acid, and two or three anti-bumping granules are added. The solution is placed into a 90°C water bath until approximately 200 μL of solvent has evaporated. The remaining solvent is evaporated under a stream of argon. The resulting methylated polysaccharide is frequently oily.

The methylated polysaccharide is hydrolyzed by suspending it in 0.3 mL of 2 M trifluoroacetic acid, which is placed into an ampule, sealed, and put into an autoclave for 1 hour at 121°C. The sample is cooled, and the ampule is opened and placed into a 40°C water bath and evaporated to dryness under a stream of nitrogen.

The dry hydrolyzate is dissolved in 1 mL of 2 M ammonium hydroxide containing 0.5 M sodium borohydride and placed in a 60°C water bath for 1 hour. Acetone (0.5 mL) is added to stop the reaction, and the solution is evaporated at 40°C under a stream of nitrogen. The residue is dissolved in 0.2 mL of 18 M acetic acid, and 1 mL of ethyl acetate, 3 mL of acetic anhydride, and 100 μL of 70% perchloric acid are added and mixed. After 5 minutes, the solution is cooled on ice and 10 mL of water is added, followed by 200 μL of 1-methylimidazole. The solution is mixed, allowed to stand for 5 minutes, and 1 mL of dichloromethane is added with shaking. The lower phase contains the methylated sugar alcohol acetates, which can be separated and quantitated by gas chromatography.

Information about the configuration of the linkages can be obtained by determining the specific optical rotation (Section 7.1.2.6), particularly if all the linkages have the same configuration. A high positive specific rotation (e.g., $+150°$ or greater) is indicative of linkages with the α-configuration, and a low positive or negative specific optical rotation (e.g., $+20$ to $-20°$) is indicative of linkages with the β-configuration. A structure having both types of configuration is more difficult to analyze but will in general have an intermediate specific rotation (e.g., $+50°$ to $+100°$), depending on the relative ratios of the two types of linkages. A definitive determination of the linkage types is obtained by analyzing the structures of the oligosaccharides obtained from partial acid hydrolysis or by ^{13}C NMR analysis.

The sequence by which the monosaccharides are attached to each other in the various linkages is also obtained by determining the structure of oligosaccharides produced from partial acid hydrolysis or specific enzyme hydrolysis. The determination of the structure of the oligosaccharides is usually easier than structural studies on the intact polysaccharide. The methods are essentially those used for oligosaccharides outlined above.

Certain structural features of polysaccharides can be more susceptible to acid hydrolysis than others. For example, furanosides are hydrolyzed many times faster than pyranosides; acetal linkages to secondary alcohols (e.g., $1 \rightarrow 2$, $1 \rightarrow 3$, $1 \rightarrow 4$ linkages in hexopyranoses) are hydrolyzed faster than acetal linkages to primary alcohols (e.g., $1 \rightarrow 6$ linkages in hexopyranoses); the α-linkage is usually hydrolyzed faster than the β-linkage. If, however, acetolysis is performed (i.e., cleavage by acidified acetic anhydride), the $1 \rightarrow 6$ linkage to a primary alcohol is cleaved faster than linkages to secondary alcohols. The use of both methods can, thus, give oligosaccharides of different types, with overlapping structures, when a mixture of linkage positions is present.

10.1.3.2 THE USE OF ENZYMES TO SPECIFICALLY HYDROLYZE LINKAGES

Specific hydrolytic enzymes can be used to determine the structure of polysaccharides. There are two general types of polymer-hydrolyzing enzymes: **endohydrolases**, which hydrolyze the acetal linkages in the *inner* part of the molecule, and **exohydrolases**, which hydrolyze the acetal linkages from the *ends* of the molecule, splitting off one or two residues at a time. Both types of enzymes are usually affected by branch linkages. That is, exohydrolases frequently stop when they reach branch linkages and give high molecular weight limit dextrins; endohydrolases usually can approach a branch linkage from both sides, but can only hydrolyze the linkages two to three monosaccharide units away from the branch linkage, thus also giving a limit dextrin. The endohydrolase limit dextrin, however, is much smaller than the exohydrolase limit dextrin.

Enzymes also have specificity for the α- or β-configurations. Furthermore, enzymes can be specific for individual monosaccharide residues or for specific linkage positions. For example, there are glucosidases, mannosidases, galactosidases, and sialidases that are specific for hydrolyzing glucose, mannose, galactose, and sialic acid linkages, respectively, and there are hydrolases that are specific for $1 \rightarrow 2$, $1 \rightarrow 3$, $1 \rightarrow 4$, and $1 \rightarrow 6$ linkages, either at specific positions within the polysaccharide molecule or in consecutive runs of the same linkage. Some enzymes have specificity for hydrolyzing branch linkages, too.

Other polysaccharide-hydrolyzing enzymes are very specific for the types of polysaccharides they will hydrolyze. For example, amylases will hydrolyze the α-1→4 linkages found in starch and glycogen but not the α-1→6 branch linkages in these molecules; pullulanase, an enzyme specific for hydrolyzing the α-1→6 linkage found in pullulan, will also hydrolyze the α-1→6 branch linkages of starch and glycogen. Dextranases will hydrolyze the α-1→6 linkages in dextrans but not the 1→6 branch linkages in starch; cellulases hydrolyze cellulose; heparinase hydrolyzes heparin, and so on. From a knowledge of the specificity of these hydrolases and the specific types of oligosaccharide products they produce, much can be learned about the structure of the parent polysaccharide.

10.1.3.3 DETERMINING THE MOLECULAR WEIGHT OF POLYSACCHARIDES

The molecular weight of a polysaccharide can be determined by two methods. The first and most accurate is the reduction of the reducing end of a known weight of polysaccharide with ^{3}H-labeled sodium borohydride. The amount of tritium incorporated into the polysaccharide after reduction is determined by liquid scintillation counting (Section 6.9). The amount of polysaccharide present can be determined by the phenol–sulfuric acid analysis (Section 7.1.2.1). Using the specific activity of the sodium borohydride divided by 4, because there are 4 moles of tritium per mole of borohydride, the number of moles of ^{3}H incorporated into the reducing end of the polysaccharide can be calculated. The number of moles of monosaccharide present (obtained from the phenol–sulfuric acid analysis) divided by the number of moles of ^{3}H incorporated into the reducing end of the polysaccharide multiplied by the anhydro molecular weight of the monomer gives the number average molecular weight of the polysaccharide. The reducing value can also be determined using alkaline copper or alkaline ferricyanide reagents (Section 7.1.2.2a,b,c) and the molecular weight obtained from the number of moles of monosaccharide divided by the number of moles of reducing group, multiplied by the anhydro molecular weight of the monomer.

The second method for determining the molecular weight of the polysaccharide involves gel filtration (Section 4.10.5). Gel filtration material such as agarose (e.g., Bio-Gel A 0.5m, A 1.5m, A 5m, A 15m, A 50m, A 150m; Sephacryl 2B) can be used to estimate the maximum molecular weight from 5×10^5 to 1.5×10^8 and the distribution of molecular weights in the sample. A more precise determination of the molecular weight by gel filtration requires specific compounds of known molecular weight that can be used as calibrating agents.

10.2 DETERMINING THE PRIMARY STRUCTURE OF PEPTIDES AND PROTEINS

The determination of the sequence of amino acids in the primary structure of a protein is a crucial step in the elucidation of the overall structure of a protein. Knowledge of the primary structure allows investigation of structure–function relationships as well as predictions about the higher order structures of a protein. Evolution of structure on a molecular level has been examined for many families of proteins by comparing the homologies of the amino acid sequences. The complete amino acid sequence of cytochrome c, a protein containing just over 100 amino acids, has been determined for more than 50 species, and the information is being used to trace the evolutionary relationships of these species.[4]

The determination of the sequence of amino acids in a protein can be divided into five distinct steps:

1. Isolation and demonstration of purity of the protein
2. Determination of the total amino acid content of the protein
3. Specific cleavages of the protein into two or more series of peptides
4. Purification of the individual peptides and determination of the amino acid sequence of each peptide
5. Deduction of the total amino acid sequence by sequence comparison of overlapping peptides

Within each of these major steps, there are important techniques to be performed, and the actual steps used will depend on the length of the peptide, the presence or absence of a specific amino acid, and the total amount of protein available for sequencing. With rapid developments in DNA sequencing techniques, it is frequently possible to have sequence data for both the protein and the DNA, and thus the structure can be proofread by comparing the results of each method. We shall first examine the process of sequencing a protein, and then (Section 10.3) the sequencing of DNA and RNA will be described in detail.

10.2.1 The Formation and Separation of Polypeptide Chains

Many proteins are oligomers, consisting of two or more identical or nonidentical polypeptides. These chains may be held together by covalent bonds, such as disulfide bonds between cysteine residues in discrete polypeptide chains, as in insulin, or by noncovalent interactions, as in the hemoglobin tetramer. Before beginning the sequencing of a multiple-chain protein complex, it is necessary to separate the individual polypeptide chains.

10.2.1.1 THE REDUCTIVE CLEAVAGE OF DISULFIDE BONDS AND ALKYLATION WITH IODOACETAMIDE

A 2% (w/v) solution of protein in Tris–HCl buffer (pH 8.2) is made 0.75 M with 2-mercaptoethanol and incubated for 1 hour at room temperature in a nitrogen atmosphere. The solution is cooled to 0°C and an equal volume of 0.75 M iodoacetamide is added. The reaction is stopped after 1 hour by the addition of excess mercaptoethanol or by dialysis of the reaction mixture against buffer without iodoacetamide.

10.2.1.2 THE SEPARATION OF POLYPEPTIDE CHAINS UNDER DISSOCIATING CONDITIONS

Treatment of the protein solution with 2% SDS or 8 M urea will dissociate oligomeric proteins into their constituent chains, which can then be separated by the methods described in Chapter 8 (Section 8.5). Human hemoglobin is dissociated into α and β chains in 50 mM 2-mercaptoethanol and 8 M urea in sodium phosphate buffer, pH 6.7.[5] The α and β-chains can be separated on carboxymethylcellulose (Section 4.2) in the presence of the same buffer.

10.2.2 Establishing the Purity of the Protein

It is essential that a protein sample be free of other contaminating proteins before sequencing is begun. There are many methods available for testing the homogeneity of a protein sample, and since no single method can reliably demonstrate homogeneity, a combination of methods generally is used. The methods commonly used are:

1. Two-dimensional analytical electrophoresis (Section 5.3.1)
2. Gel electrophoresis under dissociating conditions (Section 5.2.3.3a)
3. Ion-exchange chromatography (Section 4.2)
4. Affinity chromatography (Section 4.7)
5. Determination of the N-terminal amino acid residue (Section 10.2.4.1a,b)
6. Purification to constant specific activity (Section 9.5)

10.2.3 Determining the Amino Acid Composition

To determine the total amino acid composition, the protein must be hydrolyzed to its constituent amino acids. This is achieved by placing the sample in 6 N HCl containing 0.1% phenol and heating at 110°C in the absence of oxygen. During acid hydrolysis, some amino acids are destroyed and some peptide linkages are particularly resistant to hydrolysis. Therefore, aliquots of the hydrolysate are removed at different times (e.g., 12, 24, 48, and 72 hours) and subjected to analysis. The amounts of the labile residues (serine and threonine) are calculated by extrapolation back to zero time. Isoleucine and valine are determined at 72 hours because they form especially stable peptide linkages. Tryptophan, on the other hand, is extensively destroyed by acid hydrolysis and must be determined by other procedures, such as ultraviolet absorption spectrometry.

Once the free amino acids have been obtained, they must be separated and quantitated, using methods such as liquid–liquid, gas and ion-exchange chromatography, electrophoresis, or HPLC. The traditional procedure is liquid ion-exchange chromatography on sulfonated polystyrene cation-exchange resins (Sections 4.1 and 4.11.1). The amino acids in a buffered acidic solution have a net positive charge and are bound to the anionic sulfonate groups. As the pH and NaCl concentration of the eluting solution are increased, the amino acids are sequentially displaced from the resin and are eluted from the column. The eluant is reacted with ninhydrin (Section 7.2.2.1) and the resulting colored complex is measured by quantitative spectrophotometry (see Fig. 4-28). Thus, both the identity and the quantity of each amino acid are determined. Norleucine is added before the hydrolysis as an internal standard for quantitating the recovery of amino acids in the hydrolysate. The separation process has been improved through the use of smaller resin beads and accelerated by applying buffers under pressure. The method is sensitive to a few nanomoles of each amino acid.

Because tryptophan is destroyed during acid hydrolysis, it is measured on the intact protein. Both tryptophan and tyrosine absorb light in the ultraviolet region between 280 and 290 nm. The ratio of the number of tryptophan to tyrosine residues in a protein can be determined by measuring the protein absorbances at 280 and 288 nm in a neutral 6 M guanidinium chloride solution. The number of tyrosine residues is determined from the amino acid analysis of the protein hydrolysate. The molar extinction coefficients for tryptophan are 4815 at 288 nm and 5690 at 280 nm, and the molar extinction coefficients for tyrosine are 385 at 288 nm and 1280 at 280 nm. Using the simultaneous equation method given in Section 3.2.5, the ratio of tryptophan to tyrosine is given by:

$$\frac{\text{Trp}}{\text{Tyr}} = \frac{(A_{288} \times 1280) - (A_{280} \times 385)}{(A_{280} \times 4815) - (A_{288} \times 5690)}$$

The number of tryptophan residues is obtained by mutiplying the number of tyrosine residues times the Trp/Tyr ratio.

Proline is a secondary amine and therefore forms a different colored complex with ninhydrin (Section 7.2.2.1). The wavelength of maximum absorbance of the proline–ninhydrin product is 420 nm.

10.2.4 End-Group Analysis

End-group analysis is used both for determining the purity of the protein preparation and for characterizing the residues at the amino and carboxyl termini of the protein; the methods can also be used to determine the termini of peptides produced by specific cleavages.

10.2.4.1 DETERMINING N-TERMINAL RESIDUES

Both chemical and enzymatic methods can be used to determine the residues at the amino terminus. The chemical methods involve the formation of an acid-stable derivative of the amino group, followed by acid hydrolysis of the polypeptide and identification of the derivatized amino acid. The exoenzymatic method releases amino acids from the amino terminus in a sequential fashion.

a. The Dansyl Chloride Method. The dansyl chloride method is a technique for the sensitive detection of N-terminal residues.[6] Dansyl chloride (4-dimethylaminonaphthalene-5-sulfonyl chloride) reacts with α- and ε- amino groups to form a sulfonamide that is stable to acid hydrolysis. The sulfonamide exhibits a strong fluorescence that makes this method extremely sensitive. The reaction is pH dependent and requires an excess of dansyl chloride for complete derivatization of the protein. The side-chain groups of histidine (imidazole), cysteine (sulfhydryl), and tyrosine (phenolic hydroxyl) react with dansyl chloride, but only the tyrosine derivative is stable to acid hydrolysis.

DETERMINING THE N-TERMINAL RESIDUES USING DANSYL CHLORIDE

Protein (1–2 nmol) is dissolved in 20 μL of 1% SDS (w/v) and warmed to 40°C to effect solution of the protein. Pure N-ethyl morpholine (20 μL) is added with mixing, followed by 30 μL of 2.5% (w/v) dansyl chloride in dimethylformamide. The reaction mixture is kept at 20°C for 1 hour. The dansylated protein is precipitated by the addition of cold acetone, up to 250 μL. The suspension is held at 0°C for 10 minutes and centrifuged briefly to pellet the protein. The supernatant is discarded and the pellet is washed once with acetone containing 25% 0.1 M HCl. The precipitate is dried *in vacuo*. Next, 25 μL of 6 M HCl is added, and the tube is sealed under vacuum and placed in an oven at 105°C for 10–12 hours. The tube is cooled and broken open and the acid is removed in a vacuum desiccator over NaOH pellets. The dansyl amino acid is identified by two-dimensional chromatography on micropolyamide sheets.

The dansylated sample is dissolved in a minimum volume of 50% aqueous pyridine or pyridine–ethanol (1:1); 1–5 μL is spotted about 1 cm from the corner

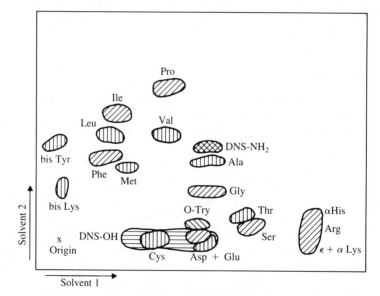

FIGURE 10-5. A typical dansyl amino acid map. Dansyl amino acid standards are separated by two-dimensional chromatography using 1.5% aqueous formic in the first dimension and toluene–glacial acetic acid (10:1) in the second. Amino acids are detected by fluorescence. DNS–OH is the hydrolysis product of dansyl chloride, and DNS–NH$_2$ is the product of dansyl chloride and ammonia.

of a 9 cm × 9 cm sheet of a double-sided polyamide thin-layer sheet (Schleicher and Schuell). The spot is dried in warm air, and a standard mixture of dansyl amino acids (2 μmol/mL) is spotted at the same position on the opposite side of the polyamide sheet. Chromatography is performed using 1.5% aqueous formic acid (v/v) until the solvent reaches the top of the sheet. The sheet is removed, dried thoroughly in warm air, rotated 90°, and placed in the second solvent, toluene–glacial acetic acid 10:1 (v/v). As soon as the solvent reaches the top, the sheet is removed and dried. The dansyl amino acids are detected by fluorescence under UV light of 254 nm. A typical dansyl-amino acid map is shown in Figure 10-5: DNS-his, DNS-arg, and ε-DNS-lys are not well resolved; Bis-DNS-lys, which is formed with N-terminal lysine, is resolved. Dansic acid (DNS-OH) is easily identified because it fluoresces a bright blue.

b. The Aminopeptidase Method. Cleavage with aminopeptidases can be used to identify residues that are unstable to acid cleavage or are not resolved by the dansylation method. These enzymes cleave residues sequentially from the amino terminus at widely differing rates, however, and therefore ambiguous results may be obtained. Leucine aminopeptidase releases hydrophobic amino acids preferentially and releases acidic residues very slowly. Aminopeptidase M has a broader specificity. However, the use of this method cannot match the sensitivity of the dansyl chloride method.

10.2.4.2 DETERMINING C-TERMINAL RESIDUES

The methods available for the determination of carboxyl-terminal residues of proteins and polypeptides are less satisfactory than those available for the amino-terminal residues and therefore are used less frequently. The most commonly used method is reaction with carboxypeptidase. Carboxypeptidases are exopeptidases that release the amino acids from the carboxyl terminus in a sequential fashion. Therefore, both the identity and the quantity of each amino acid are determined as a function of time of hydrolysis. The nature of the amino acid affects the rate of release and sometimes stops the hydrolysis. Carboxypeptidase A will not release proline or arginine and reacts slowly with lysine. Carboxypeptidase B releases both lysine and arginine. Carboxypeptidase C releases all the common amino acids from the C-terminus. A mixture of carboxypeptidases A and B is commonly used to ensure the hydrolysis of the carboxyl-terminal residues. The released amino acids can be determined by TLC, HPLC, or the amino acid analyzer.

10.2.5 Specific Methods of Peptide Bond Cleavage

The cleavage of large proteins into a mixture of smaller polypeptides, followed by the separation of the peptides and the determination of the amino acid sequence of the individual peptides, is part of the rationale for determining the entire primary structure of a protein. With current sequencing methods it is possible to obtain sequences for 30–50 residues. Limited proteolysis with highly specific enzymes or specific chemical reactions to yield a small number of large fragments (30–50 residues) is preferable to generation of many smaller fragments. Selective hydrolysis with either chemical reagents or enzymes has other advantages. From a knowledge of the amino acid composition of a protein, one can predict the number of peptides that should be produced by using reactions specific for peptide linkages of particular residues. Depending on the specificity of the cleavage reaction, the C-terminal residues of the resulting peptides will be the same for every peptide or will be limited to two or three amino acid residues. For example, cyanogen bromide cleaves only at the carboxyl group of methionine to give homoserine at the C-terminus, and trypsin cleaves only at the carboxyl groups of lysine and arginine to give these residues at the C-terminus.

10.2.5.1 ENZYMIC CLEAVAGE METHODS

a. Trypsin Hydrolysis. Trypsin hydrolyzes lysine and arginine peptide linkages, except when proline is attached to the carboxyl group of lysine or arginine. Commercially available trypsin is frequently contaminated with chymotrypsin. Therefore, trypsin preparations should be treated with tosyl-L-phenylalanine chloromethylketone (TPCK) to inhibit chymotrypsin activity. Also, calcium ions stabilize trypsin and should be present to prevent trypsin autolysis. Trypsin is usually added at 1% of the weight of the protein being hydrolyzed, and the pH of the reaction is held at 8.5 either by the addition of base during the hydrolysis or by the use of a volatile buffer, such as ammonium bicarbonate.

The specificity of trypsin can be further increased by chemically modifying the ε-amino group of the lysine residues, thereby preventing hydrolysis of the lysine peptide linkages and limiting the hydrolysis to arginine peptide linkages. In the determination of the sequence of β-galactosidase, one of the largest proteins sequenced to date, citraconic anhydride was used to modify the ε-amino groups of the lysines.[7] Large peptides result from this strategy, which aids in the fitting together of overlapping sequences. The citraconic moiety can be reversibly removed from the lysines by treatment at pH 2 for 2 hours at 20°C and the sequence of the peptides determined.

Additional trypsin susceptible linkages also can be obtained by reacting sulfhydryl groups of cysteines with ethylenimine, converting them into 5-(2-aminoethyl) cysteine,[8] thus making the cysteine carboxyl linkage susceptible to trypsin hydrolysis.

b. Chymotrypsin Hydrolysis. α-Chymotrypsin hydrolyzes at the carboxyl side of leucine, tyrosine, phenylalanine, tryptophan, and methionine. If any of these residues are attached to proline, hydrolysis does not occur. Since chymotrypsin hydrolyzes at the different residues at varying rates, a short hydrolysis time produces relatively large peptides, and longer hydrolysis times will yield smaller peptides.

c. Pepsin Hydrolysis. Pepsin does not have the high specificity that trypsin or chymotrypsin have, but it does have specificity for hydrolyzing preferentially the peptide linkage that occurs between two amino acids with hydrophobic side chains. Other sites of hydrolysis are the carboxyl side of leucine or amino acids with aromatic side chains. Pepsin is used to hydrolyze only a few especially susceptible peptide bonds at low enzyme concentration or during short reactions times. The enzyme is active at low pH, which is an advantage when working with proteins that are soluble only in acidic solutions.

d. **Other Proteinases.** Proteinases that have been used to generate peptide fragments include staphylococcal protease, thermolysin, subtilisin, and papain. Although these enzymes lack the specificity of trypsin and chymotrypsin, they are useful for generating peptides from proteins not hydrolyzed by trypsin or chymotrypsin.

10.2.5.2 CHEMICAL CLEAVAGE METHODS

Chemical cleavages are used mainly for generating large polypeptides that result from highly specific reactions. The yield of cleaved peptides varies, and the by-products of the reaction, as well as excess reagent, must be removed before further sequencing is attempted.

a. **Cyanogen Bromide Cleavage.** Cyanogen bromide cleaves specifically at the carboxyl side of methionine. The method was introduced in 1962 by Gross and Witkop[9] and is widely used. The reaction converts methionine into homoserine according to the following reactions:

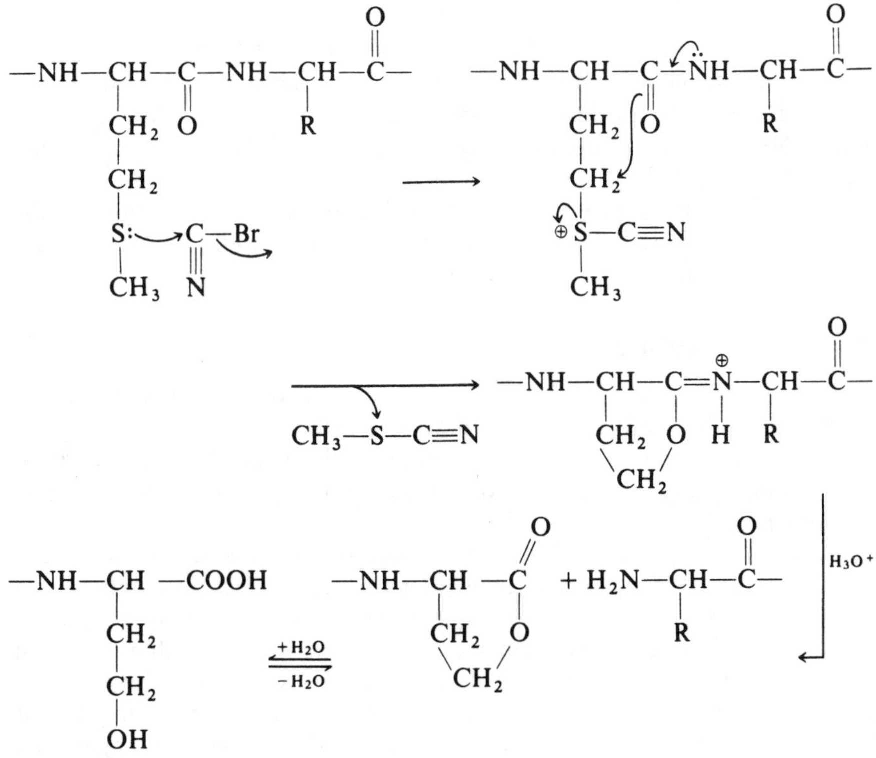

C-terminal homoserine

THE CYANOGEN BROMIDE CLEAVAGE OF PEPTIDE LINKAGES

Protein, at 5 mg/mL in 70% aqueous formic acid, is treated with a 50-fold molar excess of cyanogen bromide over methionine and is incubated in the dark, in the absence of oxygen, for 15–20 hours. The reaction mixture is then diluted with 10 volumes of water and lyophilized to remove the formic acid and excess reagent, and to concentrate the sample.

---

b. Hydrolysis of Asp-Pro Peptide Linkages with Acid. Under mild acidic conditions, the peptide linkage between the prolyl amino group and the aspartyl carboxyl group is hydrolyzed while the remaining peptide linkages remain intact. The hydrolysis is performed by treating with 70% formic acid at 40°C for 48 hours.[10,11]

c. Hydroxylamine Cleavage of Asn-Gly Peptide Linkage. The asparagine–glycine peptide linkage is more susceptible to cleavage by hydroxylamine than other peptide linkages, and approximately 70% of this linkage is cleaved.[12] The reaction is carried out at a protein concentration of 1–5 mg/mL in 6 M guanidine hydrochloride, 2 M hydroxylamine, and 0.2 M potassium carbonate, and the pH is adjusted to 9.2. The solution is incubated at 40°C for 4 hours, and the reaction is terminated by lowering the pH to 3.

d. Cleavage of Tryptophanyl Peptide Bonds with *o*-Iodosobenzoic Acid. Treatment of protein (5–10 mg/mL) with 2 mg of *o*-iodosobenzoic acid per milligram of protein in 80% acetic acid and 4 M guanidine hydrochloride for 20 hours in the dark at 24°C cleaves the peptide linkage of tryptophan but does not cleave other peptide linkages.[13]

After any cleavage protocol, it is advisable to test the effectiveness of the method by examining the number and nature of the peptides produced by gel electrophoresis, two-dimensional TLC (fingerprinting), or HPLC.

10.2.6 The Separation of Peptides

All the classical methods can be used for the preparative separation of peptides generated by the specific cleavages described in Section 10.2.5. Perhaps the most successful current techniques for separating these peptides are gel permeation chromatography, gel electrophoresis, and high performance liquid chromatography.

Peptides can be separated in reverse phase HPLC in which the stationary phase is composed of long-chain, alkyl groups bonded to silica microspheres and the mobile phase is a mixed organic–aqueous solvent. Using split-stream devices, in which only a fraction of the eluant is needed for detection, postcolumn reaction of the eluant with fluorescamine and detection by fluorescence allows the use of many peptide solvents as well as gradient elution. There are limitations, of course—for example, the concentration of the peptides for column application may cause precipitation of the peptides. Alkaline conditions cannot be used with silica-based columns, and therefore the pH of the peptide solution must be 7 or below. Gel permeation chromatography is commonly used for screening the major peptide fractions according to molecular size. Either crossed-linked dextrans or polyacrylamides can be used. The gel medium chosen will depend on the size range of the peptides produced. Sephadex G-50 or Bio-Gel P-30 will fractionate peptides in the range of 500–10,000 molecular weight. If the concentration of peptide is high, the fractions can be screened by monitoring absorbance at 280 nm. Further fractionation can be accomplished by ion-exchange chromatography, paper electrophoresis or chromatography, or two-dimensional peptide mapping.

TWO-DIMENSIONAL PEPTIDE MAPPING BY TLC[14]

A sample of the protein digest (2–5 nmol) is dissolved in 5 μL of 20% pyridine and applied to a 20 cm × 20 cm cellulose TLC plate (Machery and Nagel, or Eastman Kodak 13255 Polygram CEL 300). The plate is placed in the electrophoresis chamber, and electrophoresis buffer is applied from a moistened piece of Whatman 3MM paper placed onto the surface of the plate. Wicks made from the same paper are connected to the plate, and electrophoresis is performed for 90 minutes at 500 V, with pyridine–acetic acid–acetone–water, (20:40:150:790 by vol, pH 4.4).

At the end of the electrophoresis, the solvent is evaporated and the plate is turned 90° and placed into a chromatography tank, containing 1-butanol–acetic acid–water–pyridine (15:3:12:10, by vol). When the solvent reaches the top, the plate is removed and the solvent is evaporated by heating. The peptides can be detected by the methods listed in Chapter 7 (Section 7.2.2). Fluorescamine spray reagent is especially useful, since very low concentrations of the reagent are needed and peptides can be eluted from the stained area for further characterization. The fluorescamine reaction is pH dependent and is more intense at neutral or basic pH. Peptides with N-terminal proline or with blocked N termini are not detected.[15]

Peptides can be recovered from thin-layer plates by scraping off the silica and eluting with a suitable, volatile solvent, such as acetic acid or pyridine. Filtration or centrifugation will separate the peptide solution from the thin-layer material.

10.2.7 The Sequencing of Peptides

10.2.7.1 THE EDMAN DEGRADATION

The most important chemical method for sequencing polypeptides is the stepwise degradation process using phenylisothiocyanate devised by Edman and Begg and called the **Edman degradation**.[16] In the Edman degradation, one amino acid residue at a time is cleaved from the amino terminus of the polypeptide by reaction with phenylisothiocyanate, followed by mild acid treatment that results

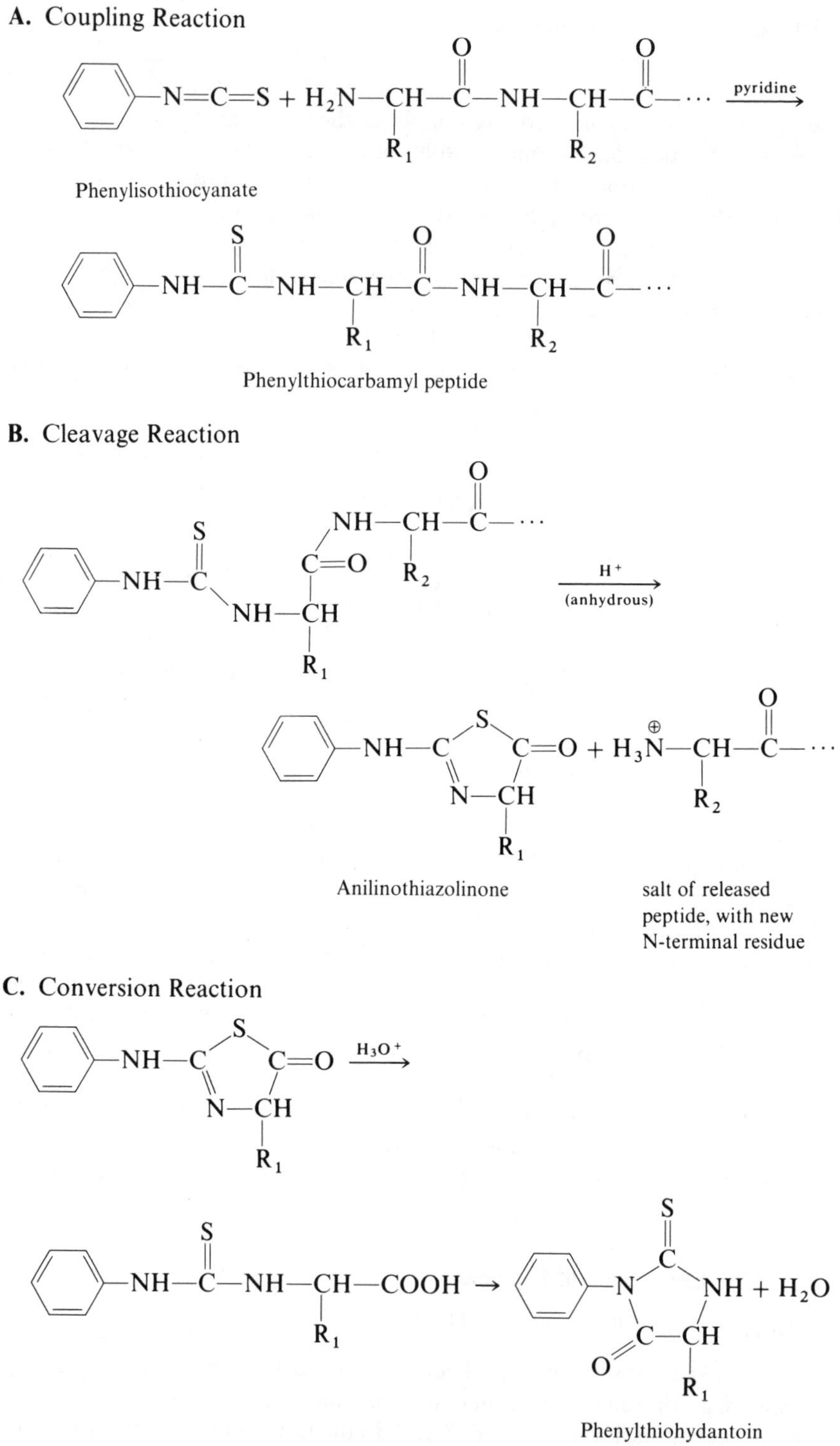
A. Coupling Reaction
pyridine
Phenylisothiocyanate
Phenylthiocarbamyl peptide
B. Cleavage Reaction
H⁺
(anhydrous)
Anilinothiazolinone
salt of released peptide, with new N-terminal residue
C. Conversion Reaction
H₃O⁺
Phenylthiohydantoin

FIGURE 10-6. Reactions of the Edman Degradation.

in the release of the derivatized N-terminal amino acid and its conversion into a stable phenylthiohydantoin derivative. The phenylthiohydantoin is identified using one of a number of chromatographic techniques, or by back-hydrolysis to the free amino acid and then quantitation of the free amino acid. An automated method using the Edman procedure was introduced in 1967, and the sensitivity has been continually improved so that the sequence of an immunoglobulin has been determined on 10 pmol.[17]

Two modifications of the Edman technique have been developed: the subtractive modification, in which the composition of the shortened peptide is measured after each step and the missing amino acid establishes the amino-terminal residue, and the dansyl–Edman modification, in which the new amino-terminal residue is identified at each degradative step by end-group analysis using dansyl chloride (Section 10.2.4.1a). A summary of the reactions involved in the Edman degradation is given in Figure 10-6.

THE EDMAN DEGRADATION: A FOUR-STEP PROCEDURE

Coupling

The peptide is dissolved in 20–50 μL of freshly distilled pyridine–water and added to an ampule. To deprotonate the α-amino group, 20 μL of trimethyl amine (25%) is added, followed by 20 μL of phenylisothiocyanate–pyridine (1:4), and the solution is flushed with nitrogen. The ampule is sealed, and the solution is mixed vigorously and heated at 50°C for 30 minutes.

Washing

Excess reagent and by-products are removed by adding 400 μL of heptane–ethyl acetate (10:1, v/v) with gentle mixing. The upper phase, containing the excess reagent and by-products, is removed. The interface, which may contain precipitated peptide, is avoided. The lower aqueous phase is washed with heptane–ethyl acetate (2:1 or 1:1) and gently mixed. The upper phase is removed, and the aqueous phase is dried in a vacuum dessicator over P_2O_5.

Cleavage and Extraction

For short peptides, 20 μL of trifluoroacetic acid is added to the dry sample and the solution is heated at 50°C for 4–6 minutes. The solution is evaporated to dryness with nitrogen *in vacuo* to evaporate the trifluoroacetic acid. Then 20 μL of 30% aqueous pyridine is added and the solution is extracted twice with 100 μL of benzene–ethyl acetate (1:2 v/v) by mixing on a vortex, followed by centrifugation. The upper layer is removed and the extracts (lower layer), containing the anilinothiazolinone are combined. The aqueous phase is dried and saved for the next Edman degradation cycle.

Conversion and Analysis

The combined benzene–ethyl acetate extracts are dried under vacuum; 40 μL of 1 N HCl in methanol is added, and the solution is heated at 50°C for 10 minutes in air. The sample is dried under vacuum and saved for analysis of the phenylthiohydantoin–amino acid.[18]

THE DANSYL–EDMAN MODIFICATION

A small aliquot of the new peptide resulting from the Edman degradation is removed for reaction with dansyl chloride. As little as 0.2 nmol is required, but it must be free of amines, ammonia, or phenol. Ten microliters of 0.2 M $NaHCO_3$ is added to the dry peptide in a glass ampule and centrifuged briefly in a microcentrifuge. The peptide is dissolved by mixing with a vortex and the contents are dried in a vacuum over P_2O_5 and NaOH. Five microliters of dansyl chloride in dimethylformamide (25 mg/mL) is added, and the ampule is covered and incubated at 37°C for 20 minutes in the dark. The reaction mixture, which should be almost colorless, is dried under vacuum. Ten microliters of 6 N HCl is added and the ampule is centrifuged and sealed. The ampule is incubated at 105°C for 16 hours. The ampule is opened and the acid evaporated under vacuum over NaOH pellets. The dansyl amino acid is identified as described in Section 10.2.4.1.

10.2.7.2 AUTOMATED LIQUID-PHASE SEQUENCE DETERMINATION

Commercially produced automated sequencing instruments, modeled on methods described by Edman and Begg,[16] are widely used. In the spinning cup automatic sequencer, the protein forms a thin film through which reagents diffuse, even if the protein is not soluble. Reliable sequence results can be obtained for polypeptides of up to 50 residues. An overview of the procedure is as follows. The protein or peptide, dissolved in a volatile acid such as trifluoroacetic or formic acid, is introduced into the spinning cup and is dried under vacuum to produce a thin film on the sides of the cup. A solution of phenylisothiocyanate is added together with the coupling buffer. Coupling proceeds for 30 minutes at 50°C. Excess reagents and solvent are removed under vacuum, and the by-products are extracted with a suitable solvent. The derivatized polypeptide is treated with acid, which causes cleavage of the amino terminal to form the anilinothiazolinone of the N-terminal residue. The solution of anilinothiazolinone is extracted and collected for identification of the phenylthiohydantoin amino acid. The dried protein film is then ready for the second cycle of degradation.

Many modifications have been made in the procedures for automated sequencing and are now available in commercial sequencing instruments. A major breakthrough has been the introduction of a polymeric quaternary ammonium salt, polybrene, that effectively anchors small quantities of both protein and peptides in the cup. This allows the sequencing of short, hydrophobic peptides that formerly were lost during extraction with organic solvents. Identification of the phenylthiohydantoins by reverse phase HPLC has extended the sensitivity of detection to subnanomolar quantities.[19]

In 1981 Hood et al. introduced the gas phase automatic sequenator.[20] The polypeptide is embedded in a matrix of polybrene and dried onto a porous glass-fiber disc and placed in a small cartridge reaction cell with a reaction volume of 50 μL. Argon gas is used to pressurize the reagent and solvent reservoirs, and all valves and tubing are miniaturized. These mechanical changes, together with greater purity of reagents and solvents, have made it possible to obtain longer sequences from smaller quantities of protein. The cycle time for each cleavage step is 45–55 minutes, which allows up to 24 cycles per day. At the same time, there have been many improvements in manual sequencing methods that do not require expensive automatic equipment and can easily handle the sequencing of relatively large numbers of smaller sized peptides.

10.2.7.3 MICROSEQUENCING

Many of the improvements in chemistry and instrumentation designed to increase the sensitivity of the analyses have led to techniques referred to as **microsequencing** in which nanomole quantities of proteins are used. There are three general approaches to microsequence analysis: (a) instrument redesign and use of ultrapure reagents in which picomole to nanomole quantities of peptides can be analyzed; (b) use of radioactive, colored, or fluorescent isothiocyanate derivatives for detection at much lower levels; and (c) radiolabeling of the protein before degradation. Using a colored derivative of phenylisothiocyanate, 4-methylaminoazobenzene-4'-isothiocyanate, it has been possible to sequence polypeptides at the 1–10 nmol level in a manual mode and identify the phenylthiohydantoins by HPLC.[21]

Another general approach relies on the radioactivity of the amino acids derived from the protein, which has been labeled during biosynthesis. Radiolabeled amino acids can be incorporated during biosynthesis in cell culture or by cell-free translation of mRNA. In most cases, a single pure protein is isolated by immunoprecipitation. The presence of an unlabeled antibody in the immunoprecipitate does not interfere with the sequence analysis of the radiolabeled protein.

10.2.7.4 PROBLEMS ENCOUNTERED BY A BLOCKED AMINO-TERMINAL RESIDUE

Many proteins have blocked amino terminal residues. Some of the blocking

groups are: formyl $H-\overset{\overset{\displaystyle O}{\|}}{C}-$ found in prokaryotic nascent proteins, acetyl

$CH_3\overset{\overset{\displaystyle O}{\|}}{C}-$ found in horse heart cytochrome c and *Neurospora* tyrosinase, pyrrolidone carboxyl $O=C$ structure found in porcine α-amylase and

immunoglobulins, and fatty acyl R—C— found in murein lipoproteins from *E. coli*.

The determination of the blocking group is generally made after studying peptides formed by digestion with proteases. The pyroglutamyl residue may be removed enzymatically by treatment with pyroglutamate aminopeptidase.[22] Acetyl groups can be removed by treatment of the protein with hydrazine, followed by reaction with dansyl chloride. *N*-Acetyl *N'*-dansyl hydrazide is formed and can be identified by chromatography on micropolyamide TLC plates. The formyl group can be removed by incubation of the peptide with 0.5 M HCl in methanol for 48 hours at 25°C, or by reaction with hydrazine.[23]

10.2.8 Deducing the Total Sequence from Overlapping Peptides

After the sequence of individual peptides resulting from various peptide bond cleavages has been determined, the sets of peptide sequences must be aligned, to obtain the total sequence of the protein. The task is easiest if at least three sets of peptides from different digests of the protein are sequenced. The difficulty of aligning small peptides increases greatly as the chain length increases, but some factors aid in the alignment process. First, amino acid residues that are fairly rare in proteins, such as tryptophan, cysteine, and methionine, can act as markers for peptide sequences. Second, information from large peptides, even if they are not completely sequenced, can be helpful in determining the location of smaller peptides. A method of rapidly increasing importance is the use of DNA sequence information. The combination of sequence information from DNA and protein techniques has the advantage of permitting the detection and correction of errors made by either method.

10.3 DETERMINING THE STRUCTURE OF NUCLEIC ACIDS

The development of sequencing methods for nucleic acids has progressed at an amazing rate. The first report of the sequence of a tRNA was published in 1965 by Holley and co-workers,[24] and the first sequence of DNA was published in 1973.[25,26] In 1975 Sanger and Coulson introduced the plus–minus method for sequencing DNA.[27] In 1977 Maxam and Gilbert reported a sequencing method involving chemical cleavage of DNA at specific bases, followed by gel electrophoretic analysis.[28] Also in 1977 Sanger et al.[29] reported an elegant DNA sequencing method using DNA polymerase, radioactive 2'-deoxyribonucleotide triphosphates, and 2',3'-dideoxyribonucleotide triphosphates to copy a DNA sequence. For their pioneering work in DNA sequencing, Gilbert and Sanger were awarded the 1980 Nobel Prize in chemistry.

Because of these developments, the time necessary to sequence DNA has been considerably shortened from the first sequence analysis, which took years to complete, to the present techniques, which can handle several hundred nucleotides in approximately 24 hours. Current developments promise even shorter times.

The importance of DNA sequencing for determining gene structure and the primary sequence of proteins has led to rapid developments in methodology. Many techniques involving end labeling with [^{32}P]phosphate and fluorescent tags, high resolution polyacrylamide gel electrophoresis, and the use of specific restriction endonucleases have been developed. We shall see how these techniques are used in the sequencing of nucleic acids.

The process of sequencing a given DNA molecule can be divided into several distinct steps:

1. Isolation and purification of DNA
2. Enzymatic cleavage of the DNA into a series of oligonucleotides that can be sequenced separately
3. Separation of DNA fragments and establishment of their purity
4. Determination of the sequences of the DNA fragments
5. Deduction of the total DNA sequence from the individual sequences of the DNA fragments.

10.3.1 The Isolation, Purification, and Cloning of DNA

The procedures for the isolation of bacterial chromosomal DNA, plasmid DNA, and phage DNA are given in Chapter 8 (Sections 8.7.2.1–8.7.2.3).

10.3.1.1 CONCENTRATING DNA ISOLATED FROM VARIOUS SOURCES

DNA can be concentrated by precipitation with cold ethanol under fixed ionic strength conditions. The salt concentration should be 0.25 M sodium acetate or 0.1 M sodium chloride. Two volumes of ice cold ethanol are added and the solution is thoroughly mixed. If the DNA is very dilute, the temperature is lowered to $-70°C$. The precipitated DNA is removed by centrifugation for 12 minutes at 12,000g. The supernatant is decanted and the final traces of ethanol are removed by placing the solution in a vacuum desiccator. The DNA is dissolved in the desired volume of buffer by incubation at 37°C for 5 minutes. The amount of DNA can be quantitated by UV absorbance (Section 7.3.2.1) or by fluorescence with ethidium bromide (Section 7.3.2.2).

10.3.1.2 DNA AMPLIFICATION BY CLONING

The technique of cloning DNA fragments from any origin into plasmids or phages and their amplification in a suitable bacterial host has made it possible to obtain increased quantities of almost any DNA for sequencing. These recombinant DNA techniques are therefore a necessary part of sequencing and are discussed here as an overview of the methods for carrying out such experiments.

Both plasmid and phage DNA can be replicated in a suitable bacterial host independently of the host chromosome. Plasmids are small, circular, self-replicating molecules that can be introduced into bacteria and will multiply to produce up to several hundred per host bacterial cell. It is possible to introduce into these vectors (plasmids or bacteriophage) foreign DNA that will also be copied and amplified. The plasmid or phage DNA can be separated from the host genome, and the inserted DNA sequences excised from the vector for sequencing, or in special cases the nucleic acids of the recombinant DNA can be sequenced while still present in the vector.

Three of the most common types of vectors are plasmids, λ bacteriophage, and the filamentous bacteriophage M13. Plasmid DNA is double stranded and circular, λ DNA is double stranded and linear, and M13 DNA is single stranded and circular. All three can be propagated in bacteria.

Bacteriophage λ DNA has a genome size of 50 kilobase (kb) pairs. In these phage particles the DNA is in a linear duplex form, but after it enters a host bacterium, the DNA circularizes and is transcribed and replicated in a circular form. During the duplication of the virus in the bacteria, the circular DNA is replicated many-fold, giving bacteriophage progeny that have double-stranded, linear DNA and are eventually released into the culture fluid when the host bacterial cell lyses. Modified λ phage have been engineered to accept and propagate foreign DNA from many diverse sources. Before foreign DNA can be inserted, however, specific regions of the λ phage must be hydrolyzed by restriction endonucleases, bacterial enzymes that hydrolyze the phosphodiester linkages of DNA at sites determined by nucleotide sequences. The amount of foreign DNA inserted cannot greatly exceed the amount of phage DNA excised. When λ phage is hydrolyzed at a single site by a restriction endonuclease, a small amount of foreign DNA can be inserted by ligation and the product is known as an **insertion vector**. When λ phage is hydrolyzed at two sites and a segment of phage DNA is released, an equivalent amount of foreign DNA can be inserted and the product is known as a **replacement** or **substitution vector**. Cloning DNA into λ vectors is similar to the process to be described for DNA insertion into plasmid vectors, with two additional steps: preparation of viable bacteriophage particles *in vivo*, and infection of the bacterial host to produce increasing quantities of the phage DNA carrying the inserted foreign DNA segment.

Many experiments with cloned DNA, in particular the primed synthesis method of sequencing (Section 10.3.4.2), require the use of single-stranded DNA, which can be obtained directly from single-stranded bacteriophage. Single-stranded bacteriophage vectors, like M13, replicate via a double-stranded circular DNA intermediate (called replicating form, RF). This RF can be isolated and manipulated *in vitro* just like a plasmid. When the modified RF is re-introduced in *E. coli*, the cells will continue to grow and divide and will release into the culture medium phage particles containing single-stranded DNA. Because M13 is a filamentous phage, there is no theoretical limit to the amount of foreign DNA that can be added to the phage DNA. In fact, modified M13 phage DNA containing foreign DNA up to five times the length of M13 has been produced. The phage particles released from the *E. coli* contain single-stranded DNA that is homologous to only one of the two strands of the cloned DNA, and therefore this DNA can be used as a template for the primed synthesis method of DNA sequencing.

The method of isolation of plasmids and their properties are given in Chapter 8 (Section 8.7.2.2).

There are several methods for inserting a DNA segment into a vector. A double-stranded vector such as a plasmid can be opened with a restriction endonuclease and joined with a DNA fragment formed from the foreign DNA by hydrolysis with the same restriction endonuclease. (For a further discussion of the specificity and action of restriction endonucleases, see Section 10.3.2). If the restriction endonuclease generates protruding ends, the ends of the vector and the DNA fragment will be complementary and will associate. DNA ligase is then added to join the DNA sequences into the vector. The association process is random, but some of the products will be circular molecules containing the inserted recombinant DNA sequence.

In some instances the molecules to be cloned will have blunt ends. These can be inserted directly into a vector, using T_4 DNA ligase, which gives blunt-end ligation, or after the addition of artificial linkers to the blunt ends to produce cohesive ends. The efficiency of incorporation of blunt-ended fragments, however, is low because the majority of the vector molecules will recircularize without the insertion of the foreign DNA.

10.3.2 The Enzymatic Cleavage of DNA and Restriction Mapping

A DNA molecule with more than 1000 base pairs cannot be completely sequenced as a single piece. Current sequencing methods can determine up to 300 nucleotides, and therefore it is necessary to cleave the native DNA that is to be sequenced into smaller fragments. Several methods are used to yield fragments for sequencing. These include mechanical shearing, sonic shearing, and enzymatic cleavage by DNAse 1 or a wide variety of restriction endonucleases. The most widely used method is hydrolysis by restriction endonucleases.

The number of restriction endonucleases available in a purified form is increasing rapidly, and the specificity and optimum conditions for action of more than 200 have been reported.[30] Many of the best characterized restriction endonucleases are available commercially, but also can be produced and purified in the laboratory. Appendix C lists commercially available restriction endonucleases and their specificities and Chapter 8, Section 8.5.4.5 contains a laboratory preparation of a restriction endonuclease.

Restriction endonucleases have the ability to recognize symmetrical sequences of four or more nucleotides and to cleave both DNA strands within the recognition sequence. For example, Eco RI from *E. coli* hydrolyzes only between

G and A in the following hexanucleotide sequence and produces fragments that protrude at their 5′ ends.

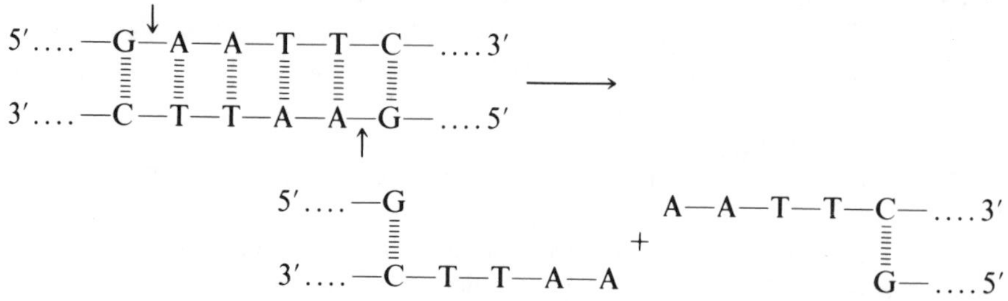

5′.... —G—A—A—T—T—C— ....3′

3′.... —C—T—T—A—A—G— ....5′   ⟶

5′.... —G   A—A—T—T—C— ....3′
                                  +
3′.... —C—T—T—A—A   G— ....5′

Pst I from *Providencia stuartii* hydrolyzes DNA between G and A in the following hexanucleotide sequence and produces fragments that protrude at their 3′ ends.

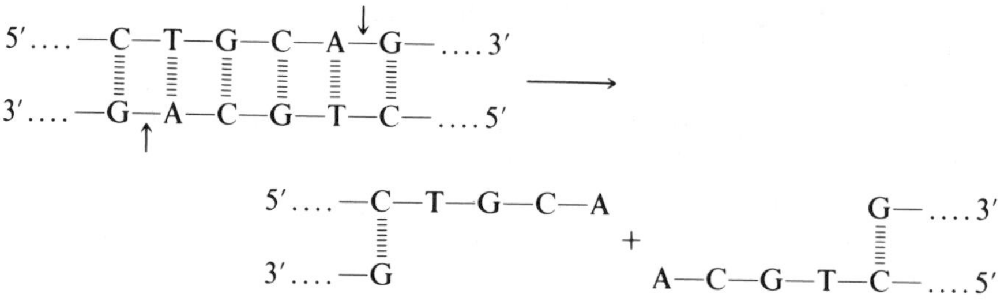

5′.... —C—T—G—C—A—G— ....3′

3′.... —G—A—C—G—T—C— ....5′   ⟶

5′.... —C—T—G—C—A   G— ....3′
                                  +
3′.... —G   A—C—G—T—C— ....5′

When the restriction fragments have protruding 5′ or 3′ ends, they are referred to as "sticky" or "cohesive" ends. Some restriction endonucleases, such as Bal I, from *Brevibacterium albidum*, hydrolyze DNA at C and G in the following hexanucleotide sequence to give "blunt ends" instead of sticky ends.

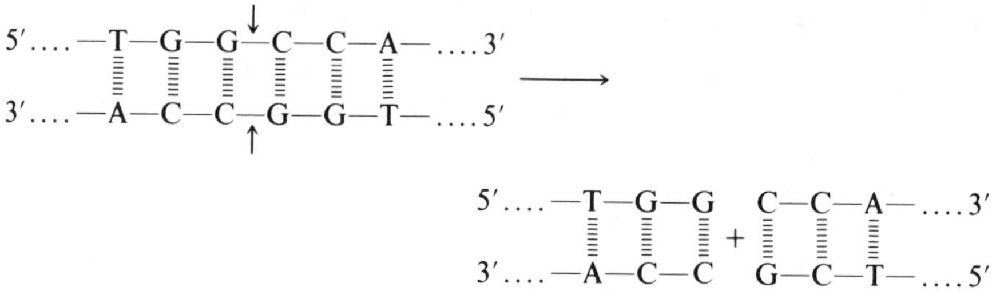

5′.... —T—G—G—C—C—A— ....3′

3′.... —A—C—C—G—G—T— ....5′   ⟶

5′.... —T—G—G   C—C—A— ....3′
                +
3′.... —A—C—C   G—G—T— ....5′

Restriction endonucleases function in bacterial cells to degrade foreign DNA (e.g., viral DNA) without damaging the bacteria's own DNA, hence the requirement for strict sequence specificity of the enzymes. Restriction endonucleases can, therefore, cleave DNA into defined sets of fragments, which can then be sequenced directly, or used for producing larger quantities of the DNA fragments to be sequenced, by cloning into vectors. An additional use of restriction endonucleases is mapping of DNA.

Before starting the sequencing of large DNAs, it is advantageous to know the cleavage sites for several restriction endonucleases. Such knowledge aids in the development of a strategy for sequencing and gives clues for obtaining overlapping fragments used in deducing the overall sequence of the DNA. The process of identifying these cleavage sites is known as **restriction mapping**.

DNA can be hydrolyzed with a series of different restriction endonucleases, and the product DNA fragments separated by gel electrophoresis. The number of DNA fragments, visualized on a gel, will reveal the number of restriction sites for a particular endonuclease, and if molecular weight marker DNAs are also electrophoresed, the approximate chain length of the fragments can be determined. For example, a 700-base-pair DNA could be treated with Hind III and then Eco RI endonucleases, yielding the following possible results:

If the individual restriction fragments are sequenced, the total sequence may be deduced by comparing the sequences of the overlapping fragments. Gel electrophoresis of the two digests, run in parallel, will indicate which fragments from the Hind III cleavage contain sites for the second enzyme (Eco RI) and vice versa (Fig. 10-7). If the hydrolyzed DNA is run in agarose, the fragments can be recovered by elution or electroelution and used for sequencing directly or for amplification by cloning.

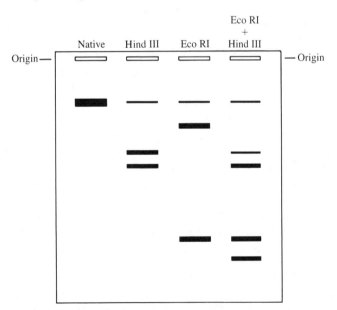

FIGURE 10-7. Slab gel electrophoretic analysis of native DNA, restriction endonuclease hydrolysis with Hind III, Eco RI, and a mixture of Hind III and Eco RI, showing the different DNA fragments in a restriction endonuclease map.

10.3.3 The Establishment of Purity and the Separation of DNA Strands by Electrophoresis

Electrophoresis of DNA in agarose minigels containing ethidium bromide provides a rapid method of measuring both the quantity of DNA and its purity. Minigels are poured on 5 cm × 8 cm glass plates and sample slots are formed with a minicomb. The gel slots will hold 3–5 μL of sample. The agarose gel (0.5–2.0%) contains 0.5 μg/mL ethidium bromide. The gel is submerged just under the surface of the buffer in the electrophoresis cell. The resistance of the gel to passage of the current is almost the same as the buffer, so a considerable fraction of the applied current is carried through the gel. Electrophoresis is carried out at high voltage (15 V/cm) for 30 minutes, during which the tracking dye, bromophenol blue, migrates almost the full length of the gel. DNA standards (0.5–50 μg/mL) can be run in the other slots of the gel and the fluorescence of the unknown can be compared to that of the standards to estimate the amount of DNA present in each fragment. This rapid electrophoresis method is especially useful for restriction endonuclease mapping of large nucleic acid molecules.

10.3.4 Sequencing Strategies

There are two procedures used for sequencing nucleic acids, and each generates a set of oligonucleotides that can be separated by electrophoresis and permits the deduction of the nucleic acid sequence.

The first is the chemical method, developed by Maxam and Gilbert,[28] which utilizes chemical reactions to modify specific bases and leads to elimination of the modified bases and cleavage of the phosphodiester linkages. Four specific chemical reactions are performed on four DNA aliquots, resulting in the formation of four distinct sets of oligonucleotides. The products are separated by electrophoresis and the sequence is read from the autoradiogram of the separated oligonucleotides. Both single- and double-stranded DNA can be sequenced. The limitation of the method is the resolving power of the sequencing electrophoresis gel, which at present is approximately 300 oligonucleotides.

The second procedure involves the primed synthesis approach, developed by Sanger. A single-stranded template, containing the sequence of interest, is enzymatically copied to produce a radiolabeled complementary strand. Using chain-terminating analogues during the copying, a series of partially copied molecules is produced that can be separated by electrophoresis on denaturing gels. The pattern of labeled bands in lanes from the four different nucleotide chain-terminating analogues is used to deduce the sequence. Although originally developed to sequence only naturally occurring single-stranded DNAs, the primed synthesis approach has been applied to single-stranded templates generated from double-stranded DNA.

Both methods require the use of a radiolabeled oligonucleotide. For the Maxam–Gilbert method, the DNA is usually labeled at the 5' terminus with ^{32}P, but can be labeled with ^{32}P at the 3' end as well. The method for 5'-end labeling depends on the phosphorylation of the 5'-hydroxyl group of the 2-deoxyribose, using $\gamma[^{32}\text{P}]$ATP as the phosphate donor and catalyzed by T_4-polynucleotide kinase.

$$5'\text{━━━━━━━}3' \quad\xrightarrow[\text{T}_4\text{-polynucleotide kinase}]{\gamma[^{32}\text{P}]\text{ATP}}\quad {}^{32}\text{P}\,5'\text{━━━━━━━}3'$$
$$3'\text{━━━━━━━}5' \qquad\qquad\qquad\qquad 3'\text{━━━━━━━}\,{}^{32}\text{P}\,5'$$

END LABELING OF DNA WITH ^{32}P

If the DNA fragment has a 5'-phosphate group, it must first be treated with alkaline phosphatase, which will catalyze the removal of the 5'-terminal phosphate group. The alkaline phosphatase is then denatured by heating and removed by centrifugation before the end labeling with T_4-polynucleotide kinase. The following are then mixed together:

1–50 pmol dephosphorylated DNA in 35 μL water

5 μL 500 mM Tris–HCl (pH 7.6), 100 mM MgCl$_2$, 50 mM DTT, 1 mM spermidine, and 1 mM EDTA

5 μL $\gamma[^{32}$P]ATP, SA > 1000 Ci/mmol, containing at least 50 pmol ATP

1 μL T_4-polynucleotide kinase containing at least 20 units

The solution is incubated at 37°C for 30 minutes and then 200 μL of 2.5 M ammonium acetate, 1 μL of tRNA (1 mg/mL), and 750 μL of 95% ethanol are added. The solution is mixed and chilled at -70°C for 5 minutes and centrifuged at 12,000g for 5 minutes at 0°C. The supernatant is removed and the pellet is dried under vacuum; the pellet is mixed with 750 μL of 95% ethanol, chilled, centrifuged, and dried as before. The final product, [5'-^{32}P]DNA, plus RNA carrier is essentially salt free and ready for strand separation.[31]

The 3' ends of DNA can be labeled by treating the 3'-recessed ends with $\alpha[^{32}$P]nucleoside triphosphates and *E. coli* DNA polymerase (Klenow fragment).

With either method, the labeling of double-stranded DNA results in label at *only one end of each strand*. Since the gel sequencing method requires one unique labeled end, it is necessary to separate the two labeled strands, or to cleave the double-stranded, end-labeled DNA with a restriction endonuclease that will give two double-stranded fragments, each with only one of the two strands labeled. These two double-stranded fragments must be separated from each other before sequencing.

SEPARATION OF DNA STRANDS BY GEL ELECTROPHORESIS[32]

The end-labeled DNA is divided into two parts:

1. 1 μg of DNA is dissolved in 40 μL of 30% DMSO containing 1 mM EDTA, 0.05% xylene cyanol, and 0.05% bromophenol blue dyes. The solution is heated at 90°C for 2 minutes to effect complete solution.

2. 0.1 μg of DNA is dissolved in 20 μL of 30% DMSO containing 1 mM EDTA and bromophenol blue dye. The gels, which are loaded as described in Section 5.3.2.1, are prepared as follows. Five grams of acrylamide and 100 mg of bisacrylamide are dissolved in 95 mL of 50 mM Tris–borate buffer (pH 8.5), containing 1 mM EDTA. The solution is degassed by applying a gentle vacuum. TEMED (30 μL) is added and 2 mL of ammonium persulfate (3% w/v). The gel is poured between glass plates and allowed to polymerize for 1–2 hours. The gel is prerun in 50 mM Tris–borate buffer (pH 8.5) for 1 hour at 8 V/cm before loading the DNA sample.

The DNA sample is loaded immediately onto the top of the gel and electrophoresed at 8 V/cm until the dye has migrated to the bottom of the gel. The DNA is located by autoradiography or ethidium bromide staining. The separated DNA fragments are obtained by cutting out the bands and placing the individual gel slices in a dialysis bag containing Tris–borate buffer. The dialysis bag is placed in a shallow layer of buffer and electrophoresed for 2–3 hours at 100 V. The polarity is reversed for 2 minutes to release the DNA from the walls of the dialysis bag. The DNA is recovered by phenol extraction and ethanol precipitation.

10.3.4.1 CHEMICAL SEQUENCING METHODS

An end-labeled nucleic acid is partially degraded using chemical reactions specific for each of the four bases, adenine, cytosine, guanine, and thymine. This is a three-step process in which the specific base is modified, eliminated, and treated to cleave the phosphodiester linkage at that position. Figure 10-8A shows the reaction sequence for the reaction of guanine with dimethylsulfate, followed by phosphodiester cleavage by piperidine. The oligonucleotide chain would be cleaved randomly at each guanine residue, producing a set of fragments all extending from the same 5′ end of the DNA strand to the base adjacent to the guanine that was modified in the methylation reaction. The 5′ end of the DNA polymer is labeled with ^{32}P (Section 6.12), so all fragments can be identified by autoradiography after separation by polyacrylamide gel electrophoresis.

If the DNA is reacted with hydrazine in the presence of 1 M NaCl, cytosine will be selectively modified and eliminated, and treatment with piperidine causes the cleavage of the phosphodiester linkage (Fig. 10-8B), producing another set of oligonucleotides, all of which will be different from the set produced by reaction of guanine.

Treatment of DNA with hydrazine in the absence of high salt, followed by piperidine, will produce cleavage at both cytosine and thymine (Fig. 10-8C). Acid treatment of DNA will cleave both the purine bases, adenine and guanine, which can be followed by treatment with piperidine to cleave the phosphodiester linkage (Fig. 10-8D).

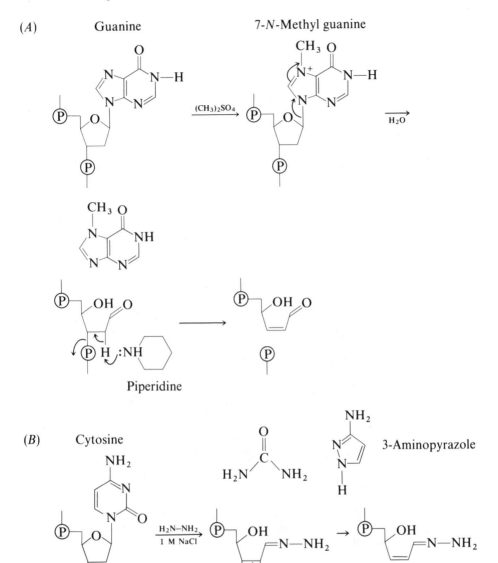

FIGURE 10-8. Maxam–Gilbert chemical modification and elimination of DNA bases, followed by phosphodiester cleavage by piperidine. (*A*) Modification of guanine by dimethylsulfate. (*B*) Modification of cytosine by hydrazine in the presence of 1 M NaCl.

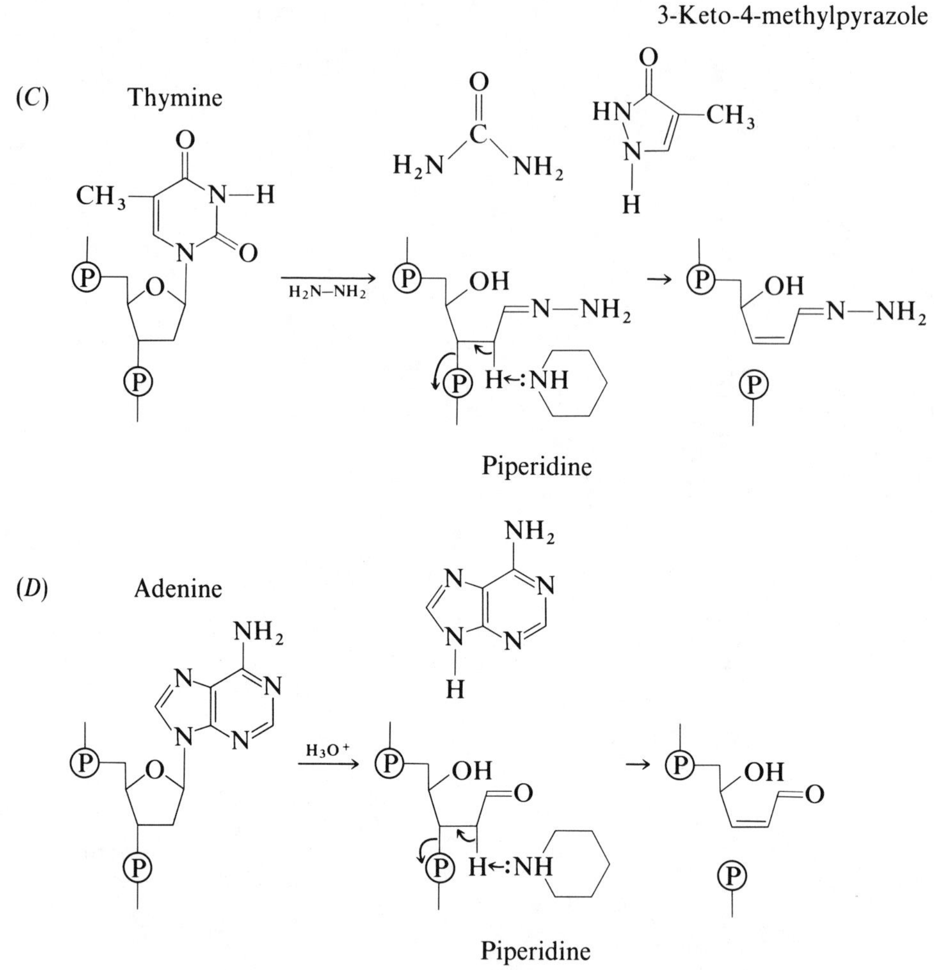

3-Keto-4-methylpyrazole

(C)   Thymine

Piperidine

(D)   Adenine

Piperidine

FIGURE 10-8 (*continued*). (*C*) Modification of thymine (and cytosine) by hydrazine in the absence of salt. (*D*) Acid hydrolysis of adenine (and guanine).

Each of these base cleavage reactions is carried out on a separate aliquot of the DNA, and the products of each reaction are separated in different lanes of an electrophoresis gel. The pattern of bands in the lanes is used to deduce the DNA sequence. Gels are loaded in the order: G reaction, A + G reaction, C + T reaction, and C reaction. If bands appear in the A + G reaction lane, but not in the G reaction lane, they are due to cleavage at adenine sites and are read as A; if bands appear in the C + T reaction lane, but not in the C reaction lane, they are due to cleavage at thymine sites and are read as T (Fig. 10-9).

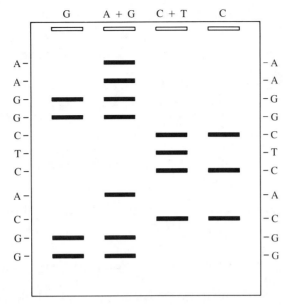

FIGURE 10-9. Sequencing gel electrophoretic analysis of four Maxam–Gilbert reactions: modification and cleavage of guanine (G) sites; hydrolysis of adenine (A) and guanine (G) sites; modification and cleavage of cytosine (C) and thymine (T) sites; and modification and cleavage of cytosine (C) sites. The different bands indicate the number and location of the DNA bases, giving the sequence of 5'—G—G—C—A—C—T—C—G—G—A—A—3' for the DNA fragment of 11 bases.

PROCEDURES FOR THE SPECIFIC DNA BASE ELIMINATIONS[34]

Modification and Cleavage at Guanines

Reaction of DNA with dimethylsulfate results in methylation at the N_7 of the guanine ring and spontaneous elimination. Treatment with piperidine cleaves the phosphodiester linkage. Adenine is also methylated but is much less reactive with dimethylsulfate than guanine.

One microgram of end-labeled DNA is dissolved in 200 μL of 50 mM cacodylate buffer, pH 8.0, containing 10 mM $MgCl_2$ and 1 mM EDTA in a 1.5 mL polyethylene snap-cap tube; 1 μL of dimethylsulfate is added and the capped tube is incubated at 23°C for 5 minutes. The reaction is stopped by adding 50 μL of 1.5 M sodium acetate, containing 1.0 M 2-mercaptoethanol, 100 μg/mL tRNA, and 750 μL of ethanol. The solution is chilled at −70°C for 5 minutes and then treated with piperidine to cleave the phosphodiester linkage.

Cleavage of Guanine and Adenine with Acid

Treatment of DNA with acid causes depurination and if followed by treatment with piperidine produces cleavage of the phosphodiester linkage. The extent of cleavage is controlled by pH and temperature.

One microgram of end-labeled DNA is dissolved in 20 μL of water; 2 μL 1.0 M piperidinium formate (4% formic acid, adjusted to pH 2 with piperidine) is added, mixed, and incubated at 23°C for 60–80 minutes. The reaction is stopped by adding 50 μL of 1.5 M sodium acetate, containing 1.0 M 2-mercaptoethanol, 100 μg/mL tRNA, and 750 μL of ethanol. The solution is chilled at −70°C for 5 minutes and treated with piperidine to cleave the phosphodiester linkage.

Cleavage at Cytosine with Hydrazine in the Presence of Salt

Treatment of DNA with hydrazine is the presence of 1 M NaCl leads to hydrazinolysis of cytosine nucleotides but not thymine nucleotides. Hydrazinolysis followed by treatment with piperidine cleaves the phosphodiester linkage of the free ribose residue.

One microgram of end-labeled DNA is dissolved in 10 μL of water and 10 μL of 5 M NaCl is added, followed by 30 μL of hydrazine (95% reagent grade, 0°C). The solution is mixed and incubated at 23°C for 5 minutes. The reaction is stopped by adding 200 μL of 0.3 M sodium acetate, 0.1 mM EDTA, and 25 μg/mL tRNA; 750 μL of ethanol is added, mixed, and chilled at −70°C for 5 minutes, followed by piperidine cleavage of the phosphodiester.

Cleavage of Cytosine and Thymine with Hydrazine in the Absence of Salt

Treatment of DNA with hydrazine, but without NaCl, leads to the hydrazinolysis of both cytosine and thymine. Treatment with piperidine cleaves the phosphodiester linkages of the free ribose residues.

The method is identical to the cleavage of cytosine with hydrazine in the presence of salt with the substitution of 10 μL of water for the 10 μL 5 M NaCl.

PIPERIDINE CLEAVAGE AND RECOVERY OF DNA FRAGMENTS

After the elimination of the base from the ribose residue, the precipitate in the modified DNA solution is centrifuged at 12,000g for 5 minutes. The supernatant is removed, 250 μL of 0.3 M sodium acetate is added, and the pellet is redissolved by vortexing; 750 μL of (0°C) ethanol is added, mixed, chilled to −70°C and centrifuged for 10 minutes. The supernatant is removed and the pellet is dried under vacuum. Next, 100 μL of fresh 1.0 M piperidine is added and mixed by vortexing. The mixture is heated to 90°C for 30 minutes in a vapor-tight tube; the solution is briefly centrifuged and then lyophilized. The solid is dissolved in 10–100 μL of water and lyophilized again to remove the last traces of piperidine. The treated DNA is dissolved in a minimum (10 μL) of loading buffer (80% de-

ionized formamide, 50 mM Tris–borate, pH 8.3, 1 mM EDTA, 0.1% w/v xylene
cyanol and bromophenol blue dyes), heated to 90°C for 1 minute, chilled in ice
water, and immediately loaded onto sequencing gel for electrophoresis.

RNA can also be sequenced directly by chemical methods similar to those
used for DNA.[33] The following specific reactions are used for cleavage of the
RNA chain: reaction of guanine with dimethylsulfate, followed by base and
aniline; reaction of adenine and guanine with diethylpyrocarbonate, followed by
aniline to cleave the phosphodiester linkage; reaction of uracil and cytosine with
hydrazine, followed by aniline; and reaction of cytosine with hydrazine in the
presence of 3 M NaCl, followed by aniline.

The RNA fragments can be separated by gel electrophoresis under conditions
identical to those for separating DNA fragments (Section 5.3.2.2).

RNA is usually end-labeled at the 3′ terminus rather than at the 5′ terminus,
which is frequently blocked and unreactive to phosphate labeling. The 3′ end is
labeled by reaction with $[5'-^{32}P]-3',5'-CDP$ and T_4-RNA ligase according to
the following reaction.

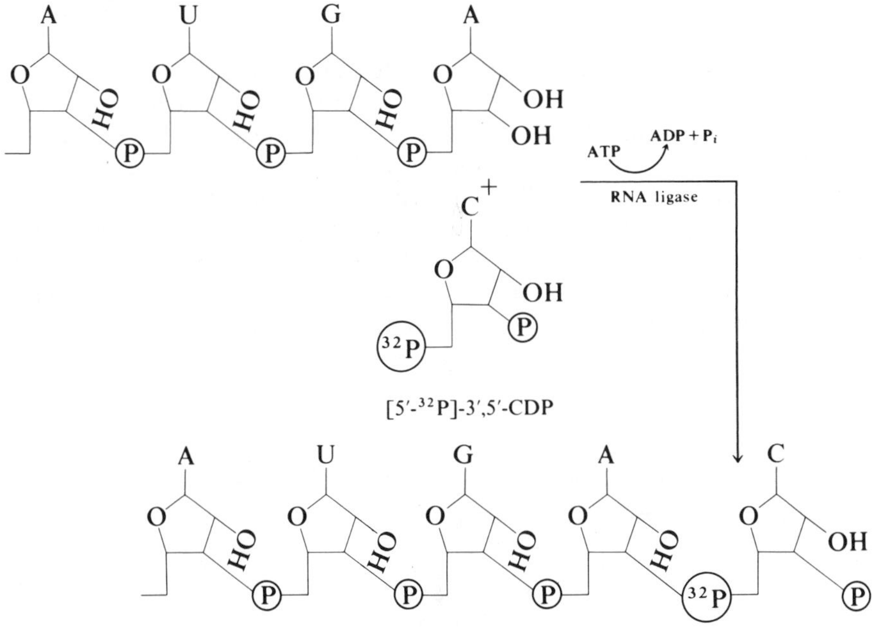

$[5'-^{32}P]-3',5'-CDP$

After each specific reaction, the samples are ready for analysis by gel electrophoresis. Large fragments are separated on 8% acrylamide gels and smaller
fragments are separated on 20% acrylamide gels. The tracking dyes bromophenol blue dye (BPB) and xylene cyanol (XC) migrate at different rates, such
that BPB reaches the bottom and XC approximately the middle of an 8% gel at
the same time as a 15-base oligonucleotide reaches the bottom; in a 20% gel

these migrations occur in the same time as a 7-base oligonucleotide. To obtain separation of the smallest fragments, therefore, the electrophoresis is terminated before the BPB has migrated to the bottom of the 20% gel. Sequencing gels include 7 M urea to prevent secondary structure formation and to ensure that all fragments travel as monomeric species. Thus the electrophoretic separation is only a function of chain length. A continuous buffer system of 90 mM Tris–borate (pH 8.3) containing 1 mM EDTA is used; 2–5 μL of sample is loaded into the gel wells and electrophoresis started immediately at a current of approximately 30 mA. The gels heat up, helping to keep the DNA denatured, but also causing the glass plates to break if too much heat is generated. Further details on pouring and running sequencing gels are given in Chapter 5 (Section 5.3.2.2).

10.3.4.2 PRIMED SYNTHESIS STRATEGY: THE DIDEOXY METHOD

In 1977 Sanger et al.[29] introduced a new method of DNA sequencing in which a single strand of DNA is copied using DNA polymerase primed by a synthetic oligonucleotide. The synthetic oligonucleotide (primer) is hybridized to a specific complementary region on the DNA strand to be copied (the template), and deoxyribonucleotides are added by the DNA polymerase to the 3′-hydroxyl end of the primer, forming a new complementary DNA strand (Fig. 10-10). Using radiolabeled nucleoside triphosphates, the new strand can be identified by radioisotope detection methods and subjected to sequencing. With the introduction of 2′,3′-dideoxynucleotide chain terminators, the older plus–minus method of Sanger and Coulson[27] was rendered obsolete. In the presence of dideoxynucleotides, the synthesis of the new DNA strand is terminated whenever a 2′,3′-dideoxynucleotide is added to the growing end of the chain because the dideoxyribose residue lacks both a 2′- and a 3′-OH group, and therefore cannot serve as a substrate for the addition of another nucleotide by DNA polymerase I.

DNA polymerase I, isolated from *E. coli*, has three distinct activities: (*a*) it is a 5′→3′ polymerase, utilizing deoxynucleoside triphosphates as substrates and requiring a template DNA strand and a primer strand with a 3′-OH group; (*b*) it is

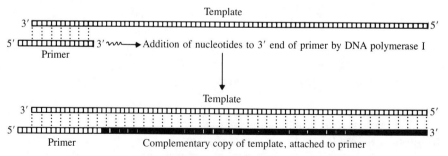

FIGURE 10-10. Schematic diagram of the complementary copying of DNA template by DNA polymerase I. The enzyme copies the template by catalyzing the addition of nucleotides to the hydroxyl at the 3′ end of the primer.

also a 3′-exonuclease, primarily degrading single-stranded DNA from the 3′ end; and (c) it can remove nucleotides from the 5′ end of double-stranded DNA.

The 3′-exonuclease activity is essentially a proofreading function, allowing the removal of any mismatched 3′-terminal residues. The 5′-exonuclease activity is detrimental to sequencing reactions because it can potentially remove residues from the 5′ end of the template strand. If the intact DNA polymerase I is treated with subtilisin, the polymerase is cleaved to produce a polypeptide chain of molecular weight 76,000 that retains the first two activities but lacks the detrimental 5′-exonuclease activity. This form of the polymerase, known as **DNA polymerase (Klenow fragment)**, is commercially available.

Four polymerization reaction digests are prepared, each containing the four deoxynucleoside triphosphates (dATP, dGTP, dCTP, dTDP) one of which is labeled with a radioisotope, normally $\alpha[^{32}P]$-dATP or α-thio$[^{35}S]$-dATP.

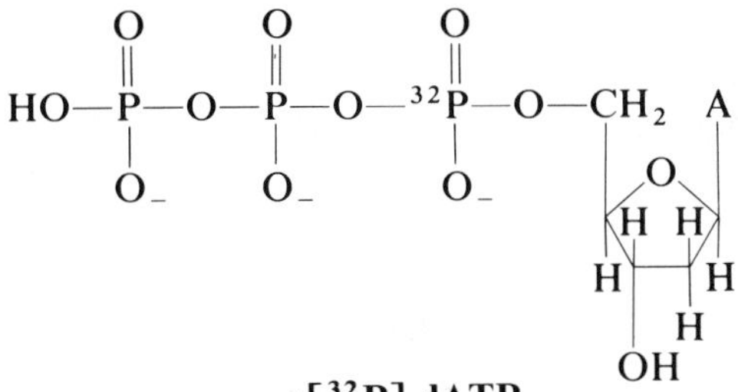

$\alpha[^{32}P]$-**dATP**

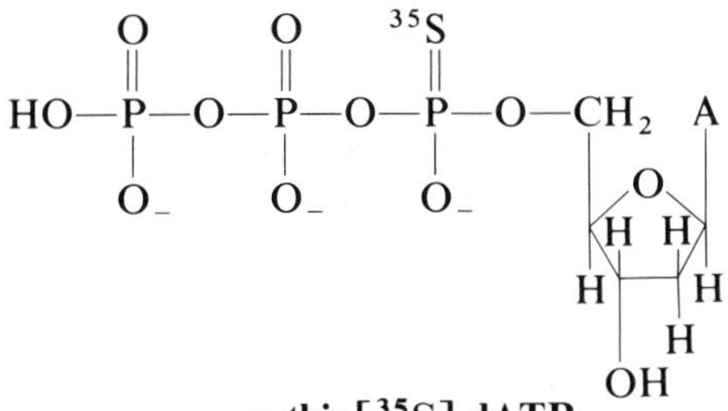

α-**thio$[^{35}S]$-dATP**

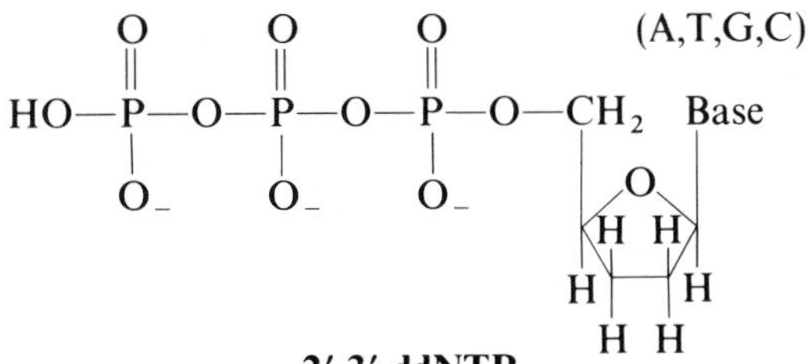

2′,3′-ddNTP

A different 2′,3′-dideoxynucleotide triphosphate (ddNTP) is added to each of the four reaction digests, and thus each reaction digest produces a mixture of oligonucleotides from a single template. The 3′ end of each oligonucleotide in the mixture is terminated with the 2′,3′-dideoxyribonucleotide.

After the polymerization reaction the mixture is heated to 100°C to denature the proteins and to separate the DNA chains. The DNA fragments are subjected to polyacrylamide gel electrophoresis under denaturing conditions to keep the DNA strands separated and to prevent formation of intramolecular secondary structures. Under these conditions the DNA fragments will separate on the basis of chain length, with the shortest fragments traveling the fastest. Each of the possible oligonucleotides theoretically will form a separate band on the gel. The oligonucleotide bands are visualized by autoradiography; ^{32}P yields the highest sensitivity, although ^{35}S gives the best resolution of the bands. Four lanes are used on a gel, corresponding to each of the reaction mixtures with the four different 2',3'-dideoxynucleotides (Fig. 10-11). In the ddT lane, each fragment ends with ddT, and therefore the position of the bands in the lane correspond to the position of thymine in the synthesized sequence. Likewise the positions of the bands in the ddA lane correspond to the positions of A in the synthesized sequence. The sequence of the DNA, starting from the primer end, can be read off the gel from the bottom (the shortest fragments) to the top (the longest fragments). This sequence represents the complement to the template, reading in a 5'→3' direction.

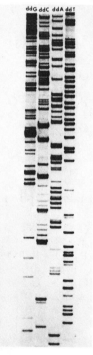

FIGURE 10-11. Autoradiogram of a Sanger dideoxy sequencing gel. Each lane corresponds to one of the four dideoxynucleotide reactions. Each fragment in the ddG lane ends with dideoxyriboguanine. The shortest fragments are at the bottom of the gel.

**DIGEST IN WHICH 2′, 3′-DIDEOXYDTHYMIDINE TRIPHOSPHATE IS
ADDED AS TERMINATOR**

Template 5′dG—T—A—A—A—A—C—G—A—C—G—G—C—C—A—G—T— . . . . 3′

ddT—C—A— . . . . 5′

ddT—G—C—C—G—G—T—C—A— . . . . 5′

copied
sequences
terminated
by ddT

ddT—G—C—T—G—C—C—G—G—T—C—A— . . . . 5′

ddT—T—G—C— . . . . 5′

ddT—T—T—G—C— . . . . 5′

ddT—T—T—T—G—C— . . . . 5′

The relative amounts of the dNTP and ddNTP in the digests determine the
frequency of the termination. A high concentration of ddNTP relative to dNTP
would result in more short chain fragments. The optimal reaction condition
would yield a fairly uniform distribution of DNA fragments up to 300+
nucleotides. The different ddNTPs are not incorporated into the DNA at the
same rates, and therefore different ratios of the different ddNTPs to dNTPs are
required.

**AUTORADIOGRAPHY OF ^{32}P-LABELED DNA FRAGMENTS ON
SEQUENCE GELS**

Thin sequencing gels will stretch or tear if handled without a backing. Gels can
be transferred from the glass electrophoresis plate to heavy chromatography
paper (Whatman 3MM) or onto old X-ray films. For ^{32}P samples, the gel need
not be dried but should be covered with plastic wrap to prevent chemical reaction
with the X-ray film. Double-coated film, such as Kodak XAR-5, and an
intensifying screen will yield the most sensitive detection of radioactivity, but
with some loss of resolution. Maximum intensity is achieved if the gel and film
are stored at −70°C during exposure. Figure A shows the proper placement of
the components.

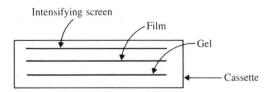

An accurate determination of a DNA sequence from a sequencing gel is not
always possible. Artifact bands may occur in one or more of the gel lanes. If the
DNA template is contaminated with traces of other DNAs, a faint background

pattern from the contaminating DNA sequence may obscure the sequencing results. Heterogeneity in the size of the primer used to initiate DNA copying will give rise to multiple bands corresponding to each position. Careful attention to the purity of both template and primer DNA is important.

Spurious or premature chain termination is sometimes observed with the dideoxynucleotides, which may be the result of particular template features. These error-termination products may be mistaken for authentic termination events. The intensity of the various bands depends on the relative concentrations of the limiting deoxy- and dideoxynucleotides in the reaction digests. Control of the intensities is largely a matter of trial and error.

With the use of α-thio[^{35}S]-dATP, sharper bands have been obtained as well as more sequence data from the upper portion of the sequencing gel, where the bands are very closely spaced.[35] To obtain the best autoradiograms with ^{35}S label, the urea, which acts as a quenching agent for ^{35}S, must be washed out of the gel: soaking in a mixture of 5% methanol and 5% acetic acid both removes the urea and fixes the DNA. All the liquid can be removed from the gel by blotting with filter paper and heating on a gel dryer. The dry gel is placed in contact with single-sided emulsion X-ray film, such as Kodak SB-5, at room temperature, without an intensifying screen. Because of the weaker β-particles emitted by ^{35}S as compared with ^{32}P, it is possible to resolve closely spaced bands in the gel.

The primed synthesis procedure of sequencing requires a single-stranded template, which can be obtained in several ways:

1. Enzymatically generated single strands from purified duplex DNA in which exonucleases digest only one strand of the duplex. Exonuclease III from *E. coli* is a $3' \rightarrow 5'$-exonuclease, specific for double-stranded DNA. It will catalyze the hydrolysis of both strands, proceeding from the 3' end of each strand.

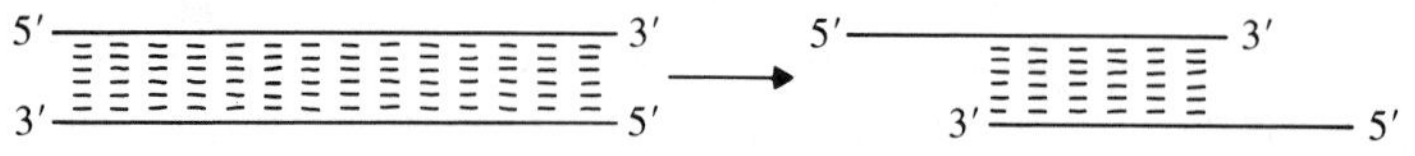

The resultant single-stranded segments can be copied by DNA polymerase I (Klenow fragment).

2. Single-stranded vectors, such as M13, in which DNA fragments are cloned into the M13, yielding single-stranded DNA in virus particles (see Section 10.3.1).

3. Single strands copied from an RNA template by reverse transcriptase, an RNA-dependent DNA polymerase.[36] For example, messenger RNA or viral RNA can be transcribed into a complementary DNA (cDNA) in the presence of deoxyribonucleotide triphosphates, Mg^{2+}, and reverse transcriptase. The enzyme requires primer strand with a 3'-OH.

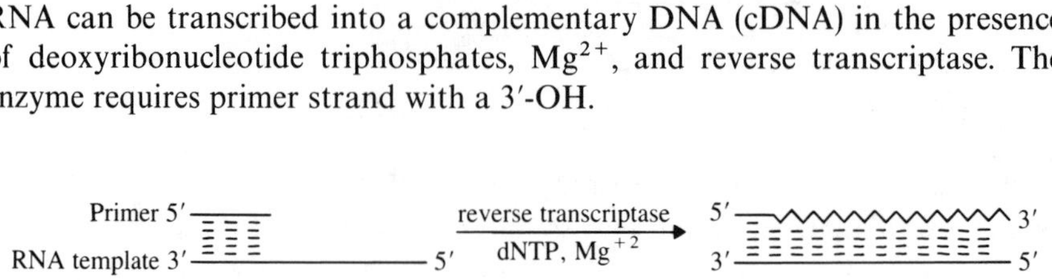

The strands are separated by base cleavage of the RNA template in NaOH at 65°C for 8 hours, leaving the DNA intact.

Synthesis of the complementary DNA sequencing strand by DNA polymerase is initiated only if a primer with a free 3'-OH is added. The primer should give specific hybridization to the template at only one position and should yield a stable complex. Three kinds of primer are used: chemically synthesized primers complementary to a known portion of the template sequence; M13 universal primer, which contains 15 nucleotides complementary to the DNA sequence adjacent to the cloning site in M13 phage; and DNA fragments obtained by restriction endonuclease hydrolysis of the double-stranded DNA template to be sequenced.

Hybridization of primer and template is accomplished by heating the mixture to 90–100°C to denature any double-stranded structures. The mixture is then cooled to 68°C for 30 minutes (for long primers) or cooled slowly to room temperature for short primers.

A method for sequencing plasmid DNA directly, without isolating single-stranded fragments, has been developed by Chen and Seeburg.[37] Plasmid DNA is denatured in 0.2 M NaOH at room temperature to expose single-stranded regions, which are then annealed with a synthetic oligonucleotide primer complementary to one strand of the plasmid DNA. The sequence determination can be carried out using the Sanger dideoxy method. The advantages of sequencing directly from the plasmid DNA include speed and the avoidance of additional cloning steps into M13 vectors to obtain the single-stranded DNA.

10.3.4.3 AUTOMATED SEQUENCING WITH FLUORESCENT-LABELED PRIMERS

Hood et al. have synthesized oligonucleotide primers containing fluorescent groups that can be used in place of radioisotopes for sensitive detection of DNA fragments in primed synthesis.[37] These modified primers are highly fluorescent, are suitable for hybridization with specific DNA sequences, and do not adversely affect the electrophoretic separation of DNA molecules in polyacrylamide gels. Four fluorescent dyes that are spectrally resolved are conjugated to the 5'-OH of the oligonucleotide primers. A fluorescent tag present on the 5' terminus of the primer oligonucleotide will also be present on the polynucleotide products of chain elongation by DNA polymerase I (Klenow fragment), and therefore all the bands on a sequencing gel will be fluorescent. By using four different fluorescent tags, and using only one of these in each of the four sequencing reactions, containing a different 2',3'-dideoxyribonucleoside triphosphate, it is possible to combine the products of the four sequencing reactions and separate the mixture by electrophoresis on a single polyacrylamide gel. An argon laser, near the bottom of the gel, is used to excite the fluorescent tags, and a detector is used to identify each fluorescent dye as it migrates to the bottom of the gel. The different fluorescent colors (emission maxima) reveal the DNA sequence because each of the fluorescent tags indicates one of the four nucleotide bases. The detection limit for the fluorescent tags is in the range of 10^{-16} mole, which is well below the amount of DNA present in a single band on sequencing gels.

10.3.5 Deducing the Complete Sequence for Large Nucleic Acids

The number of bases in a gene or in mRNA can vary from 300 to 15,000. Since the largest polynucleotide that can be sequenced in a single electrophoretic gel contains at most 300 bases, the determination of the sequence of most genes and mRNA requires that the genome or mRNA molecule be broken into smaller fragments. We have seen that restriction endonucleases will specifically hydrolyze DNA into a limited number of smaller fragments. It is these smaller fragments that must be sequenced. There are 200 known restriction endonucleases, offering a wide diversity of specificities. When different restriction endonucleases are used in different experiments and the different DNA fragments that result are sequenced a pattern emerges in which there are overlapping sequences in the different fragments, from which the total sequence can be obtained.

Because there are only four unique bases in nucleic acids compared with 20 amino acids in proteins, it is much more difficult to deduce the sequence of nucleic acids than the sequence of proteins. Especially long runs of any one nucleotide and repeated sequences of two or more nucleotides act as special markers of their location in overlapping fragments. Furthermore, when the specificities of the restriction endonucleases are known, they provide sequence markers, since these sequences appear at the ends of the fragments. The development of computer programs for comparing the sequences of different restriction endonuclease fragments has been an important contribution to obtaining the total sequence of large nucleic acids.

After establishing the sequence of DNA, the sequence of the protein gene product can be obtained using the triplet DNA code for the individual amino acids. It is first necessary, however, to establish the reading frame of the gene, that is, the proper starting point for reading the nucleotide triplets. This can be done by determining the location of the start code, or by determining a short sequence of amino acids at either the amino or carboxyl terminus of the protein gene product. With this short amino acid sequence and its triplet code, the DNA sequence is searched to find the beginning of the DNA sequence for the protein gene product. See Appendix B for the DNA code. It should be noted that the number of three-base codes for the amino acids varies from one to six, but in no case does a single triplet specify more than one amino acid.

10.4 DETERMINING THE STRUCTURE OF LIPIDS

Lipids are relatively water insoluble compounds that can be extracted from cells by nonpolar solvents such as chloroform, diethyl ether, and hexane. They are composed of several classes that appear to serve five general functions: (a) triglycerides made up of fatty acid (long-chain, C_{12}-C_{24} carboxylic acids) esters of glycerol that serve as intracellular storage depots of metabolic energy; (b) phosphoglycerides and cholesterol fatty acid esters that serve as structural components of membranes, as transport forms of metabolic energy, and in processes of sensation; (c) polyisoprenoids that serve as carriers of molecules through membranes and as intermediates in the biosynthesis of steroids and water-insoluble vitamins such as vitamins A, E, and K; (*d*) sphingosine lipids containing sphingosine combined with a variety of components such as fatty acids, amino acids, carbohydrates and are involved in nerve membranes and function in forming synapses; and (*e*) waxes made up of long-chain fatty acid esters of long-chain fatty alcohols, which serve as protective cell-wall components of the leaves of higher plants and the exoskeletons of insects. Figure 10-12 gives the structures of various kinds of lipids.

The determination of the structures of lipids provides a better understanding of their varied roles in biochemical processes. The first step in the determination is the cleavage of the lipid into its components, using both chemical and enzymatic reactions. The second step is the identification and quantitation of the components. Free fatty acids are converted to methyl or trimethyl silyl esters and separated by TLC, GLC, or HPLC; polar groups such as phosphate, choline, carbohydrate, and so forth are identified by specific chemical tests. The preparation of triglycerides, cholesterol esters, phosphatidyl choline (lecithin), phosphatidic acid, phosphatidyl glycerol, sphingomyelin, and isoprenylpyrophosphate are described in Chapter 8, Section 8.8.

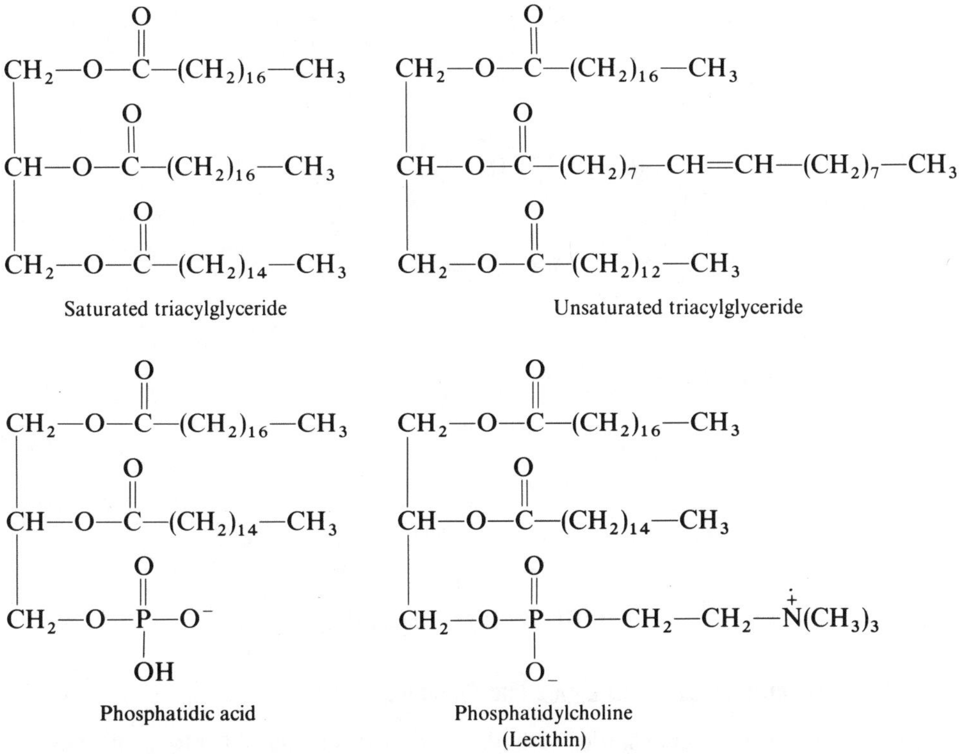

$$
\begin{array}{l}
CH_2-O-\overset{\displaystyle O}{\overset{\|}{C}}-(CH_2)_{16}-CH_3 \\[2mm]
CH-O-\overset{\displaystyle O}{\overset{\|}{C}}-(CH_2)_{16}-CH_3 \\[2mm]
CH_2-O-\overset{\displaystyle O}{\overset{\|}{C}}-(CH_2)_{14}-CH_3
\end{array}
$$

Saturated triacylglyceride

$$
\begin{array}{l}
CH_2-O-\overset{\displaystyle O}{\overset{\|}{C}}-(CH_2)_{16}-CH_3 \\[2mm]
CH-O-\overset{\displaystyle O}{\overset{\|}{C}}-(CH_2)_7-CH=CH-(CH_2)_7-CH_3 \\[2mm]
CH_2-O-\overset{\displaystyle O}{\overset{\|}{C}}-(CH_2)_{12}-CH_3
\end{array}
$$

Unsaturated triacylglyceride

$$
\begin{array}{l}
CH_2-O-\overset{\displaystyle O}{\overset{\|}{C}}-(CH_2)_{16}-CH_3 \\[2mm]
CH-O-\overset{\displaystyle O}{\overset{\|}{C}}-(CH_2)_{14}-CH_3 \\[2mm]
CH_2-O-\overset{\displaystyle O}{\overset{\|}{P}}-O^- \\[1mm]
\qquad\qquad OH
\end{array}
$$

Phosphatidic acid

$$
\begin{array}{l}
CH_2-O-\overset{\displaystyle O}{\overset{\|}{C}}-(CH_2)_{16}-CH_3 \\[2mm]
CH-O-\overset{\displaystyle O}{\overset{\|}{C}}-(CH_2)_{14}-CH_3 \\[2mm]
CH_2-O-\overset{\displaystyle O}{\overset{\|}{P}}-O-CH_2-CH_2-\overset{+}{N}(CH_3)_3 \\[1mm]
\qquad\qquad O_-
\end{array}
$$

Phosphatidylcholine
(Lecithin)

FIGURE 10-12. Representative structures of the various classes of lipids.

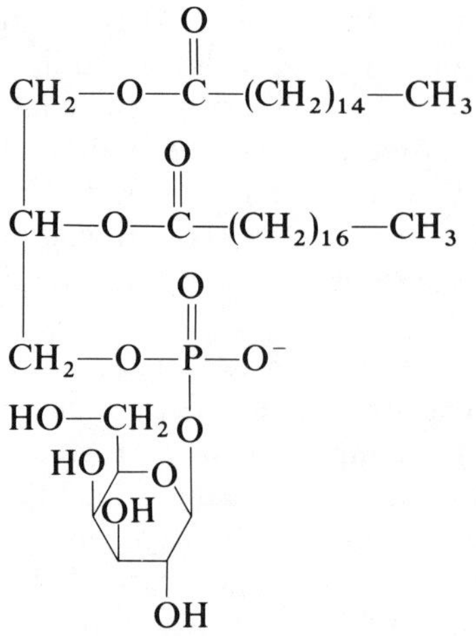

$$CH_2-O-\overset{\overset{\displaystyle O}{\|}}{C}-(CH_2)_{14}-CH_3$$

$$CH-O-\overset{\overset{\displaystyle O}{\|}}{C}-(CH_2)_{16}-CH_3$$

$$CH_2-O-\overset{\overset{\displaystyle O}{\|}}{P}-O^-$$

HO—CH₂ O

HO O

OH

OH

Phosphatidylgalactoside

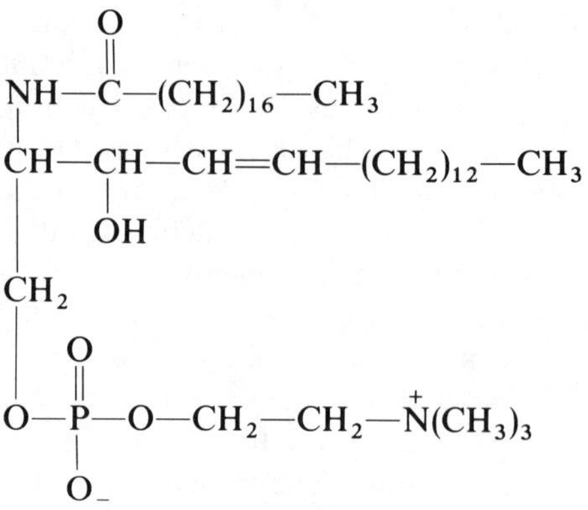

$$NH-\overset{\overset{\displaystyle O}{\|}}{C}-(CH_2)_{16}-CH_3$$

$$CH-CH-CH=CH-(CH_2)_{12}-CH_3$$

OH

CH₂

$$O-\overset{\overset{\displaystyle O}{\|}}{P}-O-CH_2-CH_2-\overset{+}{N}(CH_3)_3$$

O⁻

Sphingomyelin

$$\overset{\displaystyle O}{\underset{\displaystyle O}{\|}}{C}-(CH_2)_{14}-CH_3$$

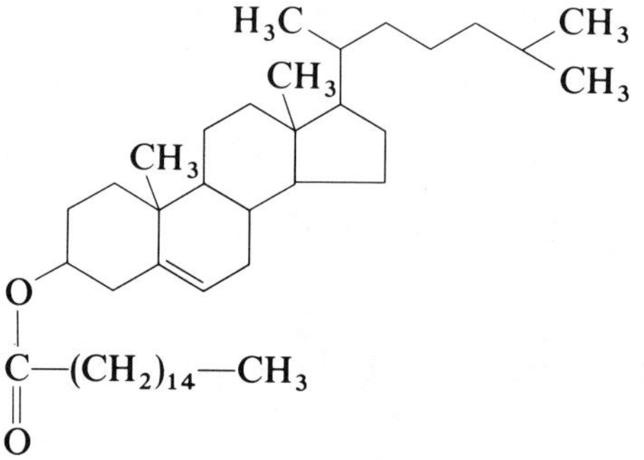

Acyl cholesterol ester

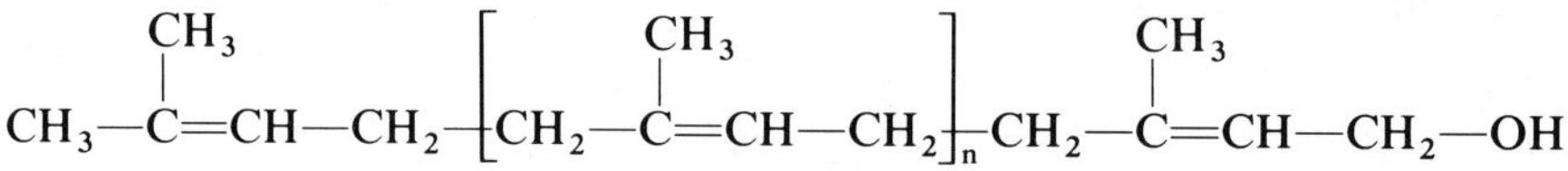

$$CH_3-\overset{\overset{\displaystyle CH_3}{|}}{C}=CH-CH_2-\left[CH_2-\overset{\overset{\displaystyle CH_3}{|}}{C}=CH-CH_2\right]_n CH_2-\overset{\overset{\displaystyle CH_3}{|}}{C}=CH-CH_2-OH$$

Polyisoprenoid

$$CH_3-(CH_2)_{16}-\overset{\overset{\displaystyle O}{\|}}{C}-O-CH_2-(CH_2)_{22}-CH_3$$

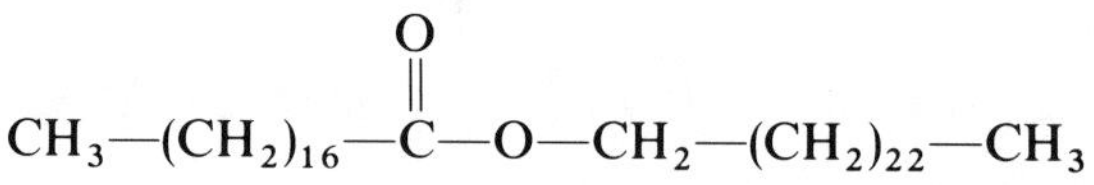

Wax

FIGURE 10-12 (*continued*)

10.4.1 Specific Cleavage of Lipids into Component Parts

Using chemical or enzymatic methods, lipids can be cleaved into their components which can then be further separated and analyzed.

10.4.1.1 SAPONIFICATION

Reaction of lipid esters, such as triglycerides, phosphodiglycerides, with NaOH releases the fatty acids as sodium salts and generates free glycerol. The lipid is refluxed under N_2 with 90% methanolic 0.3 N NaOH for 1–2 hours. The N_2 retards oxidation of unsaturated fatty acids during refluxing. After dilution of the reaction mixture with methanol, the nonsaponifiable material is extracted with petroleum ether. The alcohol phase is acidified with 6 N HCl to pH 1 and re-extracted with hexane or petroleum ether to remove the free fatty acids. This method of hydrolysis does not release fatty acids from amide linkages found in sphingolipids. Complete hydrolysis of sphingolipid esters and amides is effected by dissolving the sample in dioxane-50% saturated barium hydroxide 1:1 and refluxing for 24 hours.

10.4.1.2 ENZYMATIC CLEAVAGE

Enzymes can be used to cleave a specific class of lipids, such as phospholipids, or at a specific position in the lipid.

1. Pancreatic lipase hydrolyzes triacylglycerides to release fatty acids from sn-1 and sn-3 positions. The triglyceride, dissolved in ether or hexane, is added to the buffered enzyme solution and incubated at 40° for 5–10 minutes. The reaction is stopped by the addition of ethanol. Castor bean lipase has the same specificity as the pancreatic lipase, but is more reactive toward short chain fatty acids (39).

2. Phospholipase A_2 from snake venom specifically hydrolyzes the fatty acid ester at *sn*-2 position of *sn* 3 phosphatides such as lecithin. This enzyme is active in organic solvents such as diethyl ether and the product lysophosphatide is frequently insoluble in these solvents and thus helps to drive the reaction to completion.

3. Phospholipase C cleaves phosphatides at the sn-3 position, generating 1,2-diacylglycerols from phospholipids.

10.4.2 Determination of Lipid Components

The saponification and enzymatic cleavage of lipids yields a variety of products, such as fatty acids, mono- and diglycerides, and glycerol phosphate esters of choline, serine, ethanolamine, and inositol. In addition, glycerol and sphingosine are produced. This section describes methods for the identification of each of these components.

10.4.2.1 SEPARATION OF FATTY ACIDS BY ARGENTATION TLC

Separation on silica gel impregnated with silver nitrate is known as argentation TLC and is one of the important techniques available for separating complex lipid components. Separations are effected by the formation of weak interactions between silver ions in the stationary phase and π electrons of compounds containing double or triple carbon-carbon bonds. The argentation

TLC was suggested by Nichols in 1952.[42] Silver nitrate is added to silica gel plates by allowing a 10–20% silver nitrate solution to migrate through the silica gel prior to chromatographic separation. The advantage of Ag-TLC is that positional isomers and cis-trans isomers of unsaturated fatty acids can be separated.[43] An additional advantage of this method is that unsaturated lipids are protected against oxidation by the silver ions. The method has been used to subfractionate mixtures of sterol esters, triglycerides, di- and monoglycerides and glycerophospholipids.[44]

10.4.2.2 SEPARATION OF FATTY ACID ESTERS BY GAS-LIQUID CHROMATOGRAPHY

The most widely used method for separation and quantitation of fatty acid esters is gas-liquid chromatography, using either flame ionization or thermal-conductivity detection (see Chapter 4, Section 4.8). By using fatty acid ester standards, the individual fatty acids in the hydrolysis mixture can be identified and quantitated. The preparation of fatty acid esters is described in this section. Separation is made on the basis of chain length and the number of double bonds. The resolution and the order of elution of the fatty acid esters is dependent on the polarity of the liquid phase. For example, $C_{18:2}$ elutes before $C_{18:1}$ which elutes before $C_{18:0}$ on the relatively nonpolar liquid phases SE-30 and OV-1. If a polar liquid phase (OV-17, silar 10C, or EGSS-X) is used, the fatty acids are eluted in the order of increasing unsaturation.

Short chain fatty acid esters are separated at relatively low temperatures (160°) while higher molecular weight fatty acids require higher column temperatures and liquid phases that are stable and nonvolatile at temperatures up to 350°. Some fatty acid pairs are not easily resolved, such as $C_{18:3}$ and $C_{20:0}$ on a polar phase, so two or more separation systems may be necessary. Cis-trans isomers or double-bond positional isomers are difficult to separate by GLC. These can be resolved, however, by argentation TLC.

To quantitate the amount of fatty acid present, two factors are necessary. First, the response of the detector must be determined for authentic fatty acid ester standards of known concentration, because the detector response (area under the peak on the elution trace) may be different for each fatty acid ester. Secondly, it is necessary to add a known amount of an appropriate internal standard to the sample prior to injection into the GLC if the absolute quantity of the fatty acid is to be determined. If a standard is not available, it must be assumed that the peak area for a particular compound is proportional to the weight percent of that compound within the mixture and a semiquantitative estimate can be made.

Monoglycerides can be resolved on nonpolar GLC columns on the basis of their molecular weight after conversion to their trimethylsilyl ethers. Resolution on the basis of molecular weight and the degree of fatty acid unsaturation can be achieved using polar columns.[41]

a. **Preparation of Methyl Esters.** Boron trifluoride (14%) in methanol is added to the free fatty acid dissolved in hexane or to fatty acid samples scraped from a TLC plate. The sample is heated in a closed vial at 100°C for 5–10 minutes. An alternate method is to add 2% sulfuric acid in dry methanol to the lipid dissolved in hexane and heat to reflux in a closed vessel.

It is convenient to prepare the esters of the fatty acids in a lipid by direct transesterification. The lipid is dissolved to give 1–10% in solvent (4% NaOH in anhydrous methanol-benzene 6:4 v/v) and heated at 35°C for 1–2 minutes for glycerol esters and 15–20 minutes for cholesterol esters.

The fatty acid esters can be purified by chromatography on small silica gel columns, eluted with 1% diethyl ether in hexane. Fractions are analyzed by TLC on silica gel H plates irrigated with benzene. The fatty acid esters move in the benzene solvent, while cholesterol and other more polar compounds remain at the origin.

b. Preparation of Trimethylsilyl Esters. Trimethylsilyl esters of fatty acids are prepared by treating the dry fatty acid or acid mixture with hexamethyl-disilazane and trimethylchlorosilane 2:1. Only the free fatty acid will react.[40]

10.4.2.3 SEPARATION OF FATTY ACIDS BY HIGH PERFORMANCE LIQUID CHROMATOGRAPHY

This technique for the separation of fatty acids and lipids is discussed in Chapter 4, Section 4.14.6. Recent advances have been made in coupling the HPLC system with a mass spectrometer to separate the lipids and simultaneously obtain structural information.

10.4.2.4 DETERMINATION OF POLAR GROUPS IN LIPIDS

a. Choline. The qualitative detection of choline-containing phospholipids is described in Chapter 4, Section 4.14.4. Choline can be determined in intact phospholipid by heating with *cis*-aconitic anhydride and acetic anhydride. A green complex is formed, with a maximum absorption at 530 nm. The method is sensitive to 3–5 μg of choline.

b. Ethanolamine and Serine. Heating of phosphatidyl ethanolamine with 6 N HCl releases the ethanolamine, which can be isolated by silica gel TLC and identified by ninhydrin or fluorescamine spray. A quantitative determination can be made using the amino acid analyzer (see Section 7.2.2.1 and Section 4.11). Serine can also released from phosphatidyl serine by HCl hydrolysis and determined by quantitative amino acid analysis.[45]

c. Glycolipids. By using traditional colorimetric procedures for carbohydrates, it is possible to determine the carbohydrate content of intact glycolipids (see Section 7.1.2). Carbohydrate components of lipids can be liberated by hydrolysis of the glycosidic linkages by strong acid or by methanolysis. The free sugars can be characterized by the procedures presented in Section 10.1.

d. Glycerol and Sphingosine. Both glycerol and sphingosine can be determined by GLC. Glycerol can be converted to the triacetate by reaction with acetic anhydride or to the trimethyl silyl ether for GLC analysis.

Sphingosine can be obtained by hydrolysis of sphingolipids in HCl-methanol-water (3:29:4 v/v/v) for 18 hours at 78°C. The solution is neutralized and extracted with chloroform to obtain the sphingosine bases.[46] The solvent is evaporated under vacuum over P_2O_5, and the sphingosine is converted to the trimethylsilyl ether derivative which can be analyzed by GLC. Free sphingosine can be determined with fluorescamine (Section 7.2.2.2).

e. **Phosphorus.** Phospholipids can be eluted as described in Section 7.4.3 and phosphate quantitatively determined as described in Section 7.5.1.

f. **Cholesterol.** Free cholesterol or cholesterol esters can be quantitated by the use of cholesterol esterase-cholesterol oxidase reagent (see Section 7.4.3.2).

.

10.4.3 Characterization of Phosphatidyl Choline from Egg Yolk—An Example of Lipid Analysis[47]

Egg yolk is a rich source of lipid, containing phosphatidyl choline, phosphatidyl ethanolamine, phosphatidyl serine, sphingomyelin, cholesterol and cholesterol esters. Phosphatidyl choline compromises about 14% of the dry weight of the yolk.

Initial extraction of the egg yolk with acetone removes the nonpolar lipids, including cholesterol and its esters, triacylglycerides, and most of the yolk pigments (Fig. 10-13). Polar lipids are extracted by homogenization of the yolk solids in chloroform-methanol (2:1 v/v) and filtration on a Büchner funnel. The solid on the filter is re-extracted with chloroform-methanol. The filtrates are concentrated on a rotary evaporator at 40°C. The concentrated lipids are dissolved in a minimum volume of hexane, and the phospholipids are precipitated with 5 volumes of dry acetone. After settling, the precipitate is removed by filtration. The precipitated polar lipids are dissolved in chloroform-methanol (19:1 v/v). The composition of the extract is analyzed by silica gel G TLC using chloroform-methanol-water (65:25:4 v/v/v) as the solvent with standards of phosphatidyl choline, serine, and ethanolamine. The lipids are visualized by exposure of the TLC to iodine vapor or by sulfuric acid-methanol spray.

Pure phosphatidyl choline can be obtained by chromatography of the polar lipid mixture on alumina. The lipids are applied to the column in chloroform or chloroform-methanol 9:1, and the column is eluted with chloroform to obtain the nonpolar lipids and then with chloroform-methanol 9:1, which elutes phosphatidyl choline. The appearance of phosphatidyl choline in the eluant is monitored by TLC. The fractions containing phosphatidyl choline are pooled and concentrated in a rotary evaporator. The residue is dissolved in diethyl ether and is added to 0.01 volume of phospholipase A_2 (*Crotalus adananteus* snake venom) 5 mg/mL in 50 mM Tris-HCl, pH 8.5, containing 5 mM $CaCl_2$. The mixture is incubated at 30°C for 2–10 hours with occasional mixing to hydrolyze the fatty acids at position 2. Lysophosphatidyl choline appears as a gel like layer and the ether supernatant, containing the free fatty acids can be decanted.

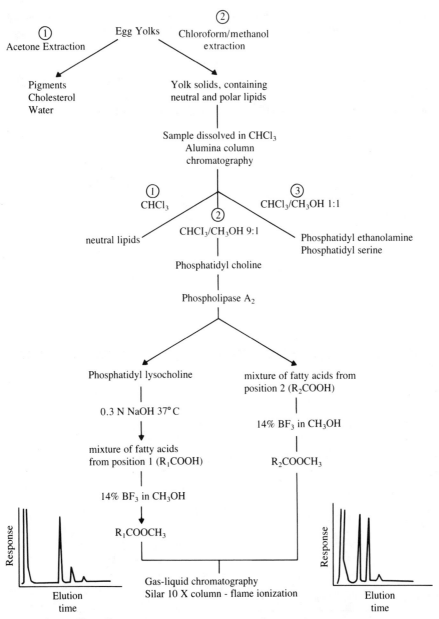

FIGURE 10-13. Scheme for determining the fatty acid composition at positions 1 and 2 of phosphatidylcholine from egg yolk.

The ether is evaporated after drying over sodium sulfate, and the fatty acids are esterified, dissolved in hexane, and analyzed by GLC.

The lysophosphatidyl choline gel is dissolved in 0.5 N KOH and incubated at 37°C for 2 hours to hydrolyze the fatty acids at position 1. The mixture is acidified to pH 1 with 6 N HCl and extracted with hexane to recover the free fatty acids. The hexane extracts are dried over sodium sulfate and then evaporated to dryness with a stream of N_2. The fatty acids obtained from the lysophosphatidyl choline are esterified, dissolved in hexane, and analyzed by GLC.

The methyl esters of the fatty acids can be separated and quantitated by GLC on a 1.8 meter column of firebrick coated with Silar 10C using a flame ionization detector. Typical GLC analyses for the fatty acids from positions 1 and 2 are shown in Figure 10-14.

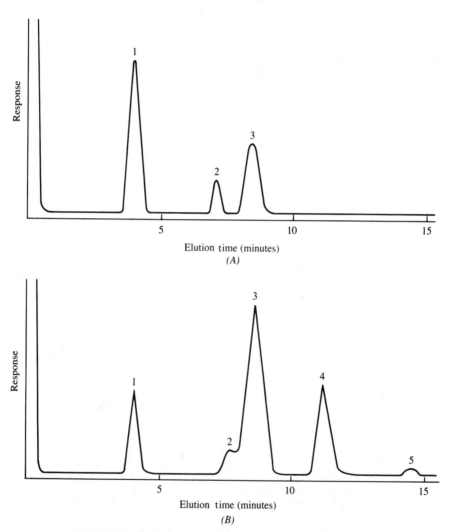

FIGURE 10-14. Separation of fatty acid methyl esters of egg yolk phosphatidylcholine. Peaks: 1 = methyl palmitate, 2 = methyl stearate, 3 = methyl oleate, 4 = methyl linoleate, 5 = methyl linolenate. (A) Position 1 fatty acids. (B) Position 2 fatty acids.

10.5 LITERATURE CITED

1. P. A. Rebers, G. E. Wessman, and J. F. Robyt, "A new thin-layer chromatographic method for analyses of amino sugars in polysaccharide hydrolyzates," *Carbohydr. Res. 153*: 132 (1986).
2. J. F. Robyt, "Paper chromatographic solvent for the separation of sugars and alditols," *Carbohydr. Res. 40*: 373 (1975).
3. P. J. Harris, R. J. Henry, A. B. Blakeney, and B. A. Stone, "An improved procedure for the methylation analysis of oligosaccharides and polysaccharides," *Carbohydr. Res. 127*: 59 (1984).
4. R. E. Dickerson, "The structure and history of an ancient protein," *Sci. Am. 226*(4): 58 (April 1972).
5. J. B. Clegg, M. A. Naughton, and D. J. Weatherall, "Abnormal human haemoglobins. Separation and characterization of the α and β chains by chromatography, and the determination of two new variants, Hb Chesapeake and Hb J (Bangkok)," *J. Mol. Biol. 19*: 91 (1966).
6. W. R. Gray, "End-group analysis using dansyl chloride," *Methods Enzymol. 25*: 121 (1972).
7. A. V. Fowler, A. J. Brake, and I. Zabin, "Amino acid sequence of β-galactosidase. VI. Limited tryptic digestion of the citraconylated protein and sequences of tryptic peptides," *J. Biol. Chem. 253*: 5490 (1978).
8. M. A. Raftery and R. D. Cole, "Tryptic cleavage at crysteinyl peptide bonds," *Biochem. Biophys. Res. Commun. 10*: 467 (1963).
9. E. Gross and B. Witkop, "Nonenzymatic cleavage of peptide bonds: The methionine residues in bovine pancreatic ribonuclease," *J. Biol. Chem. 237*: 1856 (1962).
10. R. A. Laursen, "Automatic solid-phase Edman degradation," *Methods Enzymol. 25*: 344 (1977).
11. G. Allen, "Sequencing of proteins and peptides," in *Lab. Tech. Biochem. Mol. Biol. 9*: 50 (1981).
12. P. Bornstein and G. Balian, "Cleavage at Asn-Gly bonds with hydroxylamine," *Methods Enzymol. 47*: 132 (1977).
13. M. A. Hermodson, "Cleavage of tryptophanyl bonds with *o*-iodosobenzoic acid," in *Methods in Protein Sequence Analysis*, M. Elzinga, Ed. Humana Press, Clifton, NJ, 1981, p. 313.
14. R. W. Gracy, "Two-dimensional thin-layer methods," *Methods Enzymol. 47*: 195 (1984).
15. I. Heiland, D. Brauer, and B. Wittman-Liebold, "Primary structure of protein L10 from the large subunit of *Escherichia coli* ribosomes," *Hoppe-Seylers Z. Physiol. Chem. 357*: 1751 (1976).
16. P. Edman and G. Begg, "A protein sequenator," *Eur. J. Biochem. 1*: 80 (1967).
17. M. W. Hunkapiller and L. E. Hood, "New protein sequenator with increased sensitivity," *Science, 207*: 523 (1980).
18. G. E. Tarr, "Improved manual sequencing methods," *Methods Enzymol. 47*: 335 (1975).
19. P. Edman and A. Henschen, "Sequence determination," in *Protein Sequence Determination*, S. Needleman, Ed. Springer-Verlag, New York, 1975, pp. 232–239.
20. R. M. Hewick, M. W. Hunkapiller, L. E. Hood, and W. J. Dreyer, "A gas–liquid solid phase peptide and protein sequenator," *J. Biol. Chem. 256*: 7990 (1981).
21. J. Y. Chang, D. Brauer, and B. Wittman-Liebold, "Micro-sequence analysis of peptides and proteins using 4-*N*,*N*-dimethylaminoazobenzene 4'-isothiocyanate/phenylisothiocyanate double coupling method," *FEBS Lett. 93*: 205 (1978).

22. D. N. Podell and G. N. Abraham, "A technique for the removal of pyroglutamic acid from the amino terminus of proteins using calf liver pyroglutamate amino peptidase," *Biochem. Biophys. Res. Commun. 81*: 176 (1978).

23. J. C. Sheehan and D. Yang, "The use of *N*-formylamino acids in peptide synthesis," *J. Am. Chem. Soc. 80*: 1154 (1958).

24. R. Holley, J. Apgar, G. A. Everett, J. T. Madison, M. Marquisee, S. H. Merrill, J. R. Penswick, and A. Zamir, "Structure of a ribonucleic acid," *Science, 147*: 1462 (1965).

25. H. Robertson, B. Barrell, H. Werth, and J. Donelson, "Isolation and sequence analysis of a ribosome-protected fragment from ϕX174 DNA," *Nature New Biol. 241*: 38 (1973).

26. E. Ziff, J. Sadat, and F. Galibert, "Determination of the nucleotide sequence of a fragment of ϕX174 DNA," *Nature New Biol. 241*: 34 (1973).

27. F. Sanger and A. Coulson, "A rapid method for determining sequences in DNA by primed synthesis with DNA polymerase," *J. Mol. Biol. 94*: 441 (1975).

28. A. Maxam and W. Gilbert, "A new method for sequencing DNA," *Proc. Natl. Acad. Sci. U.S.A. 74*: 560 (1977).

29. F. Sanger, S. Nickerson, and A. R. Coulson, "DNA sequencing with chain-terminating inhibitors," *Proc. Natl. Acad. Sci. U.S.A. 74*: 5463 (1977).

30. B. V. Perbal, *A Practical Guide to Molecular Cloning*. Wiley-Interscience, New York, 1984, pp. 31–59.

31. J. Hindley, "DNA sequencing," *Lab. Tech. Biochem. Mol. Biol. 10*: 258 (1983).

32. T. Maniatis, E. Fritsch, and J. Sambrook, *Molecular Cloning, A Laboratory Manual*. Cold Spring Harbor Laboratory, Cold Spring Harbor, NY, 1982, p. 164.

33. D. Peattie, "Direct chemical methods for sequencing end-labeled ribonucleic acids," in *Methods of DNA and RNA Sequencing*, S. Weisman, Ed. Praeger, New York, 1983, pp. 261–275.

34. A. Maxam and W. Gilbert, "Sequencing end-labeled DNA with base-specific chemical cleavages, *Methods Enzymol. 65*: 499–560 (1980).

35. S. A. Williams, B. E. Slatko, L. S. Moran, and S. M. De Simone, "Sequencing in the fast lane: A rapid protocol for $\alpha[^{32}P]$dATP dideoxy DNA sequencing," *Biotechniques, 4*: 138 (1986).

36. E. Retzel, M. Collet, and A. Faras, "Enzymatic synthesis of DNA by avian retrovirus reverse transcriptase *in vitro*: Optimum conditions required for transcription of large RNA templates," *Biochemistry, 19*:513 (1980).

37. E. Chen and P. Seeburg, "Supercoil sequencing: A fast and simple method for sequencing plasmid DNA," *DNA, 4*: 165 (1985).

38. L. Smith, S. Fung, M. Hunkapiller, T. Hunkapiller, and L. Hood, "The synthesis of oligonucleotides containing an aliphatic amino group at the 5'-terminus: synthesis of fluorescent DNA primers for use in DNA sequence analysis," *Nucleic Acids Res. 13*: 2399 (1985).

39. R. Ory, J. Kiser, and P. Pradel, "Studies of positional specificity of castor bean acid lipase," *Lipids, 4*: 261 (1969).

40. A. Kuksis, J. Myher, J. Marai, and K. Geher, "Estimation of plasma free fatty acids as the TMS esters," *Anal. Biochem. 70*: 302 (1976).

41. J. Myher and A. Kuksis, "Gas chromatographic resolution of homologous monoacyl and monoalkylglycerols," *Lipids, 9*: 382 (1974).

42. P. Nichols, "Coordination of silver ion with methyl esters of oleic and elaidic acids," *J. Am. Chem. Soc. 74*: 1091 (1952).

43. P. Dudley and R. Anderson, "Separation of polyunsaturated fatty acids by argentation thin-layer chromatography," *Lipids, 10*: 113 (1975).

44. A. Kuksis, "Routine chromatography of simple lipids and their constituents," *J. Chromatogr. 143*: 1 (1977).

45. J. Axelrod, J. Reichenthal, and B. Brodie, "The direct determination of phosphatidyl ethanolamine and phosphatidyl serine in plasma and red blood cells," *J. Biol. Chem. 204*: 903 (1953).

46. B. Weiss, in *Lipid Chromatographic Analysis*, Vol. 2, 2nd ed., G. Marinetti, Ed. Dekker, New York, 1979, p. 701.

47. B. J. White, C. L. Tipton, and M. Dressel, "Egg-yolk lecithin—A biochemical laboratory project," *J. Chem. Educ. 51*: 533 (1974).

APPENDIX A
LITERATURE SOURCES FOR BIOCHEMICAL ANALYSES, METHODS, AND PREPARATIONS

Literature sources consist of "series works" and journals. Sources of the first type contain reviews and summaries of methods in the research literature. They are collections of articles on general but limited topics contributed by experts in the specialties contained in the series. Sources of the second type, the various biochemical and related research journals, can be further divided into those that are exclusively devoted to methods and those that contain general biochemical studies, offering in the experimental sections of their papers many methods, some of which are new. Following is an annotated bibliography literature sources for methods and techniques used in biochemical laboratory studies.

A.1 SERIES SOURCES

Methods in Enzymology A 120-volume series introduced in 1955 and published by Academic Press, New York. There are various volume editors and many contributors (one or more different authors for each article). The series, which is still being published, covers a very wide range of subjects involving the biochemistry of enzymology, including the preparation and assay of enzymes, the preparation of substrates, enzyme structure, kinetics, and special subjects relating to enzymology (e.g., biomembranes; nucleic acids; recombinant DNA; action of hormones, drugs, and enzyme targeting).

Methods of Biochemical Analysis A 31-volume series introduced in 1954, edited by David Glick, and published by Interscience Publications of John Wiley & Sons, New York. This series, which is still being published, covers a wide variety of analytical methods involved in the qualitative and quantitative determination of compounds of interest to biochemistry. Topics include chemical, physical, and microbiological methods of analysis, and basic techniques and instrumentation for the determination of enzymes, vitamins, coenzymes, hormones, lipids, carbohydrates, proteins, nucleic acids, and products of metabolism.

Methods in Carbohydrate Chemistry An 8-volume series introduced in 1962, edited by R. J. Whistler, R. J. Smith, and J. N. BeMiller, and published by

Academic Press. This series, which is still being published, covers a wide range of general and specific methods involving all aspects of carbohydrate chemistry including analysis, isolation, synthesis, and physical chemical methods for monosaccharides, oligosaccharides, polysaccharides, and glycoproteins.

Laboratory Techniques in Biochemistry and Molecular Biology Several volumes on biochemical laboratory techniques, edited by T. S. Work and R. G. Burdon (formerly edited by T. S. Work and E. Work) covers chromatography, electrophoresis, centrifugation, isoelectric focusing, sequencing, cell culture, immunochemistry, liquid scintillation spectrometry, and related topics.

Biochemical Preparations A series published in 13 volumes from 1949 to 1971, containing the details for the preparation of many biochemical compounds, metabolites, enzymes, and coenzymes. Each preparation was reported by one laboratory and independently checked by a second laboratory. This series was particularly useful before the common biochemicals were available commercially and each laboratory had to prepare its own materials. *Biochemical Preparations* still serves as an important source for the exact methods involved in the preparation of important biochemicals.

A.2 METHODS JOURNALS*

Analytical Biochemistry [Anal. Biochem.] Published by Academic Press; serves as an outlet for the publishing of new analytical and preparative methods in biochemistry.

Analytical Chemistry [Anal. Chem.] Published by the American Chemical Society, Washington, D.C., serves as an outlet for the publishing of analytical methods of all types, some of them applicable to biochemistry.

Biotechniques Published five times per year by Eaton Publishing Co., Natick, MA; features reviews and new techniques, especially in the areas of molecular biology and biotechnology.

Preparative Biochemistry [Prep. Biochem.] Published by Marcel Dekker, New York, started in 1971 as a substitute for the discontinued *Biochemical Preparations* series. It is devoted to the rapid dissemination of information on new preparative methods and procedures in biochemistry, immunology, pharmacy, clinical chemistry, biophysics, and molecular biology.

Journal of Chromatography [J. Chromatogr.] Published by Elsevier, Amsterdam; serves as an outlet for new methods and procedures in all aspects of chromatographic separations.

Carbohydrate Research [Carbohydr. Res.] Published by Elsevier, Amsterdam. Although devoted to publishing all aspects of research on carbohydrate chemistry and biochemistry, the journal publishes a respectable number of analytical and preparative methods.

Clinical Chemistry [Clin. Chem.] Published by American Association for Clinical Chemistry, Washington, D.C. Has a high percentage of bioanalytical

---

* The abbreviation sanctioned by the International Standards Organization is given after the full name of the journal. Journal names that consist of a single word are not abbreviated.

papers, most of which are devoted to methods of analysis of medically important compounds.

Trends in Biochemical Sciences [TIBS] A monthly review journal, published by the International Union of Biochemistry and Elsevier Science Publishers. The journal features a special section on emerging techniques that presents an up-to-date review of specialized techniques used in biochemistry and molecular biology.

A.3 GENERAL JOURNALS*

Archives of Biochemistry and Biophysics [Arch. Biochem. Biophys.]
Biochemical and Biophysical Research Communications [Biochem. Biophys. Res. Commun.]
Biochemical Journal [Biochem. J.]
Biochemistry
Biochimica et Biophysica Acta [Biochim. Biophys. Acta]
Bioorganic Chemistry [Bioorg. Chem.]
Biopolymers
Canadian Journal of Biochemistry [Can. J. Biochem.]
Comparative Biochemistry and Physiology [Comp. Biochem. Physiol.]
European Journal of Biochemistry [Eur. J. Biochem.]
Federation of European Biochemical Societies Letters [FEBS Lett.]
International Journal of Biochemistry [Int. J. Biochem.]
International Journal of Peptide and Protein Research [Int. J. Peptide Protein Res.]
Journal of Applied Biochemistry [J. Appl. Biochem.]
Journal of Bacteriology [J. Bacteriol.]
Journal of Biochemistry (Tokyo) [J. Biochem. (Tokyo)]
Journal of Biological Chemistry [J. Biol. Chem.]
Journal of Chemical Education [J. Chem. Educ.]
Journal of Inorganic Biochemistry [J. Inorg. Biochem.]
Journal of Lipid Research [J. Lipid Res.]
Journal of Molecular Biology [J. Mol. Biol.]
Lipids
Macromolecules
Molecular and Cellular Biochemistry [Mol. Cell. Biochem.]
Naturwissenschaften
Nucleic Acid Research [Nucleic Acid Res.]
Proceedings of the National Academy of Sciences (U.S.A.) [Proc. Natl. Acad. Sci. U.S.A.]
Proceedings of the Society for Experimental Biology and Medicine [Proc. Soc. Exp. Biol. Med.]
Science

---

* The abbreviation sanctioned by the International Standards Organization is given after the full name of the journal. Journal names that consist of a single word are not abbreviated.

APPENDIX B
AMINO ACIDS, ABBREVIATIONS, AND DNA TRIPLET CODES

| Amino acid | Abbreviations | | DNA triplet nucleotide codes[a] | | | | | |
	3-letter	1-letter						
Methionine	Met	M	(CAT)					
Tryptophan	Trp	W	(CCA)					
Asparagine	Asn	N	(ATT)	(GTT)				
Aspartic acid	Asp	D	(ATC)	(GTC)				
Cysteine	Cys	C	(ACA)	(GCA)				
Glutamic acid	Glu	E	(CTC)	(TTC)				
Glutamine	Gln	Q	(CTG)	(TTG)				
Histidine	His	H	(ATG)	(GTG)				
Lysine	Lys	K	(CTT)	(TTT)				
Phenylalanine	Phe	F	(AAA)	(GAA)				
Tyrosine	Tyr	Y	(ATA)	(GTA)				
(Stop)	—	—	(CTA)	(TCA)	(TTA)			
Isoleucine	Ile	I	(AAT)	(GAT)	(TAT)			
Alanine	Ala	A	(AGC)	(CGC)	(GGC)	(TGC)		
Glycine	Gly	G	(ACC)	(CCC)	(GCC)	(TCC)		
Proline	Pro	P	(AGG)	(CGG)	(GGG)	(TGG)		
Threonine	Thr	T	(AGT)	(CGT)	(GGT)	(TGT)		
Valine	Val	V	(AAC)	(CAC)	(GAC)	(TAC)		
Arginine	Arg	R	(ACG)	(CCG)	(CCT)	(GCG)	(TCG)	(TCT)
Leucine	Leu	L	(AAG)	(CAA)	(CAG)	(GAG)	(TAA)	(TAG)
Serine	Ser	S	(ACT)	(AGA)	(CGA)	(GCT)	(GGA)	(TGA)

[a] In each triplet, the 5'-end nucleotide is to the left.

APPENDIX C
SOME COMMERCIALLY AVAILABLE RESTRICTION ENDONUCLEASES AND THEIR DNA SEQUENCE SPECIFICITIES

Enzyme[a,b]	Organism	Specificity
Alu I	*Arthrobacter luteus*	5'-AG↓CT-3' 3'-TC↑GA-5'
Bal I	*Brevibacterium albidum*	5'-TGG↓CCA-3' 3'-ACC↑GGT-5'
BamH I	*Bacillus amyloliquefaciens* H.	5'-G↓GATC C-3' 3'-C CTAG↑G-5'
Bcl I	*Bacillus caldolyticus*	5'-T↓GATC A-3' 3'-A CTAG↑T-5'
Bgl II	*Bacillus globigii*	5'-A↓GATC T-3' 3'-T CTAG↑A-5'
BstE II	*Bacillus stearothermophilus* ATCC 12980	5'-G↓GTNAC C-3' 3'-C CANTG↑G-5'
Cla I	*Caryophanon latum*	5'-AT↓CG AT-3' 3'-TA GC↑TA-5'
Dra I	*Deinococcus radiophilus* ATCC 27603	5'-TTT↓AAA-3' 3'-AAA↑TTT-5'
EcoR I	*Escherichia coli* RY13	5'-G↓AATT C-3' 3'-C TTAA↑G-5'
EcoR V	*Escherichia coli* B946	5'-GAT↓ATC-3' 3'-CTA↑TAG-5'
Hae III	*Haemophilus aegyptius*	5'-GG↓CC-3' 3'-CC↑GG-5'
Hha I	*Haemophilus haemolyticus*	5'-G CG↓C-3' 3'-C↑GC G-5'
Hind III	*Haemophilus influenzae* Rd	5'-A↓AGCT T-3' 3'-T TCGA↑A-5'

[a] A note on restriction endonuclease nomenclature: the first letter is the first letter of the genus, the next two letters are the first two letters of the species, the next letter or number, if present, is a strain designation, and the Roman numeral indicates the particular restriction endonuclease from the organism.

[b] Commercial sources: Sigma Chemical Co. (St. Louis, MO), Bethesda Research Laboratories (Gaithersburg, MD), Amersham Corp. (Arlington Heights, IL).

Enzyme[a,b]	Organism	Specificity
Hpa I	*Haemophilus parainfluenzae*	5'-GTT$^{\downarrow}$AAC-3' 3'-CAA$_{\uparrow}$TTG-5'
Kpn I	*Klebsiella pneumoniae*	5'-G GTAC$^{\downarrow}$C-3' 3'-C$_{\uparrow}$CATG G-5'
Nar I	*Norcardia argentinensis* ATCC 31306	5'-GG$^{\downarrow}$CG CC-3' 3'-CC GC$_{\uparrow}$GG-5'
Nco I	*Norcardia corallina* (*Rhodococcus* sp. ATCC 19070)	5'-C$^{\downarrow}$CATG G-3' 3'-G GTAC$_{\uparrow}$C-5'
Nde I	*Neisseria dentrificans* NRCC 31009	5'-CA$^{\downarrow}$TA TG-3' 3'-GT AT$_{\uparrow}$AC-5'
Nru I	*Nocardia rubra* (*Rhodococcus rhodochrous* ATCC 15906)	5'-TCG$^{\downarrow}$CGA-3' 3'-AGC$_{\uparrow}$GCT-5'
Pst I	*Providencia stuartii*	5'-C TGCA$^{\downarrow}$G-3' 3'-G$_{\uparrow}$ACGT C-5'
Pvu I	*Proteus vulgaris*	5'-CG AT$^{\downarrow}$CG-3' 3'-GC$_{\uparrow}$TA GC-5'
Pvu II	*Proteus vulgaris*	5'-CAG$^{\downarrow}$CTG-3' 3'-GTC$_{\uparrow}$GAC-5'
Sma I	*Serratia marcescens*	5'-CCC $^{\downarrow}$GGG-3' 3'-GGG$_{\uparrow}$CCC-5'
Xba I	*Xanthomonas badrii* (*X. campestris*)	5'-T$^{\downarrow}$CTAG A-3' 3'-A GATC$_{\uparrow}$T-5'
Xma III	*Xanthomonas malvacearum* M (*X. campestris*)	5'-C$^{\downarrow}$GGCC G-3' 3'-G CCGG$_{\uparrow}$C-5'

[a] A note on restriction endonuclease nomenclature: the first letter is the first letter of the genus, the next two letters are the first two letters of the species, the next letter or number, if present, is a strain designation, and the Roman numeral indicates the particular restriction endonuclease from the organism.

[b] Commercial sources: Sigma Chemical Co. (St. Louis, MO), Bethesda Research Laboratories (Gaithersburg, MD), Amersham Corp. (Arlington Heights, IL).

APPENDIX D
TABLE FOR PREPARING SOLUTIONS OF DIFFERENT CONCENTRATIONS OF AMMONIUM SULFATE

% Saturation	10	20	25	30	35	40	45	50	55	60	65	70	75	80	85	90	95	100
Grams of $(NH_4)_2SO_4$/L (0°C) to add from 0% (25°C)	52.3	107.5	137	166.4	199	230	264.5	296.8	333	371	409	451	498.7	532.2	578	625	673.5	723
	55	113	144	175	209	242	278	312	350	390	430	474	519	560	608	657	708	760
Grams of $(NH_4)_2SO_4$ to make up to 1-L solution at 0°C:	53[a]	106	133	160	187	214	240	266	294	321	345	380	404	429	456	482	510	536[a]
Grams of $(NH_4)_2SO_4$/L to bring from:																		
10		57[b]	87[b]	118[b]	149	182	215	250	287	325	365	405	448	494	530	585	634	685
20			29	59	90	121	154	188	225	260	298	337	379	420	465	512	559	610
25				29	60	91	123	157	192	228	265	304	345	386	430	475	521	571
30					30	61	93	125	160	195	232	270	310	351	394	439	485	533
35						30	62	94	128	163	199	235	275	315	358	403	449	495
40							31	63	96	131	166	205	240	280	322	365	410	458
45								31	64	98	133	169	206	245	286	330	373	420
50									32	65	100	135	172	211	250	292	335	380
55										33	66	101	138	176	214	255	298	344
60											33	67	102	140	179	219	261	305
65												34	69	105	143	182	224	267
70													34	70	108	146	187	228
75														35	72	110	149	190
80															36	73	112	152
85																37	75	114
90																	37	76
95																		38

[a] Grams of ammonium sulfate to give the desired percent by dissolving in approximately 700 mL of water and then diluting to 1 liter.

[b] Grams of ammonium sulfate to add to 1 liter of a 10% ammonium sulfate solution to make it 20%, 25%, 30%, etc.

INDEX

Periodic chart of the elements

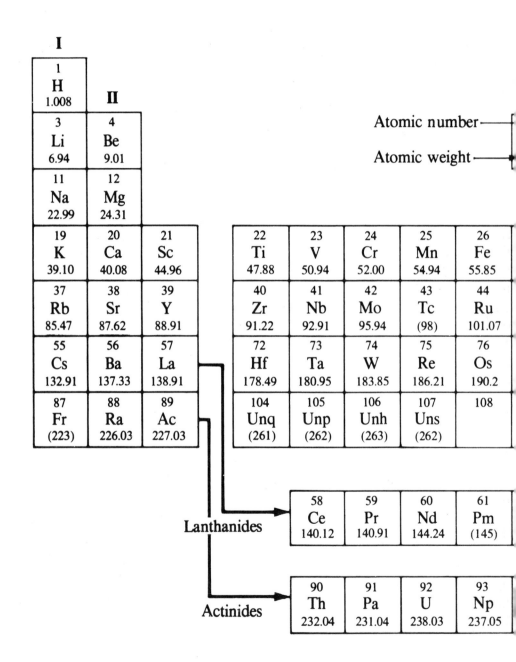